Leadership Styles, Innovation, and Social Entrepreneurship in the Era of Digitalization

Çağlar Doğru
Ufuk University, Turkey

A volume in the Advances in Business Strategy and Competitive Advantage (ABSCA) Book Series

IGI Global
DISSEMINATOR OF KNOWLEDGE

Published in the United States of America by
IGI Global
Business Science Reference (an imprint of IGI Global)
701 E. Chocolate Avenue
Hershey PA, USA 17033
Tel: 717-533-8845
Fax: 717-533-8661
E-mail: cust@igi-global.com
Web site: http://www.igi-global.com

Library of Congress Cataloging-in-Publication Data

Names: Dogru, Caglar, 1983- editor.
Title: Leadership styles, innovation, and social entrepreneurship in the era of digitalization / Çağlar Doğru, editor.
Description: Hershey, PA : Business Science Reference, [2020] | Includes bibliographical references and index. | Summary: “”This book examines leadership, social entrepreneurship, corporate social responsibility, and innovation in the digitalized era”--Provided by publisher”-- Provided by publisher.
Identifiers: LCCN 2019024182 (print) | LCCN 2019024183 (ebook) | ISBN 9781799811084 (hardcover) | ISBN 9781799811091 (softcover) | ISBN 9781799811107 (eISBN)
Subjects: LCSH: Leadership. | Technological innovations. | Entrepreneurship.
Classification: LCC HD57.7 .L4734 2020 (print) | LCC HD57.7 (ebook) | DDC 658.4/092--dc23
LC record available at https://lccn.loc.gov/2019024182
LC ebook record available at https://lccn.loc.gov/2019024183

This book is published in the IGI Global book series Advances in Business Strategy and Competitive Advantage (ABSCA) (ISSN: 2327-3429; eISSN: 2327-3437)

British Cataloguing in Publication Data
A Cataloguing in Publication record for this book is available from the British Library.

Advances in Business Strategy and Competitive Advantage (ABSCA) Book Series

ISSN:2327-3429
EISSN:2327-3437

Editor-in-Chief: Patricia Ordóñez de Pablos, Universidad de Oviedo, Spain

MISSION

Business entities are constantly seeking new ways through which to gain advantage over their competitors and strengthen their position within the business environment. With competition at an all-time high due to technological advancements allowing for competition on a global scale, firms continue to seek new ways through which to improve and strengthen their business processes, procedures, and profitability.

The **Advances in Business Strategy and Competitive Advantage (ABSCA) Book Series** is a timely series responding to the high demand for state-of-the-art research on how business strategies are created, implemented and re-designed to meet the demands of globalized competitive markets. With a focus on local and global challenges, business opportunities and the needs of society, the **ABSCA** encourages scientific discourse on doing business and managing information technologies for the creation of sustainable competitive advantage.

COVERAGE

- Balanced Scorecard
- Resource-Based Competition
- Value Chain
- Value Creation
- International Business Strategy
- Core Competencies
- Adaptive Enterprise
- Competitive Strategy
- Ethics and Business Strategy
- Entrepreneurship & Innovation

The Advances in Business Strategy and Competitive Advantage (ABSCA) Book Series (ISSN 2327-3429) is published by IGI Global, 701 E. Chocolate Avenue, Hershey, PA 17033-1240, USA, www.igi-global.com. This series is composed of titles available for purchase individually; each title is edited to be contextually exclusive from any other title within the series. For pricing and ordering information please visit http://www.igi-global.com/book-series/advances-business-strategy-competitive-advantage/73672. Postmaster: Send all address changes to above address.

Titles in this Series

For a list of additional titles in this series, please visit:
https://www.igi-global.com/book-series/advances-business-strategy-competitive-advantage/73672

Achieving Peak Sales Performance for Optimal Business Value and Sustainability
Carlton Brown (University of Plymouth, UK)
Business Science Reference • copyright 2020 • 431pp • H/C (ISBN: 9781799816393) • US $245.00 (our price)

Handbook of Research on Industrial Applications for Improved Supply Chain Performance
Jorge Luis García-Alcaraz (Universidad Autónoma de Ciudad Juárez, Mexico) George Leal Jamil (Informações em Rede Consultoria e Treinamento Ltda., Brazil) Liliana Avelar-Sosa (Universidad Autónoma de Ciudad Juárez, Mexico) and Antonio Juan Briones Peñalver (Polytechnic University of Cartagena, Spain)
Business Science Reference • copyright 2020 • 549pp • H/C (ISBN: 9781799802020) • US $345.00 (our price)

Advanced MIS and Digital Transformation for Increased Creativity and Innovation in Business
Gülay Ekren (Sinop University, Turkey) Alptekin Erkollar (ETCOP, Austria) and Birgit Oberer (ETCOP, Austria)
Business Science Reference • copyright 2020 • 251pp • H/C (ISBN: 9781522595502) • US $225.00 (our price)

Improving Business Performance Through Innovation in the Digital Economy
Ionica Oncioiu (Titu Maiorescu University, Romania)
Business Science Reference • copyright 2020 • 272pp • H/C (ISBN: 9781799810056) • US $225.00 (our price)

Ethical Consumerism and Comparative Studies Across Different Cultures Emerging Research and Opportunities
Ebtihaj Ahmed Al-A'ali (University of Bahrain, Bahrain) and Meryem Masmoudi (University of Bahrain, Bahrain)
Business Science Reference • copyright 2020 • 248pp • H/C (ISBN: 9781799802723) • US $155.00 (our price)

For an entire list of titles in this series, please visit:
https://www.igi-global.com/book-series/advances-business-strategy-competitive-advantage/73672

IGI Global
DISSEMINATOR OF KNOWLEDGE

701 East Chocolate Avenue, Hershey, PA 17033, USA
Tel: 717-533-8845 x100 • Fax: 717-533-8661
E-Mail: cust@igi-global.com • www.igi-global.com

To my mother, Özden, and my father, Zeki, for their endless support and devotion.

Editorial Advisory Board

Table of Contents

Detailed Table of Contents

Creating social value with the help of social entrepreneurship is among the most meaningful goals of ventures in the modern world. The intent of this chapter is to evaluate the effects of different leadership styles on social entrepreneurship theoretically. A detailed literature scanning was conducted in order to find correlations between different leadership styles and social entrepreneurship so as to note which of them mostly promotes social entrepreneurship. Among the leadership styles included in the study, there exists entrepreneurial leadership, responsible leadership, innovative leadership, value-based leadership, visionary leadership, strategic leadership, transformational leadership, authentic leadership, cultural leadership, paternalistic leadership, servant leadership, ambidextrous leadership, ethical leadership, and negative leadership styles. As a consequence, all of the leadership styles included in this chapter, except for negative leadership styles, were proposed to promote social entrepreneurship based on the related literature.

In this chapter, innovation, one of the most important factors for contemporary organizations and even countries, is analyzed in depth. The concept of innovation is divided into two major fields, which are social innovation and technological innovation. The relationship between these two innovation types and gaining strategic advantage are figured out in a macro perspective. To achieve this, the strongest innovators in Europe were selected. Among them are Switzerland, Sweden, Denmark, Finland, Netherlands, Germany, and Norway. To compare the countries, Social Innovation Index and European Innovation Scoreboard were employed to get secondary data. It was observed that by investing more on social and technological innovation, these European countries perform better and gain strategic advantage, which helps them build sustainable future.

This chapter describes how and why normative justification for corporate responsibility has been replaced with business case thinking. Modern organizations do link their non-financial reports to strategy because corporate responsibility is seen as a source of competitive advantage, growth, motivation, and innovation. Non-financial reporting potentiates companies to understand, learn, and redefine their social and environmental impacts and its cost; how to minimize both impact and cost; and how to capitalize on challenges that management face on a daily basis. The connection between CR strategy and non-financial reporting is described along with how it is relevant for the successful business of modern organizations. By improving CR strategies and developing unique non-financial reporting, organizations can be more effective and efficient in the local community, which will positively reflect on their business model and in achieving higher sustainability levels.

This chapter critically examines the servant leadership style of social entrepreneurs. Qualitative data from social enterprises in Palestine and Lukes third dimension power framework are used to explore the intricate forces of power, manipulation, and domination hidden within the service and follower-oriented model of servant leadership. Insights are provided on how the concept of 'service', the focal aspect in both social enterprises and servant leadership, could be another facet of soft and insidious power exercised by the social entrepreneurs over their followers. A political model of servant leadership is developed that demonstrates how social entrepreneurs could be imposing power and control over their followers under the guise of social mission, creation of social value, serving and empowering their followers and the community.

Cynthia Maria Montaudon-Tomas, UPAEP Universidad, Mexico
Ivonne M. Montaudon-Tomas, UPAEP Universidad, Mexico
Yvonne Lomas-Montaudon, Universidad Iberoamericana, Mexico

A theoretical study was conducted based on three different leadership styles: autocratic, transactional, and servant. The most relevant characteristics of each leadership style were summarized. A cross-culture study was proposed considering three countries from diverse clusters according to the Global Leadership and Organizational Behavior Effectiveness Project (GLOBE). Each culture was analyzed separately, and relevant statistics were presented as elements for comparison. Different models and tools to evaluate cultural differences were used to create a multiple perspectives overview. Information from the three leadership styles was used to further determine whether those styles fit the cultural descriptions in order to establish the most frequent and suitable leadership styles in the selected countries and to understand how leadership styles can vary from region to region.

Ravinder Jit, Maharaja Agarsen Institute of Technology, India
Anju Bharti, Maharaja Agarsen Institute of Technology, India
Urvashi Sharma, Department of Commerce, Delhi School of Economics, University of Delhi, India
P. V. Rajeev, Government of India, India

The chapter examines various strategies adopted by servant leaders to ensure discipline in the organization. The study suggests that with their inherent pro-follower orientation, servant leaders' first preference is to create an organizational environment where employees do not feel the need to indulge in indiscipline. For this, they take certain proactive initiatives such as open door policy; leading by example; clarifying goals and roles; ensuring objectivity, fairness, and transparency; etc. In case some employees still indulge in undisciplined behavior, the servant

This chapter discusses the key theoretical and empirical steps undertaken throughout the authors' previous-but-related mixed methods studies on social innovation in nonprofit organizations (NPOs) in the Australian disability sector with the aim of using the key findings of these studies to develop 'theories-in-practice' in disability NPOs. In this chapter, the authors summarize the associated theory-building processes deployed to explain how disability NPOs develop and implement social innovations and the societal 'system-level' impacts of such innovations. These theory-building processes involve two broad phases, and the culmination of these phases (grounded in the abductive logics of inquiry, complexity theorizing, and set-theoretic methods) leads to the development of several 'theories-to-practice' that not only convey the interactivity of contextual causal mechanisms leading to social innovation by NPOs, but also outline change-oriented solutions for managers who are working to address complex social challenges.

The aim of the study is to examine the relationships between ambidextrous leadership behaviors (i.e., task-oriented, relation-oriented, change-oriented, and external) in the context of leader behavior level and employee ambidextrous performance (i.e., exploitation and exploration) in the context of individual employee level by suggesting that the interaction of these two may result in organizational ambidexterity. This study aims to contribute to the immature ambidexterity literature by developing some propositions based on the previous ambidexterity researches related to ambidexterity of leaders, employees, and entire organizations. Taking into account ambidextrous leadership theory, it is proposed that while the leaders' task-oriented and relation-oriented behaviors are positively related to the employees' exploitation behaviors, the leaders' change-oriented and external behaviors are positively related to the employees' exploration behaviors. It is recommended for further studies to investigate the possible propositions with quantitative research designs.

Foreword

Both the title of this volume and its contents are extremely pertinent to our contemporary environment. The Digital Era, despite its various meanings and cultural connotations, represents yet another challenge to the twenty-first century. Academicians, business and management practitioners, and IT professionals are oftentimes confronted by a barrage of conflicting advice that take the form of media advertisements. The latter offer easy solutions to a complex set of problems---what to do with Big Data, disruptive technologies, and the like. However, little regard is allocated to the nature of the problem(s) and the toll on labour forces, national or international.

This tome sheds light on essential elements associated within the sphere of leadership, innovation, and entrepreneurialship. It addresses the complexities associated with resource management within a social and cultural context. In this sense, it examines the research question, 'are current notions and concepts of leadership relevant'. Innovation remains central to the well being of not only corporations and enterprises, but to cultures and nations. This in turn, leads to the questions of how to gauge innovation in the digital age, how one innovates, and subsequently, identify the obstacles. Lastly, by what method to evaluate the impact to the entrepreneur. To quote the Editor, *"our book has the objective to examine leadership styles to reveal how leadership impacts innovation and social entrepreneurship*".

I applaud this initiative and congratulate the editor as well as the contributing authors on their achievement. Their collaborative work contributes to the need to identify the human perspective in an era that emphasises technology over sociology, psychology, and culture.

Richard Brunet-Thornton
University of Economics, Prague

Preface

Leaders are the most important element of an organization in regards to reaching organizational goals, motivating followers to perform better, and creating an innovative work environment. To conform with successful corporate social responsibility implementations, social entrepreneurship practices have gained more importance with the development of digital technology. In today's changing business environment, managers and employees need essential capabilities in order to reach organizational goals and to survive. Among these, leadership and innovation take the lead. Leading effectively is not a option but a necessity in this globalized business life. For this reason different leadership styles gain more and more importance recently. Furthermore, a company without innovation is like a horse without legs. It cannot live unless innovation process takes place. So another issue of creating new solutions for changing problems necessitates innovation. Moreover, all types of entrepreneurship are on the agenda of even developed countries in order to create added value for global economy. For all of these reasons, the latest researches on the selected subjects of leadership, innovation and entrepreneurship will fill a critical gap in the related literature.

The concept of 'entrepreneurship' has been put in the first place among the factors affecting a country's economy since its importance was emphasized by the early economics theorists. Entrepreneurs supply an important portion of jobs especially for the young employees, while they are contributing a significant amount into real gross domestic product by their recently established firms. The concept of entrepreneurship consists of activities for establishing, developing and sustaining a business to make profit. It can also be defined as, a process of creating a new organization by taking risks for product innovation and being proactive and aggressive to compete against rivals. So as to establish a sustainable business in todays' high-competitive environment, entrepreneurs inevitably must have the vision and the experience to manage a new company. Because, nowadays the business environment is dynamic and complex, only the firms that are leaded by visionary leaders or managers who scan their environments for every opportunity can survive. That is the reason why; organizations need visionary, innovative and supportive leaders than they need before. Understanding the importance of entrepreneurship and leadership in the last century, it has been witnessed that when these two concepts are combined, then an organization can operate more effectively and only in this case, it can reach its goals and objectives more efficiently. Todays' business composition requires both leaders to act in an entrepreneurial manner and entrepreneurs to behave like a leader. Innovation in the organizations refers not only to formulation of proper ideas but also practicing them apart from the concept of creativity that refers to forming an idea primarily. Thus, innovative work behavior is defined as, generating useful ideas and solutions in favor of the organization or the group that the employee belongs to. Innovative work behavior can be taken as a process because it has a number of consecutive stages. Firstly it begins with generating new ideas, and then it is followed by disseminating this fruitful idea to other employees and in the final step, if it is accepted, the innovative idea is implemented.

Therefore in this globalization and digitalization era, leadership, entrepreneurship and innovation are getting more and more critical. In today's change-oriented and complex business environment, both entrepreneurs and leaders have to keep up with the latest developments around them. In this book, with the contributions of the potential authors, latest data and research on the issues in entrepreneurship, leadership and especially innovation will appear. Such a comprehensive book will attract both scholars and professionals.

This reference book aims to bring together the emerging topics of innovation, leadership and entrepreneurship. At the same time it provides theoretical infrastructure and moreover in order to share the latest empirical research findings within leadership, entrepreneurship and innovation. A tremendous contribution will be made to the contemporary research on these subjects by collecting the most up to date research findings and combining these with the theoretical framework. The most popular contemporary topics in leadership, entrepreneurship and innovation management fields will provide valuable insight for understanding today's business organizations operating in a global, dynamic and complex environment. Each separate chapters of the book will handle an up-to-date topic, mostly on the basis of technological advances and readers will benefit the latest management skills and techniques in this book. This scientific book will fill a critical gap by providing women leadership and social entrepreneurship issues from the perspective of innovation.

The target audience is intended to be scholars from universities all over the world, as well as the undergraduate and graduate students studying at the fields of management, organization, organizational behavior, human resources management, leadership studies and management information systems. Furthermore, the professional managers, specialists, consultants, educationalists and the employees of especially knowledge and technology intensive companies are in the scope of the target audience of this reference book. In this context;

Chapter 1 figures out the concepts of social entrepreneurship, social value creation and social innovation. The intent of this chapter is to evaluate the effects of different leadership styles on social entrepreneurship theoretically. A detailed literature scanning was conducted in order to find correlations between different leadership styles and social entrepreneurship so as to note which of them mostly promote social entrepreneurship. Among the leadership styles included in the study, there exist entrepreneurial leadership, responsible leadership, innovative leadership, value-based leadership, visionary leadership, strategic leadership, transformational leadership, authentic leadership, cultural leadership, paternalistic leadership, servant leadership, ambidextrous leadership, ethical leadership and negative leadership styles.

In Chapter 2, the concept of innovation is divided into two major fields, which are social innovation and technological innovation. The relationship between these two innovation types and gaining strategic advantage is figured out in a macro perspective. To achieve this, the strongest innovators in Europe were selected. Among them are, Switzerland, Sweden, Denmark, Finland, Netherlands, Germany and Norway. To compare the countries, Social Innovation Index and European Innovation Scoreboard were employed to get secondary data. It was observed that, by investing more on social and technological innovation, these European countries perform better and gain strategic advantage, which help them build sustainable future.

Chapter 3 describes how and why normative justification for corporate responsibility has been replaced with business case thinking. In this chapter, the connection between CR strategy and non-financial reporting is described and how it is relevant for the successful business of modern organizations. By improving CR strategies and developing unique non-financial reporting, organizations can be more effective and efficient in the local community, which will positively reflect on their business model and in achieving higher sustainability levels

Chapter 4 figures out the relationship between social entrepreneurship and servant leadership which is a special type of leadership. In order to achieve this, in this chapter, Qualitative methodology encompassing the case study design was used to investigate power dimensions within servant leadership style of social entrepreneurs. Six in-depth, semi-structured interviews were conducted. The two enterprises were contacted for in-person interviews with their owners and other senior executives by the secondary author.

In Chapter 5, a cross-cultural theoretical analysis is conducted. This was done by including autocratic, transactional and servant leadership in three countries which are, Japan, France and Mexico. After examining the relationship between leadership and social entrepreneurship, this chapter provides valuable insights for the effects of different leadership styles in three continents.

Chapter 6 identifies the relationship between servant leadership and organizational discipline. This is achieved by conducting qualitative analysis. Research consists fifteen semi-structured interviews with leaders in education sector, corporate sector and public sector.

Chapter 7 investigates the relationship between strategic leadership, innovation and design practices as manufacturing small-medium enterprises (SMEs) begin to transition to advanced manufacturing technology. This chapter presents a case study of a prototyping method.

Chapter 8 aims at revealing themes in this research field throughout the years between 1970 and 2016In this study, by using the terms collaboration and social innovation together. In this chapter co-citation analysis was applied to find out theoretical foundations of this recently emerged field. Six clusters with different attributes, such as cross-sector partnerships in social innovation, the definition of social innovation, transition studies, social entrepreneurship, innovation studies, and inter-organizational relations *were obtained.*

Chapter 9 discusses the key theoretical and empirical steps undertaken throughout the authors' previous-but-related mixed methods studies on social innovation in nonprofit organizations (NPOs) in the Australian disability sector. The aim of the chapter is to use the key findings of these studies to develop 'theories-in-practice' in disability NPOs. In this chapter, the authors summarize the associated theory-building processes deployed to explain how disability NPOs develop and implement social innovations and the societal 'system-level' impacts of such innovations.

Chapter 10 explains the interactions between leader ambidextrous behavior, employee ambidexterity and organizational ambidexterity. These interactions are subdivided into the interactions between task-oriented leadership behavior, relation oriented leadership behavior, change- oriented leadership behavior and external leadership behavior and employees' ambidextrous behaviors.

Chapter 11 is a theoretical chapter in which authentic leaders and their effects on innovation is figured out. While achieving this, a social entrepreneurship perspective is used. After giving the definitions of the concepts of social entrepreneurship, authentic leadership and innovation, a theoretical synthesis is provided in this chapter.

In Chapter 12, a brief discussion is made in order to provide a better understanding of the desired organizational outcomes of this concept and its role in the authentic leadership and organizational commitment behavior of employees. It is stated that under authentic leadership style followers with low psychological capital are thought to be more inclined to show organizational commitment than their peers with high psychological capital.

Chapter 13 examines the effects of authentic leadership and leader optimism on follower affective commitment. This research was undertaken in the format of an experiment. Four 2*2 experiments were carried out in separate groups consisting of at least 10 persons. Each experiment was composed of two paragraphs. The first paragraph described a fictional authentic leader or a fictional inauthentic leader.

In Chapter 14, an analysis of the electronic decision making system, which is thought to benefit from the heavy bureaucratic system which does not take into account the expectations of the citizen in the public administration system, is used to make a more flexible structure. The focus of this chapter is on the need to design the decision-making mechanisms of the state according to the expectations of the citizen. For this purpose, requests and complaints from the citizens through the electronic environment should be taken into consideration in the decision-making process.

Leadership Styles, Innovation, and Social Entrepreneurship in the Era of Digitalization is a refereed, scientific and pivotal reference book that provides vital research on the application of business organizations operating in a global, complex environment. While highlighting topics such as social entrepreneurship, innovation, and leadership, this publication explores recent technological advances and the methods of the latest management skills and techniques. This book is ideally designed for human resources professionals, managers, leaders, executives, CEOs, specialists, consultants, researchers, students, and professors seeking current research on human resources management and management information systems in a digital society.

Consequently, as it will be seen throughout the whole book that there is an intense devotion from chapter authors to complete the project. At this point, I appreciate their diligent work. Moreover, it was impossible to bring this reference book to an end without the help of IGI Global employees. I wish everyone to benefit from this peer-reviewed scientific book.

Çağlar Doğru
Ufuk University, Turkey

Chapter 1
Leadership Styles That Mostly Promote Social Entrepreneurship:
Towards a Conceptual Framework

Çağlar Doğru
https://orcid.org/0000-0002-4215-8979
Ufuk University, Turkey

ABSTRACT

Creating social value with the help of social entrepreneurship is among the most meaningful goals of ventures in the modern world. The intent of this chapter is to evaluate the effects of different leadership styles on social entrepreneurship theoretically. A detailed literature scanning was conducted in order to find correlations between different leadership styles and social entrepreneurship so as to note which of them mostly promotes social entrepreneurship. Among the leadership styles included in the study, there exists entrepreneurial leadership, responsible leadership, innovative leadership, value-based leadership, visionary leadership, strategic leadership, transformational leadership, authentic leadership, cultural leadership, paternalistic leadership, servant leadership, ambidextrous leadership, ethical leadership, and negative leadership styles. As a consequence, all of the leadership styles included in this chapter, except for negative leadership styles, were proposed to promote social entrepreneurship based on the related literature.

DOI: 10.4018/978-1-7998-1108-4.ch001

INTRODUCTION

From the early works of Bowen (1953) on the social responsibilities of entrepreneurs, research on social entrepreneurship has come a long way until now. Unfortunately, because of the importance of the topic still there is need for extra work. This is due to the absence of both a consensus on the definition of the term and lack of studies revealing the antecedents and consequences of social entrepreneurship. As Short, Moss and Lumpkin (2009) set forth in their article, beginning with 1990s, there has been an increasing interest on the concept of social entrepreneurship among scholars. Throughout this period, it has been noted that, the concept of 'social entrepreneurship' has been linked to 'social value creation' (Austin, Stevenson and Wei-Skillern, 2006), 'social goal' (Weerawardena and Mort, 2001), and 'social enterprise' (Campbell, 1998) from a narrow perspective of non-profit organizations or philanthropists (Moss and Lumpkin, 2009). Beyond this, in accordance with the emerging literature about social entrepreneurship, the concept will be held in a broader meaning throughout this chapter. Because according to the existing literature, it can easily be seen neither that all the nonprofit organizations are social entrepreneurs (Weerawardena and Mort, 2006) nor all the profit organizations are not social entrepreneurs (Thompson, Alvy and Lees, 2000).

To demonstrate entrepreneurship behaviors, individuals must possess leadership characteristics (Cunningham and Lischeron, 1993; Gupta, MacMillan and Surie, 2004). Precisely, a very special type of entrepreneurship, which is social entrepreneurship, can be best exercised by individuals' certain leadership behaviors (Cogliser and Brigham, 2004; Litzky, Godshalk and Walton-Bongers, 2010). These leadership behaviors are called as leadership styles in general in the literature.

In this chapter, it will be discussed which leadership styles promote social entrepreneurship mostly or do they really affect social entrepreneurship behaviors by scanning the existing literature. This theoretical chapter has the intention of putting forth a conceptual framework for leadership styles' enhancing social entrepreneurship. The rationale behind this idea has been embodied by the future research direction from the study of Short, Moss and Lumpkin (2009). Hence, to attain this, the concept of social entrepreneurship will be exhibited in the following background section before explaining the leadership styles, which boost social entrepreneurship most.

BACKGROUND

The Concept Of Social Entrepreneurship

Before, figuring out the concept of social entrepreneurship, it should be put forth what the concept of 'entrepreneurship' is. Based on the Schumpeterian Model, entrepreneurship necessitates the presentation of a new good or service, adding a quality or improvement, a new technique for production, a new market for selling goods and services, or a new source inputs (i.e. raw materials or half-manufactured goods) (Schumpeter, 1934). This type of entrepreneurship can be considered as the allocation of entrepreneurship between the groups of tasks mentioned above (Baumol, 1990). In addition, this type of entrepreneurship emphasizes on innovation and creativity (Carland, Hoy, Boulton and Carland, 1984). According to the related literature, among the characteristics of entrepreneurs, there exist their being risk-taking (McClelland, 1961), ambitious, self-confident (Timmons, 1978), hardworking (Beugelsdijk and Noorderhaven, 2005) and achievement oriented (Kessler, 2008).

After touching briefly on the concept of entrepreneurship, a more precise concept which is social entrepreneurship can be emphasized. Social entrepreneurship has been on the agenda of both practitioners and scholars since 1980s. On the basis of literature scrunity it is observed that the concept of social entrepreneurship has been attracted more attention from governments and institutions rather than from scholars until now (i.e. Mair, Robinson and Hockerts, 2006; Short, et al., 2009). But in the last decades, there has been a growing interest on social entrepreneurship by the scholars.

Social entrepreneurship can be simply defined as, a specific type of entrepreneurship within or across the organizations operating in private, public or not-for-profit sectors, in which innovation and social value creation are the key issues (Austin, et al., 2006). Additionally, social entrepreneurship can be defined by Johnson (2000) as an entrepreneurial movement that consists of social oriented intentions.

The components of social entrepreneurship which can be seen on Figure 1. should be understood in order to distinguish the concept of social entrepreneurship from related constructs.

First of all, social value is at the center of the social entrepreneurship since entrepreneurship cannot be taken as social, unless there is social value creation. In this chapter, social value is defined as, a response to a social need or a solution to a social problem which creates better conditions for the society. These needs and problems of societies may vary from educational, economic, political and cultural (Alvord, Brown and Letts, 2002), to environmental, equality and health needs and problems. In order to create social value, entrepreneurs try to fulfill their social

Figure 1. Components of social entrepreneurship

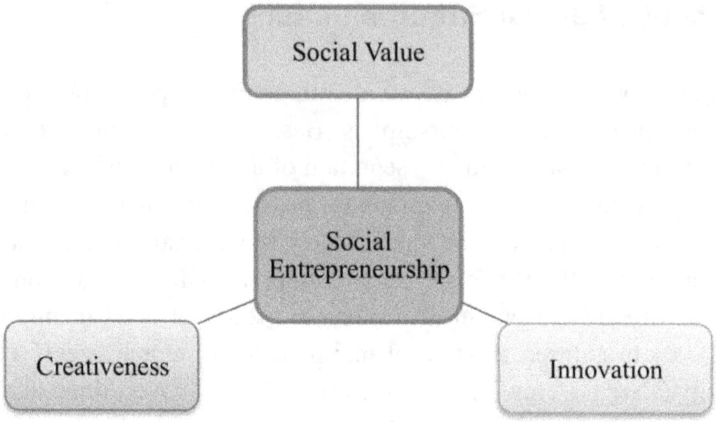

missions. These social missions are creative, innovative and satisfactory for the society (Mort, Weerawardena and Carnegie, 2003).

Furthermore, creativeness is obligatory for social entrepreneurship (Perrini and Vurro, 2006). Here, creativeness refers to unique ideas and solutions of entrepreneurs to unique social problems (Gibb, 2006). Trying to solve a social problem or to reach a social aim, entrepreneurs should generate unique and effective projects.

The last main component of social entrepreneurship comes to be the concept of innovation. That is why social entrepreneurs are innovative entrepreneurs (Prabhu, 1999). Together with the social value creation and creativeness, innovation makes difference for social entrepreneurship from other types of entrepreneurship especially commercial entrepreneurship. Social entrepreneurs are the ones who bring about innovative solutions to social issues (Nicholls, 2006).

After figuring out the main components of social entrepreneurship, here it is suitable to touch on the concept on the basis of sectors like, not-for-profit, private and public sectors. There is still a controversy on the organizations' types on being socially entrepreneur. Such that, social entrepreneurs may be not-for-profit organizations or privately owned or public owned organizations. Not all the not-for-profit organizations are social entrepreneurs when they have not a social mission, they don't conduct innovation and they don't create social value by their activities. Besides, not all the private owned organizations have only commercial missions (Weerawardena and Mort, 2006). The ones with social missions, creativeness and innovation can be called as social entrepreneurs. In public sector, while being a more complicated situation, governments may support public organizations to be active on some chosen social movements or problems. In the literature, it has been put forth that, since private companies chase for innovation, they are a bit more suitable for social entrepreneurship (i.e. Canadian Centre for Social Entrepreneurship, 2001; Thompson, Alvy and Lees, 2000). After them, today's modern not-for-profit organizations, which are faced with a fierce competition (Weerawardena and Mort, 2006), are called to be successful social entrepreneurs (Boschee, 1998). Following these two types of organizations, publicly held organizations are more strengthened than yesterday to find solutions to social problems in an innovative way (Roper and Cheney, 2005). These social entrepreneurial organization models are shown at Figure 2.

MAIN FOCUS OF THE CHAPTER

The Relationship Between Leadership and Social Entrepreneurship

All of the social enterprises from private, not-for-profit or public sectors need effective leadership (Thompson, 2008). This is because, according to Dym and Hutson (2005), today social enterprises need a more integrated and comprehensive directions to reach their aims with the help of leadership more than yesterday due to changing demands of shareholders. Leadership refers to a social process (Barker, 2001) in which an individual or a group influences one or more followers who have different skills, abilities and characteristics in order to fulfill the organization's goals and objectives (Winston and Patterson, 2006). Here, in this chapter we are interested in a leader's role in motivating followers to reach especially social goals and objectives.

Furthermore, before stepping into the subject of 'effects of leadership styles to social entrepreneurship', a research question like '*Is there a relationship between leadership and social entrepreneurship?*' should be answered. To accomplish this, the related literature was scanned in detail. Unfortunately, it has been witnessed that there is a scarcity on empirical results of researches which links leadership and social entrepreneurship. Additionally, there are only a few theoretical studies concentrating on the importance of leadership for social entrepreneurship (i.e. Alvord, et al., 2002; Prabhu, 1999; Short, et al., 2009). Still, it can be proposed that, like in every social institutions, organizations or companies, leadership is also vital for social entrepreneurship. Motivating group members for a social mission in an effective way will foster social entrepreneurship. There is no doubt that, demonstrating proper leadership by being visionary, inclusive, responsible, inspiring, creative, innovative and an efficient communicator, social entrepreneurs will be effective.

Figure 2. Sectorial based social entrepreneurship

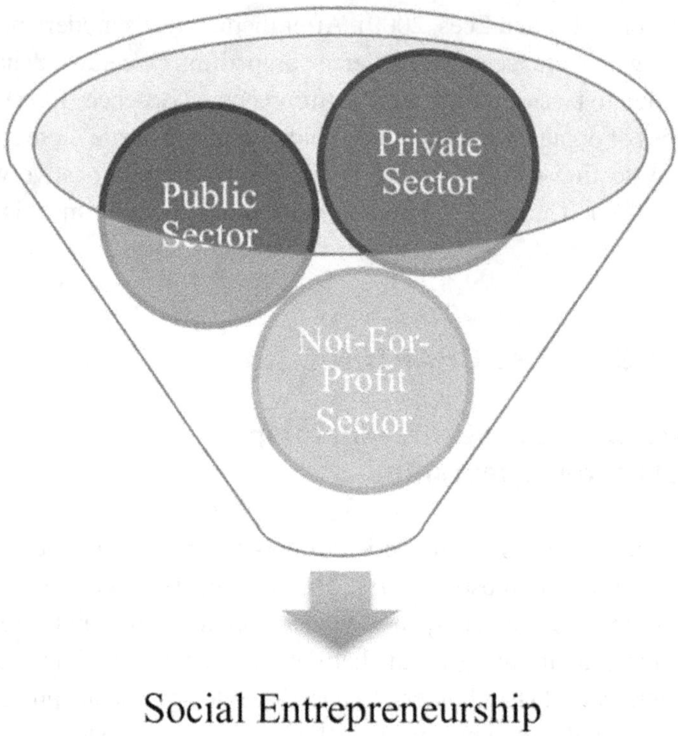

This is adequate for accepting that there is a positive correlation between leadership and social entrepreneurship.

Leadership Styles That Mostly Promote Social Entrepreneurship

In this section, the subject is being more specific in a way that exploring which leadership styles are more correlated with social entrepreneurship in todays context. There have been promising attempts to figure out the relationship between leadership styles and social entrepreneurship. Among these substantial studies existing in the literature, some examples are given at Table 1.

Only the most related ones among many alternative leadership styles have been selected to form a theoretical framework below.

Table 1. Examples of leadership studies linked to social entrepreneurship in the literature

Leadership style	Linked to	Method	Sample	Author(s)
Entrepreneurial Leadership	Nonprofit organization	Interviews, observation, primary and secondary	Board members of non profit organization	Santora, Seaton and Sarros (1999)
Responsible Leadership	Social entrepreneurship	Case study	Fundación Paraguaya	Maak and Stoetter (2012)
	Responsible Corporations	Theoretical	N/A	Pruzan and Miller (2006)
	Social responsibility	Case studies	24 German Companies	Székely and Knirsch (2005)
Transformational Leadership	Social entrepreneurship	Quantitative method	107.738 individuals from 27 countries	Muralidharan and Pathak (2018)
	Social entrepreneurship	Case studies	Influencers	Bhutiani, et.al, (2012)
	Social responsibility	Quantitative method	Managers from 112 companies	Waldman et.al, (2006)
Authentic Leadership	Cultural entrepreneurship Social entrepreneurship	Theoretical	N/A	Dacin et.al, (2010)
Paternalistic Leadership	Social capital Philanthropy	Theoretical	N/A	Casson and Della Giusta (2007)
Servant Leadership	Social responsibility	Theoretical	N/A	Van Dierendonck (2011)
Strategic Leadership	Social entrepreneurship Nonprofit organization	Theoretical	N/A	Mort, et.al, (2003)
Visionary Leadership	Social entrepreneurship	Theoretical	N/A	Thompson, et. al., (2000)
Ambidextrous Leadership	Social capital Entrepreneurial orientation	Quantitative method	Companies in Vietnam	Luu (2017)
Ethical Leadership	Social entrepreneurship	Theoretical	N/A	Harris, et.al., (2009)
Value-Based Leadership	Social justice Support for community	Theoretical	N/A	Prilleltensky (2000)

Entrepreneurial Leadership

The concept of entrepreneurial leadership is the integration of two important constructs which are entrepreneurship and leadership. As entrepreneurs are the ones who seek opportunities and new ideas in order to establish a new venture (Ward, 2004) and leaders are the ones who influence and motivate their followers (Hersey, Blanchard, and Natemeyer, 1979), entrepreneurial leadership refers to special type of leaders who focus on innovativeness and new opportunities to bring an idea to life successfully.

Social entrepreneurship, which is a special type of entrepreneurship, is made of entrepreneurial activities which consists of innovation and new opportunities to create social value. Based upon this, entrepreneurial leadership is understood to be highly correlated to social entrepreneurship. Social entrepreneurship is in need of leaders demonstrating entrepreneurial behaviors. In the literature also there are some studies emphasizing on this relationship (i.e. Prabhu, 1999; Santora, et.al., 1999; Smith, Besharov, Wessels and Chertok, 2012) Because of this:

Proposition 1: Entrepreneurial leadership is positively related to social entrepreneurship.

Innovative Leadership

Nowadays, it is impossible for organizations to survive without innovation. Innovation can be defined as a set of effective reactions from organizations, organizational members or their alliances to technological demands in order to adapt and survive in the business eco-system (Archibugi and Iammarino, 2002; Baregheh, Rowley and Sambrook, 2009). Innovative leaders are the ones, who foster innovation in their organizations and innovative work behavior among employees or followers.

As social entrepreneurship puts innovation in the center of activities, innovative leadership is expected to promote social entrepreneurship. Furthermore, since social entrepreneurship is an integration of innovation and social change (Perrini and Vurro, 2006), innovative leadership is among the most suitable leadership styles for social entrepreneurship. Hence;

Proposition 2: Innovative leadership is positively related to social entrepreneurship.

Responsible Leadership

Responsible leadership refers to the degree to which a leader acts responsibly to all of the stakeholders of the organization in order to benefit all the parties (Voegtlin, Patzer and Scherer, 2012). Responsible leaders play key roles in transforming their organizations into sustainable and socially responsible organizations (Székely and Knirsch, 2005).

The essence of social entrepreneurship that is to serve society by bringing solutions to social problems fits into responsible leadership. Responsible leaders are expected to be successful at social change and social entrepreneurship. For this reason;

Proposition 3: Responsible leadership is positively related to social entrepreneurship.

Value-Based Leadership

Values are vital for organizations. Leaders who possess encouraging values are more effective at motivating followers (Graber and Kilpatrick, 2008). Being related to responsible leadership and ethical leadership (Pless, 2007), value-based leadership is a type of leadership in which leaders share their values with followers (Graber and Kilpatrick, 2008), promote shared values with them (Prilleltensky, 2000) and emphasize on morality (Garg and Krishnan, 2003).

Value-based leadership is a special style of leadership in which leaders also share social values with their followers and try to fulfill socially meaningful missions. Being honest, ethical, inclusive and participative (Garg and Krishnan, 2003), value-based leadership can foster social entrepreneurship as the two constructs intersect at these points. So;

Proposition 4: Value-based leadership is positively related to social entrepreneurship.

Visionary Leadership

Being visionary is among the basic elements of entrepreneurship. A successful social entrepreneur easily anticipates the social need and its nature by being visionary (Thompson, et. al., 2000). Furthermore, social entrepreneurship mostly relies on visionary leadership, as this type of leadership relies on an inspiring and encouraging vision (Kirkpatrick and Locke, 1996). Visionary leaders are capable of indicating the conditions and best solutions to social problems, with the intention of creating social value. The requirement of being visionary for a social entrepreneur can be fulfilled by visionary leadership. Thus;

Proposition 5: Visionary leadership is positively related to social entrepreneurship.

Strategic Leadership

Based on the 'upper echelons theory' of Hambrick and Mason (1984), the concept of strategic leadership differs from other leadership styles by emphasizing on the top management leadership due to the nature of strategy formulation. Strategic leadership refers to formulate and implement strategies suitable with the organizational mission and vision (Davies and Davies, 2004). Either in private or in public and not-for-profit sectors, all types of social entrepreneurship require the most effective strategies for both social welfare and survival. With the help of strategic leadership, social ventures will survive and social mission will be fulfilled and social value will be created as well. Apparently;

Proposition 6: Strategic leadership is positively related to social entrepreneurship.

Transformational Leadership

Transformational leadership is a special type of leadership in which the leader develops the followers and behaves like a mentor to respond their needs and wants (Bass, 1990) and also redefines the problems and supports the followers to be innovative and creative (Bass and Avolio, 1994).

At this point, transformational leadership is among the most fruitful leadership styles for social entrepreneurship. This is because, the nature of transformational leadership supports the innovation and creativeness side of social entrepreneurship while bringing new solutions to existing problems at the same time. For this reason;

Proposition 7: Transformational leadership is positively related to social entrepreneurship.

Authentic Leadership

Having roots in positive psychology (Fredrickson, 2001; Sheldon and King, 2001), authentic leadership is a type of leadership in which the leader is supportive for followers to develop themselves while he/she behaves appropriately (Luthans and Avolio, 2003). Authentic leaders affect their followers by being moral, enthusiastic and optimistic and cheerful (Avolio, Gardner, Walumbwa, Luthans and May, 2004). According to Jensen and Luthans (2006), today's business ventures and entrepreneurship needs authentic leadership more than ever. It is because today's employees are affected from authencity of their entreprenuers or leaders when they are more transparent and eager to empower them.

Authentic leadership is among the most relevant leadership styles with social entrepreneurship. The reason is that, social entrepreneurship will be more effective for creating solutions under the conditions which authentic leaders set for the employees. It is evident that when the leader is trustable and participative, the followers are more eager to perform better in innovation and social entrepreneurship. Thus;

Proposition 8: Authentic leadership is positively related to social entrepreneurship.

Cultural Leadership

Cultures are strengthened by their members' exchanging shared values, myths and rituals (Trice and Beyer, 1984). Cultural leadership refers to the degree to which leaders play key role in innovating the culture. While innovating the culture, the personal characteristics, behaviors, use of cultural forms and traditions of cultural leaders are important. Cultural leaders innovates culture and create impression of success and competence (Beyer and Browning, 1999; Trice and Beyer, 1984). Since this type of innovation either social or cultural is in the center of social entrepreneurship. With the help of effective cultural leadership, social entrepreneurship will be more strengthened. Hence;

Proposition 9: Cultural leadership is positively related to social entrepreneurship.

Paternalistic Leadership

A paternalistic employer, feel obliged to care for his/her employees, to support them and create solutions for their problems in order to prevent a wider negative impacts for the society (Casson and Della Giusta, 2007). Demonstrating paternalistic leadership is a very important characteristic of an entrepreneur (Dedahanov, Lee, Rhee and Yoon, 2016) especially a social entrepreneur since paternalistic leaders are sensitive about both the problems of followers and the other shareholders of their organizations. But, from another point of view paternalistic leadership possesses authority to constrain followers' activities, which may cause negative perceptions for them. But with a balance of integrity and helpfulness, paternalistic leaders are successful at affecting their followers in order to do a favor to other parties (Farh and Cheng, 2000). Except for the authoritarian style, paternalistic leadership provides a basis for social entrepreneurship because of its benevolent dimension. So,

Proposition 10: Paternalistic leadership is positively related to social entrepreneurship.

Servant Leadership

Being introduced to the literature by Greenleaf (1977), the concept of servant leadership consists of two opposite words of 'servant' and 'leader'. This is because servant leadership requires service first, and then leadership (Greenleaf 1977). An important proportion of studies on servant leadership concentrate on the traits of servant leadership (Joseph and Winston, 2005). Generally they indicate that servant leaders are; initiative, persuasive, and aware about themselves and their environment and they listen to the others as well (Greenleaf 1977; Spears, 1995). Due to the characteristics of servant leaders, it can be noted that servant leadership promotes social entrepreneurship. Because serving to others and especially to society comes first in social entrepreneurship. That is why;

Proposition 11: Servant leadership is positively related to social entrepreneurship.

Ambidextrous Leadership

The concept of ambidexterity refers to capability of using both left and right hands with equal comfort (Rosing, Frese and Bausch, 2011). Ambidextrous leadership can be defined as leadership style which consists of both leadership behaviors on exploration and exploitation and the ability of moving from one another due to the conditions (Benner and Tushman, 2003; Zacher and Rosing, 2015). Having been imported from organization science to leadership studies (Rosing, et al., 2011), ambidextrous leadership is said to be positively correlated with both innovation and entrepreneurship (Tuan Luu, 2017). Creating an innovative climate and easing employees' readiness to change and generate a social impact, ambidextrous leadership is also among the leadership styles which promote social entrepreneurship. For this reason;

Proposition 12: Ambidextrous leadership is positively related to social entrepreneurship.

Ethical Leadership

Ethical leaders are the ones, who set ethical standards, discuss about ethics with the followers and be fair and make rational decisions (Brown and Treviño, 2006). Ethical leaders exert effective leadership and followed by others because of their ethical traits which followers perceive important. Ethical leaders are sensitive about ethical problems and they take initiative to solve them. From this point of view, they are important factors for a society to be transformed into a society with high ethical standards. This mission of ethical leaders makes themselves a vital ingredient for social entrepreneurship. Ethical leaders are expected to affect social entrepreneurship positively (Harris, Sapienza and Bowie, 2009). Hence it is stated below;

Proposition 13: Ethical leadership is positively related to social entrepreneurship.

Negative Leadership Styles

Not all the leadership styles are constructive for the followers and the organization. In the literature a growing number of studies concentrate on the concept of 'dark side of leadership' (i.e. Clements and Washbush, 1999; Conger, 1990; Mathieu, Neumann, Hare and Babiak, 2014). Based on the classification made in the research of Schilling (2009), among the contents of negative leadership behaviors there exist;

'insincere leadership', 'despotic leadership', 'exploitative leadership', 'restrictive leadership', 'failed leadership', 'avoiding leadership' and 'laissez-faire leadership'. Naturally all these negative leadership styles have negative effects on followers in terms of their behaviors and performances. Furthermore, negative leadership neither eases social entrepreneurship nor reinforces it. It may be destructive for social entrepreneurship. For this reason;

Proposition 14: Negative leadership styles are negatively related to social entrepreneurship.

FUTURE RESEARCH DIRECTIONS

Despite of the fact that there are a growing number of researches on the social entrepreneurship, because of the value of the subject, still there is a need for further studies. These may be concentrated on different dimensions of social entrepreneurship which need to be developed. For example, the research questions of this study may be tested by qualitative, quantitative or mixed research. Successful social entrepreneurs

Figure 3. Conceptual framework of the relationship between leadership styles and social entrepreneurship

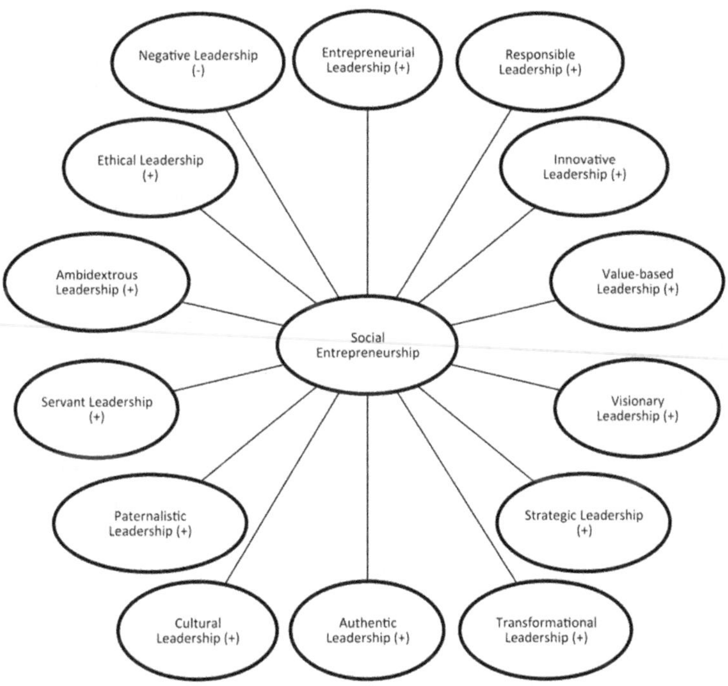

from different countries or regions may be examined on the basis of their leadership styles in order to figure out which leadership style is more correlated with social entrepreneurship based on the conceptual framework of this chapter.

Moreover, the question of which leadership styles are more effective on creating social value may also be examined by taking also innovation into account. Furthermore, the circumstances which enable successful social entrepreneurships in various sectors or countries should be figured out. Also, the followers' affection by different leadership to perform better for social ventures should be analyzed. Consequently, as the social entrepreneurship phenomenon is getting more and more importance each day, the factors which affect social entrepreneurship, social value creation and social innovation should be carefully discussed in the future studies.

CONCLUSION

Since entrepreneurship begins with innovative ideas and value creation, a special kind of entrepreneurship that is social entrepreneurship puts social value creation and social innovation in the center. By achieving social innovation and coming up with solutions to the societal problems, social entrepreneurs are the ones who create value in their society. In our society, social problems are getting more important more than ever. So, there are many benefits of concentrating on the concept of social entrepreneurship and discover the factors easing social entrepreneurship. A detailed literature scan showed that there have being growing numbers of published articles and books on social entrepreneurship, but still there is a need for improvement. As our chapter focused on the relationship between leadership styles and social entrepreneurship, it was noted that there were not sufficient studies related to these constructs figuring out this relationship. So the intent of this chapter which was to help expand the existing literature, by adding a theoretical study has been achieved. This is because, a future research direction from Short et.al (2009) about discussing the leadership styles' effects on social entrepreneurship were discussed theoretically based on literature scanning.

Throughout this chapter, in accordance with the related literature, entrepreneurial leadership, responsible leadership, innovative leadership, value-based leadership, visionary leadership, strategic leadership, transformational leadership, authentic leadership, cultural leadership, paternalistic leadership, servant leadership, ambidextrous leadership, ethical leadership and negative leadership styles were related to social entrepreneurship. Entrepreneurial leadership was proposed to promote social entrepreneurship because this leadership style is characterized by value creation and innovativeness. Responsible leadership is about taking care of the entire stakeholders, which in turn proves the positive correlation with social entrepreneurship. Innovative leadership is also positively correlated with social entrepreneurship because of the innovative characteristic of the leaders. Moreover, value-based leadership is beneficial for social entrepreneurship since it is based on the leaders' values, communicating these values with the followers. Visionary leadership is also important for social entrepreneurs because these leaders have the clear vision to serve their society. Likewise, strategic leadership is vital for social entrepreneurs to formulate and implement strategies. Transformational leaders are the ones who support their followers to change and develop themselves while being innovative and creative which are crucial for social entrepreneurship. Furthermore, leaders' being ethical, transparent, supportive, empowering and caring are fruitful for social entrepreneurship that all these characteristics are available in authentic, cultural, ethical, servant and paternalistic leadership. On the contrary, based upon the negative results for social entrepreneurship, negative leadership style which are insincere leadership, despotic leadership, exploitative leadership, restrictive leadership, failed leadership, avoiding leadership and laissez-faire leadership styles are negatively correlated to social entrepreneurship. Consequently, leaders should be more participative, supportive, empowering, visionary, democratic, responsible, ethical, trustable to foster social entrepreneurship.

REFERENCES

Alvord, S. B., Brown, L. D., Christine, W. D., & Letts, C. (2002). *Social entrepreneurship and social transformation: an exploratory study.* Hauser Center for Nonprofit Organizations and The Kennedy School of Government, Harvard University, Working Paper 15, 1-26.

Archibugi, D., & Iammarino, S. (2002). The globalization of technological innovation: Definition and evidence. *Review of International Political Economy*, *9*(1), 98–122. doi:10.1080/09692290110101126

Austin, J., Stevenson, H., & Wei-Skillern, J. (2006). Social and commercial entrepreneurship: Same, different, or both? *Entrepreneurship Theory and Practice*, *30*(1), 1–22. doi:10.1111/j.1540-6520.2006.00107.x

Avolio, B. J., Gardner, W. L., Walumbwa, F. O., Luthans, F., & May, D. R. (2004). Unlocking the mask: A look at the process by which authentic leaders impact follower attitudes and behaviors. *The Leadership Quarterly*, *15*(6), 801–823. doi:10.1016/j.leaqua.2004.09.003

Bareghe h, A., Rowley, J., & Sambrook, S. (2009). Towards a multidisciplinary definition of innovation. *Management Decision, 47*(8), 1323–1339. doi:10.1108/00251740910984578

Barker, R. A. (2001). The nature of leadership. *Human Relations, 54*(4), 469–494. doi:10.1177/0018726701544004

Bass, B. M. (1990). From transactional to transformational leadership: Learning to share the vision. *Organizational Dynamics, 18*(3), 19–31. doi:10.1016/0090-2616(90)90061-S

Bass, B. M., & Avolio, B. J. (Eds.). (1994). *Improving organizational effectiveness through transformational leadership*. Sage Publications.

Baumol, W. J. (1996). Entrepreneurship: Productive, unproductive, and destructive. *Journal of Business Venturing, 11*(1), 3–22. doi:10.1016/0883-9026(94)00014-X

Benner, M. J., & Tushman, M. L. (2003). Exploitation, exploration, and process management: The productivity dilemma revisited. *Academy of Management Review, 28*(2), 238–256. doi:10.5465/amr.2003.9416096

Beugelsdijk, S., & Noorderhaven, N. (2005). Personality characteristics of self-employed; an empirical study. *Small Business Economics, 24*(2), 159–167. doi:10.100711187-003-3806-3

Beyer, J. M., & Browning, L. D. (1999). Transforming an industry in crisis: Charisma, routinization, and supportive cultural leadership. *The Leadership Quarterly, 10*(3), 483–520. doi:10.1016/S1048-9843(99)00026-0

Bhutiani, D., Flicker, K., Nair, P., & Groen, A. (2012). Is Social Entrepreneurship transformational Leadership in action? In J. Kickul & S. Bacq (Eds.), *Patterns in Social Entrepreneurship Research*. Edward Elgar Publishing.

Boschee, J. (1998). *Merging mission and money: A board member's guide to social entrepreneurship*. The National Center for Nonprofit Boards.

Bowen, H. R. (1953). *Social responsibilities of the businessman*. New York: Harper & Row.

Brown, M. E., & Treviño, L. K. (2006). Ethical leadership: A review and future directions. *The Leadership Quarterly, 17*(6), 595–616. doi:10.1016/j.leaqua.2006.10.004

Campbell, S. (1998). Social entrepreneurship: How to develop new social-purpose business ventures. *Health Care Strategic Management, 16*(5), 17–18. PMID:10179045

Canadian Centre for Social Entrepreneurship (CCSE). (2001). *Social entrepreneurship discussion paper*, (1). Author.

Carland, J. W., Hoy, F., Boulton, W. R., & Carland, J. A. C. (1984). Differentiating entrepreneurs from small business owners: A conceptualization. *Academy of Management Review*, *9*(2), 354–359. doi:10.5465/amr.1984.4277721

Casson, M., & Giusta, M. D. (2007). Entrepreneurship and social capital: Analysing the impact of social networks on entrepreneurial activity from a rational action perspective. *International Small Business Journal*, *25*(3), 220–244. doi:10.1177/0266242607076524

Clements, C., & Washbush, J. B. (1999). The two faces of leadership: Considering the dark side of leader-follower dynamics. *Journal of Workplace Learning*, *11*(5), 170–176. doi:10.1108/13665629910279509

Cogliser, C. C., & Brigham, K. H. (2004). The intersection of leadership and entrepreneurship: Mutual lessons to be learned. *The Leadership Quarterly*, *15*(6), 771–799. doi:10.1016/j.leaqua.2004.09.004

Conger, J. A. (1990). The dark side of leadership. *Organizational Dynamics*, *19*(2), 44–55. doi:10.1016/0090-2616(90)90070-6

Cunningham, J. B., & Lischeron, J. (1993). Entrepreneurship: Some popular impressions and new information. *Journal of Small Business and Entrepreneurship*, *10*(2), 46–50. doi:10.1080/21692610.1993.11683545

Dacin, P. A., Dacin, M. T., & Matear, M. (2010). Social entrepreneurship: Why we don't need a new theory and how we move forward from here. *The Academy of Management Perspectives*, *24*(3), 37–57.

Davies, B. J., & Davies, B. (2004). Strategic leadership. *School Leadership & Management*, *24*(1), 29–38. doi:10.1080/1363243042000172804

Dedahanov, A. T., Lee, D. H., Rhee, J., & Yoon, J. (2016). Entrepreneur's paternalistic leadership style and creativity: The mediating role of employee voice. *Management Decision*, *54*(9), 2310–2324. doi:10.1108/MD-11-2015-0537

Dym, B., & Hutson, H. (2005). *Leadership in nonprofit organizations: Lessons from the third sector*. Thousand Oaks, CA: Sage Publications.

Farh, J. L., & Cheng, B. S. (2000). A cultural analysis of paternalistic leadership in Chinese organizations. In *Management and organizations in the Chinese context* (pp. 84–127). London: Palgrave Macmillan. doi:10.1057/9780230511590_5

Fredrickson, B. L. (2001). The role of positive emotions in positive psychology: The broaden-and-build theory of positive emotions. *The American Psychologist*, *56*(3), 218–226. doi:10.1037/0003-066X.56.3.218 PMID:11315248

Garg, G., & Krishnan, V. R. (2003). Transformational leadership and organizational structure: The role of value-based leadership. *Transformational leadership: Value-based management for Indian organizations*, 82-100.

Gibb, A. A. (2006). *Entrepreneurship. Unique Solutions for Unique Environments. Can this be achieved with the existing paradigm?* Paper as background to Plenary presentation ICSB World Conference, Melbourne, Australia.

Graber, D. R., & Kilpatrick, A. O. (2008). Establishing values-based leadership and value systems in healthcare organizations. *Journal of Health and Human Services Administration*, 179–197. PMID:18998522

Greenleaf, R. K. (1977). *Servant leadership: A journey into the natüre of legitimate power and greatness*. New York: Paulist Press.

Gupta, V., MacMillan, I. C., & Surie, G. (2004). Entrepreneurial leadership: Developing and measuring a cross-cultural construct. *Journal of Business Venturing*, *19*(2), 241–260. doi:10.1016/S0883-9026(03)00040-5

Hambrick, D. C., & Mason, P. A. (1984). Upper echelons: The organization as a reflection of its top managers. *Academy of Management Review*, *9*(2), 193–206. doi:10.5465/amr.1984.4277628

Harris, J. D., Sapienza, H. J., & Bowie, N. E. (2009). Ethics and entrepreneurship. *Journal of Business Venturing*, *24*(5), 407–418. doi:10.1016/j.jbusvent.2009.06.001

Harris, J. D., Sapienza, H. J., & Bowie, N. E. (2009). Ethics and entrepreneurship. *Journal of Business Venturing*, *24*(5), 407–418. doi:10.1016/j.jbusvent.2009.06.001

Hersey, P., Blanchard, K. H., & Natemeyer, W. E. (1979). Situational leadership, perception, and the impact of power. *Group & Organization Studies*, *4*(4), 418–428. doi:10.1177/105960117900400404

Jensen, S. M., & Luthans, F. (2006). Entrepreneurs as authentic leaders: Impact on employees' attitudes. *Leadership and Organization Development Journal*, *27*(8), 646–666. doi:10.1108/01437730610709273

Johnson, S. (2000). Literature review on social entrepreneurship. *Canadian Centre for Social Entrepreneurship*, *16*(23), 96–106.

Joseph, E. E., & Winston, B. E. (2005). A correlation of servant leadership, leader trust, and organizational trust. *Leadership and Organization Development Journal*, *26*(1), 6–22. doi:10.1108/01437730510575552

Kessler, A. (2008). Success factors for new businesses in Austria and the Czech Republic. *Entrepreneurship and Regional Development*, *19*(5), 381–403. doi:10.1080/08985620701439959

Kirkpatrick, S. A., & Locke, E. A. (1996). Direct and indirect effects of three core charismatic leadership components on performance and attitudes. *The Journal of Applied Psychology*, *81*(1), 36–51. doi:10.1037/0021-9010.81.1.36

Litzky, B. E., Godshalk, V. M., & Walton-Bongers, C. (2010). Social entrepreneurship and community leadership: A service-learning model for management education. *Journal of Management Education*, *34*(1), 142–162. doi:10.1177/1052562909338038

Luthans, F., & Avolio, B. J. (2003). Authentic leadership development. *Positive organizational scholarship,* 241-258.

Maak, T., & Stoetter, N. (2012). Social entrepreneurs as responsible leaders: 'Fundación Paraguaya' and the case of Martin Burt. *Journal of Business Ethics*, *111*(3), 413–430. doi:10.100710551-012-1417-0

Mair, J., Robinson, J., & Hockerts, K. (2006). Introduction. In J. Mair, J. Robinson, & K. Hockerts (Eds.), *Social entrepreneurship.* New York: Palgrave Macmillan. doi:10.1057/9780230625655_1

Mathieu, C., Neumann, C. S., Hare, R. D., & Babiak, P. (2014). A dark side of leadership: Corporate psychopathy and its influence on employee well-being and job satisfaction. *Personality and Individual Differences*, *59*, 83–88. doi:10.1016/j.paid.2013.11.010

McClelland, D. C. (1961). *Achieving society*. New York: The Free Press. doi:10.1037/14359-000

Mort, S. G., Weerawardena, J., & Carnegie, K. (2003). Social entrepreneurship: Towards conceptualisation. *International Journal of Nonprofit and Voluntary Sector Marketing*, *8*(1), 76–88. doi:10.1002/nvsm.202

Muralidharan, E., & Pathak, S. (2018). Sustainability, transformational leadership, and social entrepreneurship. *Sustainability*, *10*(2), 567. doi:10.3390u10020567

Nicholls, A. (2008). *Social entrepreneurship: New models of sustainable social change*. Oxford, UK: Oxford University Press.

Perrini, F., & Vurro, C. (2006). Social entrepreneurship: Innovation and social change across theory and practice. In J. Mair, J. Robinson, & K. Hockerts (Eds.), *Social entrepreneurship* (pp. 57–85). New York: Palgrave Macmillan. doi:10.1057/9780230625655_5

Pless, N. M. (2007). Understanding responsible leadership: Role identity and motivational drivers. *Journal of Business Ethics*, *74*(4), 437–456. doi:10.100710551-007-9518-x

Prabhu, G. N. (1999). Social entrepreneurial leadership. *Career Development International*, *4*(3), 140–145. doi:10.1108/13620439910262796

Prilleltensky, I. (2000). Value-based leadership in organizations: Balancing values, interests, and power among citizens, workers, and leaders. *Ethics & Behavior*, *10*(2), 139–158. doi:10.1207/S15327019EB1002_03

Pruzan, P., & Miller, W. C. (2006). Spirituality as the basis of responsible leaders and responsible companies. In *Responsible leadership* (pp. 88–112). Routledge.

Roper, J., & Cheney, G. (2005). Leadership, learning and human resources management: The meanings of social entrepreneurship today. *Corporate Governance*, *5*(3), 95–104. doi:10.1108/14720700510604733

Rosing, K., Frese, M., & Bausch, A. (2011). Explaining the heterogeneity of the leadership-innovation relationship: Ambidextrous leadership. *The Leadership Quarterly*, *22*(5), 956–974. doi:10.1016/j.leaqua.2011.07.014

Santora, J. C., Seaton, W., & Sarros, J. C. (1999). Changing times: Entrepreneurial leadership in a community-based nonprofit organization. *The Journal of Leadership Studies*, *6*(3-4), 101–109. doi:10.1177/107179199900600308

Schilling, J. (2009). From ineffectiveness to destruction: A qualitative study on the meaning of negative leadership. *Leadership*, *5*(1), 102–128. doi:10.1177/1742715008098312

Schumpeter, J. A. (1934). *The theory of economic develeopment*. Cambridge, MA: Harvard University Press.

Sheldon, K. M., & King, L. (2001). Why positive psychology is necessary. *The American Psychologist*, *56*(3), 216–217. doi:10.1037/0003-066X.56.3.216 PMID:11315247

Short, J. C., Moss, T. W., & Lumpkin, G. T. (2009). Research in social entrepreneurship: Past contributions and future opportunities. *Strategic Entrepreneurship Journal*, *3*(2), 161–194. doi:10.1002ej.69

Smith, W. K., Besharov, M. L., Wessels, A. K., & Chertok, M. (2012). A paradoxical leadership model for social entrepreneurs: Challenges, leadership skills, and pedagogical tools for managing social and commercial demands. *Academy of Management Learning & Education, 11*(3), 463–478. doi:10.5465/amle.2011.0021

Spears, L. C. (1995). *Reflections on leadership: How Robert K. Greenleaf's theory of servant-leadership influenced today's top management thinkers*. John Wiley & Sons.

Székely, F., & Knirsch, M. (2005). Responsible leadership and corporate social responsibility: Metrics for sustainable performance. *European Management Journal, 23*(6), 628–647. doi:10.1016/j.emj.2005.10.009

Thompson, J., Alvy, G., & Lees, A. (2000). Social entrepreneurship–a new look at the people and the potential. *Management Decision, 38*(5), 328–338. doi:10.1108/00251740010340517

Thompson, J. L. (2008). Social enterprise and social entrepreneurship: Where have we reached? A summary of issues and discussion points. *Social Enterprise Journal, 4*(2), 149–161. doi:10.1108/17508610810902039

Timmons, J. A. (1978). Characteristics and role demands of entrepreneurship. *American Journal of Small Business, 3*(1), 5–17. doi:10.1177/104225877800300102

Trice, H. M., & Beyer, J. M. (1984). Studying organizational cultures through rites and ceremonials. *Academy of Management Review, 9*(4), 653–669. doi:10.5465/amr.1984.4277391

Tuan Luu, T. (2017). Ambidextrous leadership, entrepreneurial orientation, and operational performance: Organizational social capital as a moderator. *Leadership and Organization Development Journal, 38*(2), 229–253. doi:10.1108/LODJ-09-2015-0191

Tuan Luu, T. (2017). Ambidextrous leadership, entrepreneurial orientation, and operational performance: Organizational social capital as a moderator. *Leadership and Organization Development Journal, 38*(2), 229–253. doi:10.1108/LODJ-09-2015-0191

Van Dierendonck, D. (2011). Servant leadership: A review and synthesis. *Journal of Management, 37*(4), 1228–1261. doi:10.1177/0149206310380462

Voegtlin, C., Patzer, M., & Scherer, A. G. (2012). Responsible leadership in global business: A new approach to leadership and its multi-level outcomes. *Journal of Business Ethics, 105*(1), 1–16. doi:10.100710551-011-0952-4

Waldman, D. A., Siegel, D. S., & Javidan, M. (2006). Components of CEO transformational leadership and corporate social responsibility. *Journal of Management Studies, 43*(8), 1703–1725. doi:10.1111/j.1467-6486.2006.00642.x

Ward, T. B. (2004). Cognition, creativity, and entrepreneurship. *Journal of Business Venturing, 19*(2), 173–188. doi:10.1016/S0883-9026(03)00005-3

Weerawardena, J., & Mort, G. S. (2001). Learning, innovation and competitive advantage in not-for-profit aged care marketing: A conceptual model and research propositions. *Journal of Nonprofit & Public Sector Marketing, 9*(3), 53–73. doi:10.1300/J054v09n03_04

Weerawardena, J., & Mort, G. S. (2006). Investigating social entrepreneurship: A multidimensional model. *Journal of World Business, 41*(1), 21–35. doi:10.1016/j.jwb.2005.09.001

Winston, B. E., & Patterson, K. (2006). An integrative definition of leadership. *International Journal of Leadership Studies, 1*(2), 6-66.

Zacher, H., & Rosing, K. (2015). Ambidextrous leadership and team innovation. *Leadership and Organization Development Journal, 36*(1), 54–68. doi:10.1108/LODJ-11-2012-0141

ADDITIONAL READING

Argyris, C., & Argyris, C. (1976). *Increasing leadership effectiveness*. New York: Wiley.

Certo, S. T., & Miller, T. (2008). Social entrepreneurship: Key issues and concepts. *Business Horizons, 51*(4), 267–271. doi:10.1016/j.bushor.2008.02.009

Chen, M. H. (2007). Entrepreneurial leadership and new ventures: Creativity in entrepreneurial teams. *Creativity and Innovation Management, 16*(3), 239–249. doi:10.1111/j.1467-8691.2007.00439.x

Dacin, M. T., Dacin, P. A., & Tracey, P. (2011). Social entrepreneurship: A critique and future directions. *Organization Science, 22*(5), 1203–1213. doi:10.1287/orsc.1100.0620

Defourny, J., & Nyssens, M. (2010). Conceptions of social enterprise and social entrepreneurship in Europe and the United States: Convergences and divergences. *Journal of Social Entrepreneurship, 1*(1), 32–53. doi:10.1080/19420670903442053

Dulewicz, V., & Higgs, M. (2005). Assessing leadership styles and organisational context. *journal of Managerial Psychology, 20*(2), 105-123.

Egri, C. P., & Herman, S. (2000). Leadership in the North American environmental sector: Values, leadership styles, and contexts of environmental leaders and their organizations. *Academy of Management Journal, 43*(4), 571–604.

Hunt, J. G. (1991). *Leadership: A new synthesis.* Sage Publications, Inc.

London, M. (2010). Understanding social advocacy: An integrative model of motivation, strategy, and persistence in support of corporate social responsibility and social entrepreneurship. *Journal of Management Development, 29*(3), 224–245. doi:10.1108/02621711011025768

Mair, J., & Marti, I. (2006). Social entrepreneurship research: A source of explanation, prediction, and delight. *Journal of World Business, 41*(1), 36–44. doi:10.1016/j.jwb.2005.09.002

Peredo, A. M., & McLean, M. (2006). Social entrepreneurship: A critical review of the concept. *Journal of World Business, 41*(1), 56–65. doi:10.1016/j.jwb.2005.10.007

Phills, J. A., Deiglmeier, K., & Miller, D. T. (2008). Rediscovering social innovation. *Stanford Social Innovation Review, 6*(4), 34–43.

Santos, F. M. (2012). A positive theory of social entrepreneurship. *Journal of Business Ethics, 111*(3), 335–351. doi:10.100710551-012-1413-4

Shapero, A., & Sokol, L. (1982). The social dimensions of entrepreneurship. Encyclopedia of entrepreneurship, 72-90.

Thompson, J. L. (2002). The world of the social entrepreneur. *International Journal of Public Sector Management, 15*(5), 412–431. doi:10.1108/09513550210435746

Todorovic, Z. W., & Schlosser, F. K. (2007). An entrepreneur and a leader!: A framework conceptualizing the influence of leadership style on a firm's entrepreneurial orientation—Performance relationship. *Journal of Small Business and Entrepreneurship, 20*(3), 289–307. doi:10.1080/08276331.2007.10593401

Vroom, V. H., & Jago, A. G. (1988). *The new leadership: Managing participation in organizations.* Prentice-Hall, Inc.

Zahra, S. A., Gedajlovic, E., Neubaum, D. O., & Shulman, J. M. (2009). A typology of social entrepreneurs: Motives, search processes and ethical challenges. *Journal of Business Venturing, 24*(5), 519–532. doi:10.1016/j.jbusvent.2008.04.007

KEY TERMS AND DEFINITIONS

Entrepreneurial Leadership: A leadership style in which leaders are concentrated on innovation and new ideas in order to establish a venture.

Entrepreneurship: Bringing a creative idea to life by organizing and managing factors of production.

Leadership: The process of affecting and motivating followers to reach organizational goals and objectives.

Leadership Styles: Different ways of leading followers in different circumstances.

Responsible Leadership: Leading in a way that acting responsibly to followers, their organization, society and environment.

Social Entrepreneurship: Coming up with permanent solutions to social problems by using entrepreneurial tools and techniques.

Transformational Leadership: A special leadership style in which leaders encourage followers to change and develop themselves.

Chapter 2
Gaining Strategic Advantage through Social and Technological Innovation:
Evidence from European Countries

Çağlar Doğru
https://orcid.org/0000-0002-4215-8979
Ufuk University, Turkey

ABSTRACT

In this chapter, innovation, one of the most important factors for contemporary organizations and even countries, is analyzed in depth. The concept of innovation is divided into two major fields, which are social innovation and technological innovation. The relationship between these two innovation types and gaining strategic advantage are figured out in a macro perspective. To achieve this, the strongest innovators in Europe were selected. Among them are Switzerland, Sweden, Denmark, Finland, Netherlands, Germany, and Norway. To compare the countries, Social Innovation Index and European Innovation Scoreboard were employed to get secondary data. It was observed that by investing more on social and technological innovation, these European countries perform better and gain strategic advantage, which helps them build sustainable future.

DOI: 10.4018/978-1-7998-1108-4.ch002

INTRODUCTION

Today's business world has many common aspects with the wild nature. The members of both environments face similar factors while trying to survive. Gaining competitive advantage and adaptation capability to new conditions in the long run are the key factors for companies in the business life and for all the creatures on earth. It is explained by a single word of 'survival', which has been imported from physical sciences to social sciences. Survival of organizations is vital in this knowledge era where constant change is ready (Agarwal, Sarkar and Echambadi, 2002; Christensen, Suárez and Utterback, 1998; Klepper and Simons, 2000). Due to the fast changes and developments in technology and other factors in the environment, both organizations and even countries seek ways to gain competitive advantage. According to resource-based view (Wernerfeldt, 1984), firms control their resources and capabilities so as to survive. Firms can possess competitive advantage over other firms by their tangible and intangible assets (Barney, 1986; Rumelt, 1984). These may vary from human resources, managerial abilities (Isik, Arditi, Dikmen and Birgonul, 2009) and financial resources to effective marketing capabilities, building customer royalty or even effective service recovery strategies (Eren, 2018).

Although there are so many ways to gain competitive advantage and to survive for organizations, due to modern business world circumstances, innovation is accepted to be more effective than many of them (Buddelmeyer, Jensen and Webster, 2009; Cefis and Marsili, 2006; Dervitsiotis, 2010). All the organizations and even countries invest in innovation more than ever. In today's globalized world, not only technological innovation but also social innovation ease organizations to be more effective. Although the importance of innovation for an organization is well understood in the literature (i.e. Baumol, 2002; Schumpeter, 1942), there is still lack of both empirical and theoretical researches. Besides, many of the existing papers concentrate on technological innovation of companies. This chapter is bringing social innovation and technological innovation together to gain strategic advantage for firms and their home countries in turn. After figuring out the theoretical basis of strategic advantage and social/technological innovation, secondary data analysis of innovation taking place in selected European countries will be analyzed.

BACKGROUND

The Concept of Strategic Advantage

Before explaining the concept of strategic advantage, a related concept of competitive advantage should be clearly identified. Afterwards, in order to make distinction between the two closely related concepts, examples will be given.

First of all, competitive advantage is the internal resources, values and abilities of firms or nations which are creating advantages over firms or nations (Peteraf, 1993; Porter, 1985; Porter, 2011). These are the advantages that mostly rival institutions do not have and will not have in the short term. The competitive advantage gains importance when there is a competition in the sector or market (Foss, 1999). The organizations having competitive advantage over rivals have the superiority based on their resources (Oliver, 1997). These resources are inimitable easily (Lubit, 2001; Wright, Mcmahan and Mcwilliams, 1994) and organizations reach their goals and objectives effectively based on the resource-based view (Barney, 2001). For example Japan has a competitive advantage of achieving cooperation and specialization better than western countries based on self-managing work teams and long-term supplier relations (Hill, 1995).

The concept of strategic advantage has been chosen over the generally accepted concept of competitive advantage, because as Bettencourt, Lusch and Vargo (2014) state, the term of 'strategic advantage' transform the understanding of outplaying the rival elements in the environment to a more cooperative way. This is achieved by organizations' developing strategic capabilities to adapt and survive in the constantly changing environments. Gaining strategic advantage does not necessiate, having only capabilities which rivals do not possess. Strategic advantage gains importance even there is not rivalry in the market. This is the difference between strategic advantage and competitive advantage. But it doesn't mean that strategic advantage is not for competitive markets at all. Strategic advantage may be any advantage that is useful for an organization in the long run and beneficial to implement strategic plans which are made for reaching strategic goals and objectives whether there is a competition or not. Strategic advantage is apperant when any of the firms' resources are valuable and have contribution fort he long term success of the organization in the market (Barney, 1991). This is the case for the norwegian salmon farming industry in which companies use differantiation on the basis of 'country of origin' and gain strategic advantage in sustainable norwegian salmon farming (Toft, 2018).

MAIN FOCUS OF THE CHAPTER

Social Innovation

Social innovation broadly refers to the degree to which humanitarian solutions are found to critical social problems (Fairweather, 1967). In other words, social innovation can be defined as innovative activities and solutions which have the aim of bringing solutions to social issues (Mulgan, 2006). In the literature most of the studies note that social innovation is achieved mostly by social enterprises (i.e. Defourny and Nyssens, 2010; Harrisson, Klein and Browne, 2011; Weber, 2012). On the contrary, social innovation may be practiced by commercial ventures. There are also some studies on this subject (i.e. Herrera, 2015; Scholl, 2013).

Although social innovation can be achieved by both social and commercial ventures, social innovation should be distinguished from business innovation. In social innovation, there is no aim of profit maximization as it occurs for business innovation. The core idea behind social innovation is to create social value rather than economic value as it is in business innovation (Van der Have and Rubalcaba, 2016). Social innovation can be achieved by nonprofit organizations, public institutions and private ventures. The basic requirements for the social innovation process are, social value creativeness and responsiveness to social issues (Mumford, 2002). Without creative and original ideas and solutions which generate social value, social innovation is not possible. According to Mulgan (2006), the most popular social issues waiting for solutions are; ageing populations, chronic diseases, difficulty in transitions to adulthood, immigrant and refugee issues and climate change. Besides all, although there has been a short history of the concept of social innovation, there is an increasing attention from organizations, communities, countries and scholars. There are some examples of studies on social innovation on Table 1.

As it is seen on Table 1., almost every study is theoretical. Publications may vary from whole books, book chapters, articles and review papers. Clearly, it can easily be noted that there is a need in practical studies on social innovation.

Technological Innovation

Unlike social innovation, companies and governments have come a long way with technological innovation since they invest more time and money on technological innovation in order to maximize their efficiency and gain strategic advantage. This is because organizations and states face more complexity, uncertainty and competition than ever based on the developments in communication, information and computer technologies (Hobday, 1998; Newman, 2005). And they see technological innovation

Table 1. Examples of social innovation studies in the literature

Author (s)	Subject	Publication Type	Theoretical / Practical
Mulgan (2006)	Social Innovation	Article	Theoretical
Pol and Ville (2009)	Social Innovation	Article	Theoretical
Fairweather (1967)	Social Innovation	Book	Theoretical
Nicholls & Murdock (2011)	Social Innovation	Book	Theoretical
Mumford (2002)	Social Innovation	Article	Case Study
Moulaert (2013)	Social Innovation	Book	Theoretical
Cajaiba-Santana (2014)	Social Innovation	Article	Theoretical
Taylor (1970)	Social Innovation	Article	Theoretical
Dawson & Daniel (2010)	Social Innovation	Article	Theoretical
Phillips, Lee, Ghobadian, O'Regan & James (2015)	Social Innovation	Article	Review
Marcy & Mumford (2007)	Social Innovation	Article	Causal analysis
Mumford & Moertl (2003)	Social Innovation	Article	Case Study

as a key instrument to catch up with the latest developments and gain strategic advantage in the market.

Technological innovation which is vital for gaining strategic advantage and for survival in turn (Baumol, 2002), can be defined as the generating and using a new process or a new product by the help of computer and information technologies (Pennings and Buitendam, 1987; Teece, 1986). In another comprehensive definition technological innovation is 'a repetitive process compromising development of an invention to end-users through adoption and diffusion' (Garcia and Calantone, 2002).

Technological innovation can be classified as radical innovation or incremental innovation based on the level of newness of innovation (Dewar and Dutton, 1986). In radical innovation, there is no or little know-how on new technologies and inventions whereas in incremental innovation, there is absolutely more knowledge on the new process or product (Green, Gavin and Aiman-Smith, 1995).

Even radical or incremental, there is now a consensus on being innovative is advantageous for every organization, even for countries. Innovativeness helps organizations to perform better and to reach their goals and objectives easier and more effective. Moreover, it also eases countries to increase their gross domestic product and also give them opportunity of having comparative advantages over other countries.

RESEARCH METHODOLOGY

In this chapter in order to fulfill the aim of figuring out evidences from European countries about social and technological innovation on gaining strategic advantage, seven European countries are selected. These seven countries are selected on the basis of the classification of European Commission's 'European Innovation Scoreboard'. According to their classification, there are five countries in the 'innovation leader' profile. These are; Switzerland, Sweden, Denmark, Finland and Netherlands. Since they are among the most important social innovator countries, Norway and Germany are selected from 'strong innovator' segment and added to the research sample, reaching up to seven countries.

Moreover, analysis conducted in this chapter is based on secondary data obtained from various sources. Social innovation data was obtained from the Economist Intelligence Unit report published in 2016. The methodology of that research is as follows: As stated in the Economist Intelligence Unit report (2016), the social innovation index included seven quantitative and ten qualitative scores and divided into four pillars which were 'policy and institutional framework', 'financing', 'entrepreneurship' and 'society'.

Furthermore, to figure out the technological innovation capacity of sample countries, the interactive tool of European innovation scoreboard was used. Again using secondary data, the performance of countries was indicated performance relative to European Union in 2010. Dimension was selected as 'summary innovation index'. To constrain the time interval, a five-year period of 2014-2018 was chosen.

ANALYSIS

Social Innovation

The data obtained from the social innovation index based on the year 2016 is shown on Figure 1.

As it can be seen on Figure 1, Denmark practiced the best performance on social innovation in 2016 with a score of 71.2. After Denmark, Germany performed 66.0, Sweden performed 65.7, Switzerland performed 61.6, Finland and Norway performed both 59.2, and Netherlands performed 57.7. When looking at the chosen seven countries, the performances of countries are close to each other ranging from 57.7 to 71.2. Denmark paying attention to the selected criteria achieved social innovation more than other countries and created more social value. But results show that still other selected countries are the pioneers of the social innovator countries since, there are countries like Bangladesh, Saudi Arabia, Paraguay and Philippines performing

Figure 1. Social innovation score for selected countries
Source: The Economist Intelligence Unit (2016). Old problems, new solutions: Measuring the capacity for social innovation across the world (Social innovation index 2016). London, UK: The Economist Intelligence Unit.

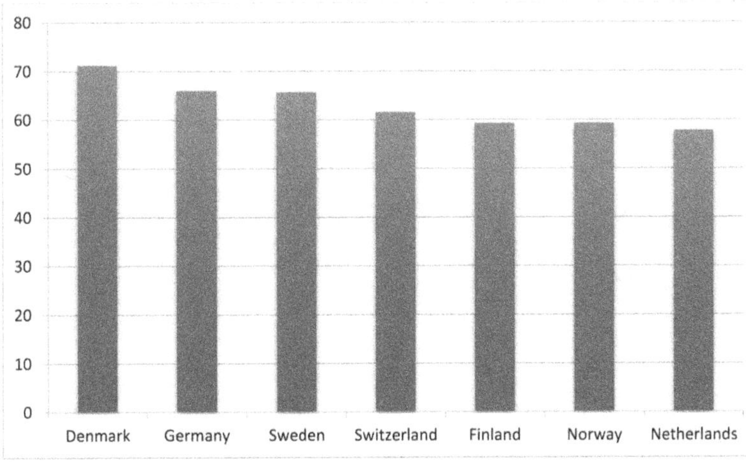

30.6, 30.2, 28.1 and 27.6 respectively. The best performers in the social innovation index were chosen to be United States, United Kingdom and Canada performing 79.4, 77.3 and 75.7 respectively.

According to social innovation index 2016, the seven European countries are among the most influential countries, which invest on social innovation. Also as seen on Figure 1, northern European countries dominate the whole group. They invest more in sustainable resources and social projects that create social value.

Technological Innovation

Switzerland

The innovation scores of Switzerland in the five-year period of 2014-2018 can be seen on Figure 2 below.

Switzerland performed best in the period of 2014-2018 among the other selected countries. Its scores were between 160.65 and 170.57. Switzerland has an increasing steady trend from 2014 to 2018. This means, Switzerland invests more in innovation and the most innovative country in Europe. As it invests more, it also harvests more by gaining strategic advantage over other European and non-European countries.

Figure 2. Innovation scores of Switzerland (2014-2018)
Source: European Innovation Scoreboard 2019, https://www.interactivetool.eu

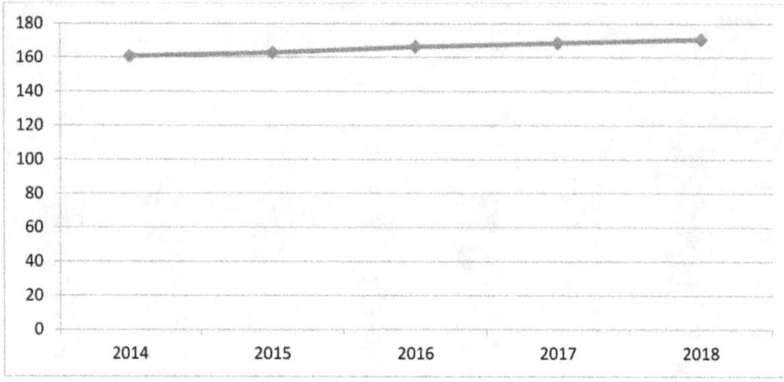

Sweden

The innovation scores of Sweden in the five-year period of 2014-2018 are demonstrated on Figure 3.

After Switzerland, Sweden turned out to be second country in the period of 2014-2018 among the other selected countries. The innovation scores of Sweden occurred between 144.10 and 148.47. Sweden has a steady trend from 2014 to 2018. The highest value occurred in 2017. Sweden is one of the strongest innovator countries both in the world and in Europe. Sweden is gaining strategic advantage with innovative products and innovation processes. As there is a significant interest in social and technological innovation, Sweden and Swedish companies invest in the future.

Figure 3. Innovation scores of Sweden (2014-2018)
Source: European Innovation Scoreboard 2019, https://www.interactivetool.eu

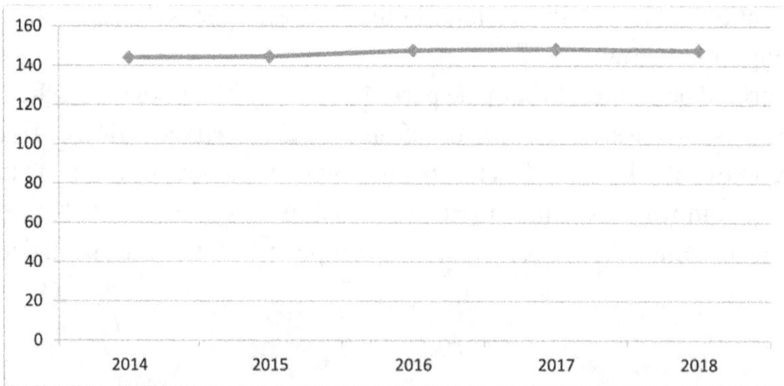

Denmark

The innovation scores of Denmark in the five-year period of 2014-2018 are shown on Figure 4.

Denmark turned out to be third country in the period of 2014-2018 among the other selected countries. The innovation scores of Denmark occurred between 139.97 and 143.20. Like Sweden, Denmark has also a stable trend from 2014 to 2018. It is important to note that Denmark was the leader in the Social Innovation Index 2016 among European countries. It is understood that Denmark also performs well in technological innovation.

Finland

Figure 4. Innovation scores of Denmark (2014-2018)
Source: European Innovation Scoreboard 2019, https://www.interactivetool.eu

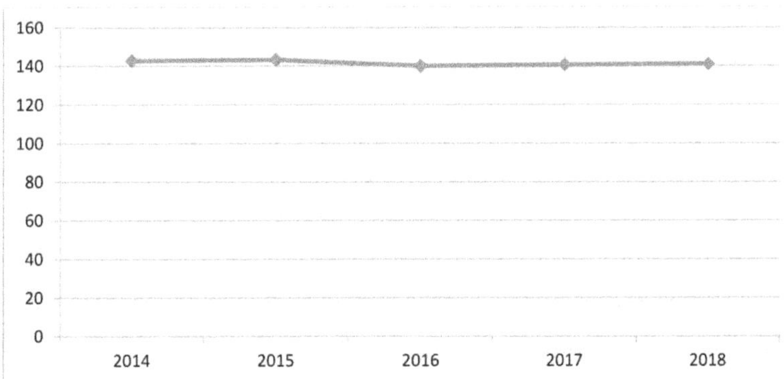

The innovation scores of Finland in the five-year period of 2014-2018 can be observed on Figure 5 below.

Finland is the forth country among the other selected countries in the period of 2014-2018. The scores took place between 129.39 and 145.87. It can be observed from Figure 5 that Finland has an increasing trend 2014 to 2018. Finland has accelerated in 2018, increasing from the value of 134.77 to 145.87.

Netherlands

The innovation scores of Netherlands in the five-year period of 2014-2018 are demonstrated on Figure 6.

Figure 5. Innovation scores of Finland (2014-2018)
Source: European Innovation Scoreboard 2019, https://www.interactivetool.eu

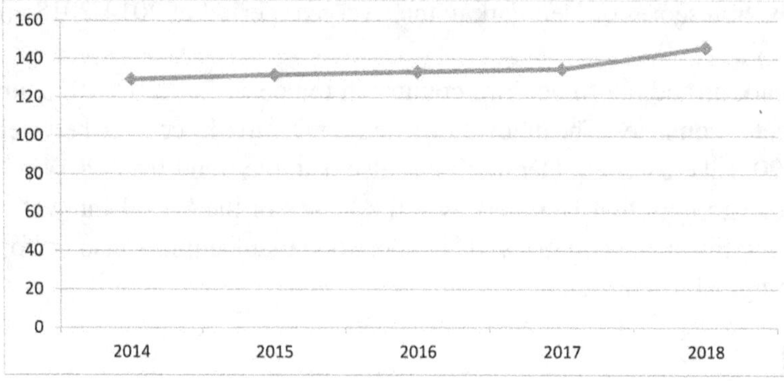

Netherlands is the fifth country in the period of 2014-2018 among the other selected countries. The innovation scores of Netherlands occurred between 125.57 and 134.99. The important movement took place in 2017 as it can be observed on Figure 6. The trend is expected to increase in the near future, since Netherlands is among the most innovative countries.

Germany

The innovation scores of Germany in the five-year period of 2014-2018 can be seen on Figure 7 below.

Germany turned out to be the sixth country in the period of 2014-2018 among the selected European countries. The innovation scores of Germany were between

Figure 6. Innovation scores of Netherlands (2014-2018)
Source: European Innovation Scoreboard 2019, https://www.interactivetool.eu

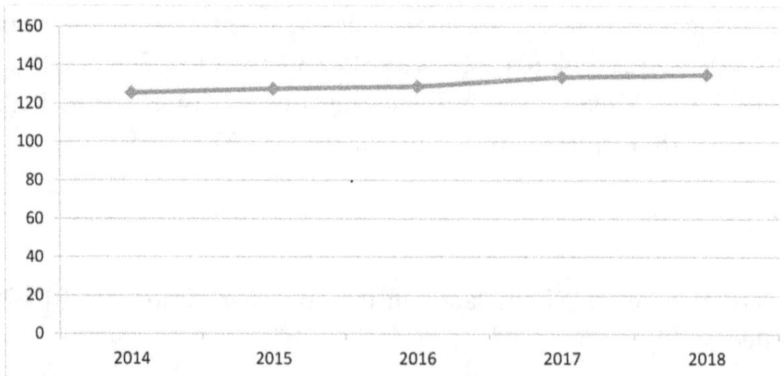

Figure 7. Innovation scores of Germany (2014-2018)
Source: European Innovation Scoreboard 2019, https://www.interactivetool.eu

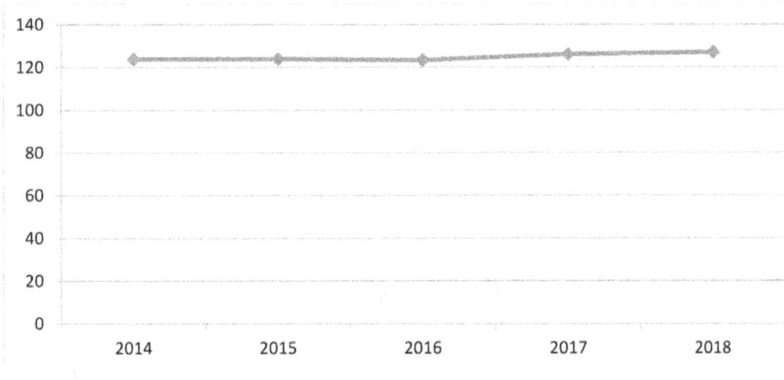

123.78 and 126.88. Germany performed a more stable trend like Sweden and Denmark than other countries. It is important to note that in 2016, Germany is the second country in social innovation and the sixth country in technological innovation among European countries.

Norway

The innovation scores of Norway in the five-year period of 2014-2018 are shown on Figure 8.

Being a strong innovator, Norway is the last country in the research group. The innovation scores of Norway occurred between 100.92 and 127.71. With this result it turned out to be that Norway's scope was the largest one. There is a jump from

Figure 8. Innovation scores of Norway (2014-2018)
Source: European Innovation Scoreboard 2019, https://www.interactivetool.eu

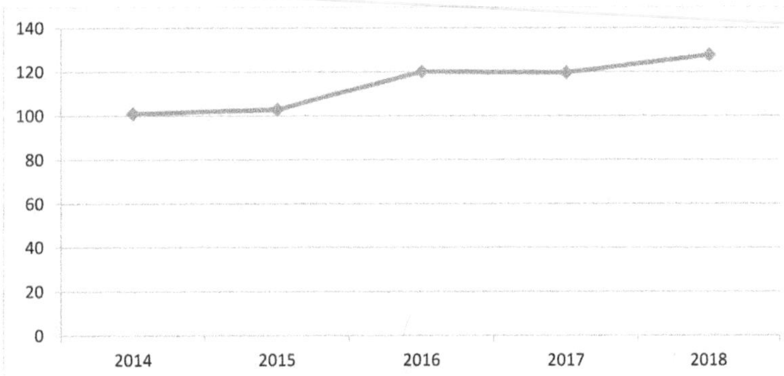

102.76 to 120.09 from 2015 to 2016. Norway's grade is increasing year-to-year reaching its maximum in 2018.

On Figure 9, it is seen that Sweden, Netherlands and Germany have a stable trend in scoring innovation. Furthermore, whereas Switzerland has a moderate increase in innovation, Denmark has a slight decrease. The notable countries are Finland and Norway. They are jumping parallel from bottom values to top values of their own in the period of 2014-2018.

FUTURE RESEARCH DIRECTIONS

Social innovation researches are growing but still there is a need for further studies. In the future, both quantitative and qualitative method of research on social innovation should be conducted. So as to practice quantitative research method, reliable and valid scales for social innovation or social innovativeness for organizations or managers should be generated. Also as well as developed countries, developing countries need to emphasize social innovation projects in order to build more sustainable future. Furthermore, since it is precise that technological innovation is a must in production processes, efficiency and effectiveness can be sustained by the help of innovation. At this point, innovativeness and innovative climate is vital. In the future, the innovative capacity and innovation orientation of organizations and even countries should be examined. Since it is understood that all types of innovation are vital for success, some other minor innovation types should also be questioned.

Figure 9. Innovation scores of all selected countries (2014-2018)
Source: European Innovation Scoreboard 2019, https://www.interactivetool.eu

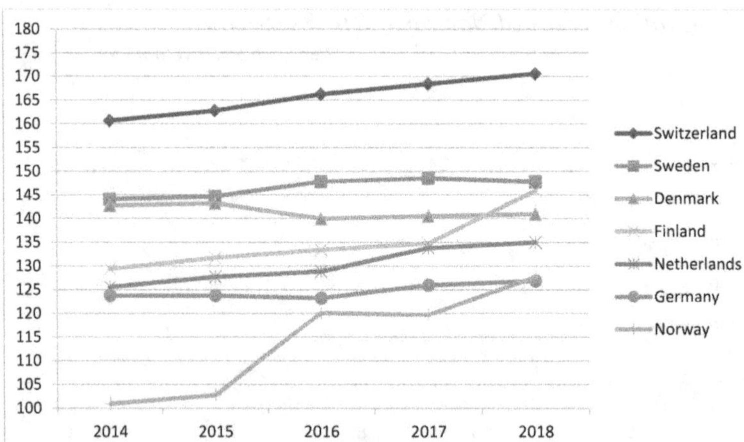

CONCLUSION

During the whole of this chapter, gaining strategic advantage by social and technological innovation was discussed. Gaining strategic advantage, eases organizations both to survive in the long-term and to operate effectively. In the present business context, according to researches and professional experience innovation turns out to be the most influential factor to help an organization gain strategic advantage. Due to the complexity of social and economic life, innovation has two main dimensions. One of the dimensions is social innovation, while the other one is technological innovation literally. According to the related literature some organizations focus on social innovation for creating social value and benefit the society they live in. On the other side, the rest focus on technological innovation on producing goods and services.

In this context, the host countries of those organizations mentioned above benefit from social and technological innovation generated by companies, social ventures or public ventures. In this chapter, seven innovative leaders European countries were selected to analyze their innovation trends. These countries were, Switzerland, Sweden, Denmark, Finland, Netherlands, Germany and Norway.

Looking at the Social Innovation Index 2016, all of these seven countries performed well at generating social innovation and creating social value. Among them Denmark took the lead. Their performances were close to each other but significantly better than the rest of the world except for United States, United Kingdom and Canada.

According to European Innovation Scoreboard, within the five-year period of 2014-2018, Switzerland was the innovator leader of the selected group. It was noted that Switzerland, Sweden, Denmark and Germany had a relatively stable innovator performance. Notably, Finland, Norway and Netherlands had an increasing trend for innovation. Among these three, Finland and Norway had a jump parallel to each other in 2017.

As a result, these chosen countries show the way how social and technological innovations help countries gain strategic advantage. With the help of innovation, countries maintain sustainable economic and social advantages both for their host companies and their nations. Both social and technological innovations maintain guarantees for a better life for humanity in the future.

REFERENCES

Agarwal, R., Sarkar, M. B., & Echambadi, R. (2002). The conditioning effect of time on firm survival: An industry life cycle approach. *Academy of Management Journal*, *45*(5), 971–994.

Barney, J. (1991). Firm resources and sustained competitive advantage. *Journal of Management*, *17*(1), 99–120. doi:10.1177/014920639101700108

Barney, J. B. (1986). Strategic factor markets: Expectations, luck, and business strategy. *Management Science*, *32*(10), 1231–1241. doi:10.1287/mnsc.32.10.1231

Barney, J. B. (2001). Resource-based theories of competitive advantage: A ten-year retrospective on the resource-based view. *Journal of Management*, *27*(6), 643–650. doi:10.1177/014920630102700602

Baumol, W. J. (2002). *The free-market innovation machine: Analyzing the growth miracle of capitalism*. Princeton University Press. doi:10.1515/9781400851638

Bettencourt, L. A., Lusch, R. F., & Vargo, S. L. (2014). A service lens on value creation: Marketing's role in achieving strategic advantage. *California Management Review*, *57*(1), 44–66. doi:10.1525/cmr.2014.57.1.44

Buddelmeyer, H., Jensen, P. H., & Webster, E. (2009). Innovation and the determinants of company survival. *Oxford Economic Papers*, *62*(2), 261–285. doi:10.1093/oep/gpp012

Cajaiba-Santana, G. (2014). Social innovation: Moving the field forward. A conceptual framework. *Technological Forecasting and Social Change*, *82*, 42–51. doi:10.1016/j.techfore.2013.05.008

Cefis, E., & Marsili, O. (2006). Survivor: The role of innovation in firms' survival. *Research Policy*, *35*(5), 626–641. doi:10.1016/j.respol.2006.02.006

Christensen, C. M., Suárez, F. F., & Utterback, J. M. (1998). Strategies for survival in fast-changing industries. *Management Science, 44*(12-part-2), S207-S220.

Dawson, P., & Daniel, L. (2010). Understanding social innovation: A provisional framework. *International Journal of Technology Management*, *51*(1), 9–21. doi:10.1504/IJTM.2010.033125

Defourny, J., & Nyssens, M. (2010). Conceptions of social enterprise and social entrepreneurship in Europe and the United States: Convergences and divergences. *Journal of Social Entrepreneurship*, *1*(1), 32–53. doi:10.1080/19420670903442053

Dervitsiotis, K. N. (2010). Developing full-spectrum innovation capability for survival and success in the global economy. *Total Quality Management*, *21*(2), 159–170. doi:10.1080/14783360903549865

Dewar, R. D., & Dutton, J. E. (1986). The adoption of radical and incremental innovations: An empirical analysis. *Management Science, 32*(11), 1422–1433. doi:10.1287/mnsc.32.11.1422

Eren, B. A. (2018). Şikayete yol açan hataların ve müşterilerin bankadan ayrılmasını engelleyen faktörlerin belirlenmesi: Kamu ve özel mevduat banka müşterilerine yönelik bir araştırma. *Uluslararası Sosyal Araştırmalar Dergisi, 11*(57), 586–600.

European Union. (2019). *European Innovation Scoreboard*. Retrieved from https://www.interactivetool.eu

Fairweather, G. W. (1967). *Methods for experimental social innovation*. Oxford, UK: Wiley and Sons.

Foss, N. J. (1999). Networks, capabilities, and competitive advantage. *Scandinavian Journal of Management, 15*(1), 1–15. doi:10.1016/S0956-5221(98)00030-X

Garcia, R., & Calantone, R. (2002). A critical look at technological innovation typology and innovativeness terminology: A literature review. *Journal of Product Innovation Management, 19*(2), 110–132. doi:10.1111/1540-5885.1920110

Green, S. G., Gavin, M. B., & Aiman-Smith, L. (1995). Assessing a multidimensional measure of radical technological innovation. *IEEE Transactions on Engineering Management, 42*(3), 203–214. doi:10.1109/17.403738

Harrisson, D., Klein, J. L., & Browne, P. L. (2011). Social innovation, social enterprise and services. In F. Gallouj & F. Djellal (Eds.), *The handbook of innovation and services: A multi-disciplinary perspective* (pp. 197–218). Edward Elgar Publishing.

Herrera, M. E. B. (2015). Creating competitive advantage by institutionalizing corporate social innovation. *Journal of Business Research, 68*(7), 1468–1474. doi:10.1016/j.jbusres.2015.01.036

Hill, C. W. (1995). National institutional structures, transaction cost economizing and competitive advantage: The case of Japan. *Organization Science, 6*(1), 119–131. doi:10.1287/orsc.6.1.119

Hobday, M. (1998). Product complexity, innovation and industrial organisation. *Research Policy, 26*(6), 689–710. doi:10.1016/S0048-7333(97)00044-9

Isik, Z., Arditi, D., Dikmen, I., & Birgonul, M. T. (2009). Impact of corporate strengths/weaknesses on project management competencies. *International Journal of Project Management, 27*(6), 629–637. doi:10.1016/j.ijproman.2008.10.002

Klepper, S., & Simons, K. L. (2000). The making of an oligopoly: Firm survival and technological change in the evolution of the US tire industry. *Journal of Political Economy, 108*(4), 728–760. doi:10.1086/316100

Lubit, R. (2001). The keys to sustainable competitive advantage. *Organizational Dynamics, 29*(3), 164–178. doi:10.1016/S0090-2616(01)00026-2

Marcy, R. T., & Mumford, M. D. (2007). Social innovation: Enhancing creative performance through causal analysis. *Creativity Research Journal, 19*(2-3), 123–140. doi:10.1080/10400410701397180

Moulaert, F. (Ed.). (2013). *The international handbook on social innovation: collective action, social learning and transdisciplinary research*. Edward Elgar Publishing. doi:10.4337/9781849809993

Mulgan, G. (2006). The process of social innovation. *Innovations: Technology, Governance, Globalization, 1*(2), 145–162. doi:10.1162/itgg.2006.1.2.145

Mumford, M. D. (2002). Social innovation: Ten cases from Benjamin Franklin. *Creativity Research Journal, 14*(2), 253–266. doi:10.1207/S15326934CRJ1402_11

Mumford, M. D., & Moertl, P. (2003). Cases of social innovation: Lessons from two innovations in the 20th century. *Creativity Research Journal, 15*(2-3), 261–266. doi:10.1080/10400419.2003.9651418

Newman, L. (2005). Uncertainty, innovation, and dynamic sustainable development. *Sustainability: Science. Practice and Policy, 1*(2), 25–31.

Nicholls, A., & Murdock, A. (Eds.). (2011). *Social innovation: Blurring boundaries to reconfigure markets*. Springer. doi:10.1057/9780230367098.0006

Oliver, C. (1997). Sustainable competitive advantage: Combining institutional and resource-based views. *Strategic Management Journal, 18*(9), 697–713. doi:10.1002/(SICI)1097-0266(199710)18:9<697::AID-SMJ909>3.0.CO;2-C

Pennings, J. M., & Buitendam, A. (1987). *New technology and organizational innovation: The development and diffusion of micro-electronics*. Cambridge: Ballinger Publishing.

Peteraf, M. A. (1993). The cornerstones of competitive advantage: A resource-based view. *Strategic Management Journal, 14*(3), 179–191. doi:10.1002mj.4250140303

Phillips, W., Lee, H., Ghobadian, A., O'Regan, N., & James, P. (2015). Social innovation and social entrepreneurship: A systematic review. *Group & Organization Management, 40*(3), 428–461. doi:10.1177/1059601114560063

Pol, E., & Ville, S. (2009). Social innovation: Buzz word or enduring term? *Journal of Socio-Economics*, *38*(6), 878–885. doi:10.1016/j.socec.2009.02.011

Porter, M. E. (1985). Technology and competitive advantage. *The Journal of Business Strategy*, *5*(3), 60–78. doi:10.1108/eb039075

Porter, M. E. (2011). *Competitive advantage of nations: creating and sustaining superior performance*. New York: Simon and Schuster.

Rumelt, R. P. (1984). Competitive strategic management. In R. Lamb (Ed.), Toward a Strategic Theory of the Firm, (pp. 556-570). Prentice Hall.

Scholl, J. (2013). Inclusive Business Models as a Key Driver for Social Innovation. In T. Osburg & R. Schmidpeter (Eds.), *Social Innovation. CSR, Sustainability, Ethics & Governance*. Berlin: Springer. doi:10.1007/978-3-642-36540-9_9

Schumpeter, J. A. (1942). *Capitalism, socialism and democracy*. New York: Harper and Row.

Taylor, J. B. (1970). Introducing social innovation. *The Journal of Applied Behavioral Science*, *6*(1), 69–77. doi:10.1177/002188637000600104

Teece, D. J. (1986). Profiting from technological innovation: Implications for integration, collaboration, licensing and public policy. *Research Policy*, *15*(6), 285–305. doi:10.1016/0048-7333(86)90027-2

The Economist Intelligence Unit. (2016). *Old problems, new solutions: Measuring the capacity for social innovation across the world. (Social innovation index 2016)*. London, UK: The Economist Intelligence Unit.

Toft, B. J. (2018). *Competitive advantage in a global market: The Case of the Norwegian Salmon Farming Industry A comparative study of the Chilean salmon farming industry* (Master's thesis). University of Agder.

Van der Have, R. P., & Rubalcaba, L. (2016). Social innovation research: An emerging area of innovation studies? *Research Policy*, *45*(9), 1923–1935. doi:10.1016/j.respol.2016.06.010

Weber, J. M. (2012). Social innovation and social enterprise in the classroom: Frances Westley on bringing clarity and rigor to program design. *Academy of Management Learning & Education*, *11*(3), 409–418. doi:10.5465/amle.2011.0403

Wernerfelt, B. (1984). A resource-based view of the firm. *Strategic Management Journal*, *5*(2), 171–180. doi:10.1002mj.4250050207

Wright, P. M., McMahan, G. C., & McWilliams, A. (1994). Human resources and sustained competitive advantage: A resource-based perspective. *International Journal of Human Resource Management*, *5*(2), 301–326. doi:10.1080/09585199400000020

ADDITIONAL READING

Archibugi, D., & Iammarino, S. (2002). The globalization of technological innovation: Definition and evidence. *Review of International Political Economy*, *9*(1), 98–122. doi:10.1080/09692290110101126

Brown, T., & Wyatt, J. (2010). Design thinking for social innovation. *Development Outreach*, *12*(1), 29–43. doi:10.1596/1020-797X_12_1_29

Garvin, D. A. (1995). Leveraging processes for strategic advantage. *Long Range Planning*, *28*(6), 126–126. doi:10.1016/0024-6301(95)99980-E

Howell, J. M., & Higgins, C. A. (1990). Champions of technological innovation. *Administrative Science Quarterly*, *35*(2), 317–341. doi:10.2307/2393393

Kickul, J., & Gundry, L. (2002). Prospecting for strategic advantage: The proactive entrepreneurial personality and small firm innovation. *Journal of Small Business Management*, *40*(2), 85–97. doi:10.1111/1540-627X.00042

Lettice, F., & Parekh, M. (2010). The social innovation process: Themes, challenges and implications for practice. *International Journal of Technology Management*, *51*(1), 139–158. doi:10.1504/IJTM.2010.033133

McElroy, M. W. (2002). Social innovation capital. *Journal of Intellectual Capital*, *3*(1), 30–39. doi:10.1108/14691930210412827

Pennings, J. M., & Harianto, F. (1992). The diffusion of technological innovation in the commercial banking industry. *Strategic Management Journal*, *13*(1), 29–46. doi:10.1002mj.4250130104

Pol, E., & Ville, S. (2009). Social innovation: Buzz word or enduring term? *Journal of Socio-Economics*, *38*(6), 878–885. doi:10.1016/j.socec.2009.02.011

Price, R. M. (1996). Executive forum: Technology and strategic advantage. *California Management Review*, *38*(3), 38–56. doi:10.2307/41165842

Schuler, D. A., Rehbein, K., & Cramer, R. D. (2002). Pursuing strategic advantage through political means: A multivariate approach. *Academy of Management Journal*, *45*(4), 659–672.

Slater, S. F., & Mohr, J. J. (2006). Successful development and commercialization of technological innovation: Insights based on strategy type. *Journal of Product Innovation Management*, *23*(1), 26–33. doi:10.1111/j.1540-5885.2005.00178.x

Voorberg, W. H., Bekkers, V. J., & Tummers, L. G. (2015). A systematic review of co-creation and co-production: Embarking on the social innovation journey. *Public Management Review*, *17*(9), 1333–1357. doi:10.1080/14719037.2014.930505

Wang, C. H., Lu, I. Y., & Chen, C. B. (2008). Evaluating firm technological innovation capability under uncertainty. *Technovation*, *28*(6), 349–363. doi:10.1016/j.technovation.2007.10.007

KEY TERMS AND DEFINITIONS

Innovation: Implementing new ideas, new products or new processes.

Social Innovation: Practicing new ideas on society in order to create social value.

Strategic Advantage: A kind of advantage in which an organization gains by having core capabilities or superiorities.

Strategy: A long-term road map that is an organization follows in order to attain its goals.

Technological Innovation: Generating new techniques and processes in producing goods and services.

Chapter 3
Leadership for Sustainability: Connecting Corporate Responsibility Reporting and Strategy

Nikolina Markota Vukić
RRIF d.o.o., Croatia

Mislav Ante Omazić
Faculty of Economics and Business, University of Zagreb, Croatia

Mirjana Pejic-Bach
https://orcid.org/0000-0003-3899-6707
Faculty of Economics and Business, University of Zagreb, Croatia

Ana Aleksić
https://orcid.org/0000-0002-9702-6703
Faculty of Economics and Business, University of Zagreb, Croatia

Jovana Zoroja
Faculty of Economics and Business, University of Zagreb, Croatia

ABSTRACT

This chapter describes how and why normative justification for corporate responsibility has been replaced with business case thinking. Modern organizations do link their non-financial reports to strategy because corporate responsibility is seen as a source of competitive advantage, growth, motivation, and innovation. Non-financial reporting potentiates companies to understand, learn, and redefine their social and environmental impacts and its cost; how to minimize both impact and

DOI: 10.4018/978-1-7998-1108-4.ch003

cost; and how to capitalize on challenges that management face on a daily basis. The connection between CR strategy and non-financial reporting is described along with how it is relevant for the successful business of modern organizations. By improving CR strategies and developing unique non-financial reporting, organizations can be more effective and efficient in the local community, which will positively reflect on their business model and in achieving higher sustainability levels.

INTRODUCTION

In a globalized and interconnected world, the distant becomes close, and time and space are no longer constraints for business interaction. In a modern world, organizations are encouraged by stakeholders to balance between the need for profitability and need to contribute to the local community, social equity, and environmental protection (Leon, 2013). The beginnings of the concept of corporate (social) responsibility (hereafter CR) in the academic literature are found in the articles from the 1930s where for the first time, in addition to the profit function, social service function of business is also mentioned (Okoye, 2009). During the 1960s, the growing refinement of consumer demands helped to spread this concept, followed in the 1970s by the environmental movement and the growing concern for the impact of the economy on society in the 1990s. Since the 1990s CR has been seen as an integral part of business activities and strategy (Dahlsrud, 2008) with a balance between social, environmental, and economic impact becoming a crucial element of all organizational activities. As a consequence the *raison d'être* of contemporary CR is very much a product of general developments and trend in modern society.

Although the majority of academics still use term corporate social responsibility, in this chapter, we will use term corporate responsibility to emphasize the balance between the economic, social, and environmental impact of an enterprise. CR is multidimensional and can benefit business organizations and their stakeholders in a multitude of ways; many driving increased profitability and resource efficiency. Still leading to corporate responsibility requires leaders to pay attention to how the concept is framed and introduced into the organization (Quinn & Dalton, 2009).

On its road to sustainable business models, organizations need to implement CR in their strategic processes (Bonn & Fisher, 2011). In order to achieve this integration, the strategy itself should be aligned with organizational values, and stakeholders needs and wants have to be integrated into organizational past, present, and future.

As such, CR strategy is focused on the successful integration of organization into the overall network of society. This requires full integration of societal needs, the natural environment, and corresponding business imperatives (Quinn & Dalton, 2009) into the organization environment, its process, structure, and culture. Thus sustainability efforts need to become an integral part of business strategy if they are to provide long-term value to both the organization and society (Galpin & Whittington, 2012). In that sense, leaders are encouraged to change the organization's views of success from an economic-based to a sustainability-based measurement (Quinn & Dalton, 2009, p. 22). Companies are motivated to include social and environmental activities in their business to be considered legitimate following national, local, and industrial norms and regulations. Also, as stated by Husted and Allen (2011), companies will take part in responsible and sustainable activities to imitate their competitors and thus preserve their social legitimacy, build a reputation, avoid negative perception, and ensure the long-term survival of the organization.

Countries around the world, because of the diversity of legislative mechanisms, are not homogeneous in the implementation of laws and regulations, and the formal institutional frameworks, involved in creating an environment for corporate responsibility. According to Campbell (2007), instead of legally prescribed non-financial reporting, some countries have adopted voluntary initiatives, primarily to help businesses while analyzing and compiling reports on their sustainability. Furthermore, more recently, Adams (2016) highlighted digitization produced increased transparency, raising the issues of credibility and emphasizing more the need for non-financial assurance. Also, digitization, technological and communicational revolutions have created a global community of active and involved stakeholders. That's especially true for investors and customers, who have increased their expectations for transparent business operations (Fung, 2014). Moreover, digitization has made information more available and accessible. Accordingly, the increasing number of non-financial reports published by companies around the world are moving from black-box to glass-box business models where transparency is rather important. Transparency is something that many managers see as a prerequisite as they know that companies with hard to understand and complex business structures are riskier and less valuable for investments.

The trend of increasing the number of non-financial reports is confirmed by the fact that over the last decade, reporting on non-financial information such as the impact on the environment and corporate society has become widespread. This trend is confirmed by the GRI statistics that publishes the number of organizations and reports on corporate responsibility since 1999. Only 11 companies were reported in the first year of research, while according to the latest statistics at May 2019, the number of organizations that reported was 13,701, and there were 53,675 non-financial reports (GRI, 2019).

Intending to provide more insights into CR strategy and non-financial reporting, this chapter explores more deeply their interrelation and provides an empirical analysis of their connection. In addition to theoretical examination on their relation, empirical part of the chapter examines the connection between the level of non-financial reporting and type of CR strategy implemented by the company on a sample of 154 companies originating from Austria, Belgium, Croatia, France, Germany, Italy, Netherlands, Sweden, Switzerland and United Kingdom. We used the online questionnaire, as well as the GRI database, to collect data on the level of the non-financial reporting and CR strategy.

The level of the non-financial reporting is described by using the GRI framework, reporting outside the framework, and the intent of reporting. Five levels (types) of CR strategy are examined: (1) Innovative strategy; (2) Proactive strategy; (3) Accommodative strategy; (4) Defensive strategy; and (5) Obstructionist strategy. These strategies are described by using three dimensions (1) Companies' relationship towards the impact of non-financial reporting on their reputation; (2) Interactive connection with their stakeholders and (3) updated version of the theoretical framework according to Lee (2011).

The contribution and objective of this chapter are in providing additional insights into CR strategy and reporting, especially in the light of modern technological developments and digitalization of business activities, processes, and communication. Besides, the chapter provides results of empirical research examining the connection of the level of non-financial reporting with the type of CR strategy implemented.

The chapter consists of the following sections. Following the introduction, the second section presents the background and rationale of CR and CR strategy, its context and connection with institutional framework and stakeholders. The next section outlines the concept of CR or non-financial reporting, together with elements related to the quality of reporting. This is followed with the section where empirical research is presented, including the methodology used and results of our analysis. After discussion, the last section brings a summary of results and presents future research directions and limitations of the chapter.

CORPORATE RESPONSIBILITY AND STRATEGY

Nowadays, it is widely recognized that corporations need to act in a responsible way to contribute to social well-being, environmental protection, and competitiveness as well as to ensure financial success (Moneva & Ortas, 2010). Activities oriented towards economic and financial achievements need to be supplemented by those oriented towards "minimization of ecological footprints and increased attention to social aspects" (Lee, 2009, p. 3274). Also, some organizations perceive sustainability initiatives as a branding and image strategy, while others engage in these activities for additional reasons, such as to cut cost or as a differentiation strategy to help them be distinct from their competitors (Galpin & Whittington, 2012).

If you were a newcomer in the field of role of business in society, and you started browsing specialized literature, surely you would be confused by a number of different terms and definitions that imply similar or identical meanings: corporate social responsibility, public responsibility, responsible business conduct, corporate societal responsibility, corporate social performance, corporate citizenship, business citizenship, stakeholding company, do-good-company, business ethics, sustainable company, and triple-bottom-line approach. That only proves how fast and varied this area is developing. CR is a slippery phrase, and the concept has often been accused of complexity, contradictory, vagueness, lack of clarity, ambiguity, and uncertainty (Pedersen, 2015). There it is important to define it bearing on our minds that every definition is an abstraction that offers the value of a panorama at the expense of detail.

The World Business Council for Sustainable Development defines CR as the "continuing commitment by business to behave ethically and contribute to economic development while improving the quality of life of the workforce and their families as well as of the local community and society at large" (Fontaine, 2013; WBCSD, 2000). CR can be seen as a framework that provides short-term performance and long-term organizational sustainability (Pejic Bach et al., 2018). It also presents a "context-specific organizational actions and policies that take into account stakeholders' expectations and the triple-bottom-line of economic, social, and environmental performance" (Aguinis, 2011; Aguinis & Glavas, 2012). It changes the role and utility of businesses in its environment, moving it above the core function of producing and selling goods (Wang et al., 2016). European Commission defines corporate responsibility as the "the responsibility of enterprises for their impacts on society" (EC, 2019, p. 3). According to them to fully meet their social responsibility, companies "should have in place a process to integrate social, environmental, ethical, human rights and consumer concerns into their business operations and core strategy in close collaboration with their stakeholders, with the aim of maximizing the creation of shared value for their owners/shareholders and civil society at large and identifying, preventing and mitigating possible adverse impacts." To summarize, CR is a form of corporate social and environmental self-regulation integrated into its business model or, how companies manage the business processes to produce an overall positive impact on society.

Still, one of the major barriers to CR implementation and fulfilment is how to design and coordinate broad set of decisions and actions on the local and global level (Mikolajuk & Yeh, 2002), capturing multiple shareholder interests and values. How organization balances and arranges priorities related to different aspects of CR is often seen as a specific challenge for many organizations (Wang et al., 2016). In that sense, organizations have developed a variety of strategies for dealing with this intersection of various dimensions (Quinn & Dalton, 2009).

In general, CR strategy presents a framework for managing a company's current and future impacts on society and the environment. It represents the social dimension of an organization that can be associated with the economic dimension to create economic and social values (Husted & Allen, 2011). Consequently, an organization that understands its social responsibility and incorporates elements of socially responsible business into its business strategy is likely to have better business results and influence the development of industry, state and entire society (Galbreath, 2009).

CR strategy implemented by the organization can be interpreted using two complementary, systemic theoretical frameworks: institutional and shareholder theory. The institutional theory examines business models that make the organization legitimate and viable within a certain institutional and social environment. While shareholder theory, according to Suchman (1995), studies the way organizations respond to external and internal expectations and impacts to maintain their reputation, and thus viability. Institutions, by providing prescribed, normative, and cognitive forms of social behavior, enable the stability of social behavior. Stakeholders, on the other hand, form individual or collective mechanisms that directly affect the company's strategy and are aimed at stakeholder or social interests.

To achieve a successful social strategy, organizations must establish a balance between the business community, local community, and the regulators. As Hawkins (2006) states, balance between these three groups is primarily based on the conflict that needs to be resolved. Furthermore, according to Hawkins (2006), understanding the needs of the local community, can bring new values for the organization and achieve market competitiveness. This creates a dynamic relationship with business strategy, as competition is likely to exploit new market opportunities. The role of regulators is to support the creation of general wealth and prosperity through the constant balancing of all the consequences that it creates.

Different authors have categorized different CR strategies. For instance, Heikkurinen (2010) makes a distinction between five levels of strategy: passive, reactive, proactive, entrepreneurial, and creative. According to Van Bommel (2011), it is possible to differentiate between resigning, defensive, and offensive strategy. Lee (2011) categorizes different social strategies according to the intensity of institutional and stakeholder pressures and differs among obstructionist, defence, accommodative, and proactive strategy. Interesting classification of CR strategies of other authors include approaches such as "do minimum harm/do maximum good" (Wheeler, Colbert & Freeman, 2003) or "responsive/strategic" (Porter & Kramer, 2006).

When defining and implementing CR into business strategy, six dimensions need to be considered in the context of CR and business strategy, including: (1) corporate mission, (2) strategic activities, (3) market, (4) consumer needs, (5) resources, and (6) competitive advantage (Galbreath, 2007). Additionally, a strategy that contains elements of sustainability needs to build in dialogue with suppliers and distributors to ensure a sustainable business strategy along the value chain (Bonn & Fisher, 2011).

Although managers believe sustainability is increasingly important to business strategy (Galpin & Whittington, 2012) research (e.g., Porter & Kramer, 2006) has shown that often management teams fail to incorporate sustainability and CR into business strategy. Quinn and Dalton (2009) therefore state that to ensure success in pursuing a sustainability agenda leaders must be careful how the concept is designed and implemented in the organization; they should build capacity in their system and culture to support sustainability, while at the same time ensure deep stakeholders engagement.

Besides, Smith and Sharicz (2011, p. 76-77) based on previous studies provide a list of basic leadership qualities and skills needed for integrating social and sustainability issues into an organization:

- "questioning business as usual by being open to new ideas and challenging others to adopt new ways of working;
- understanding the role of each player in society (government, unions, business, social partners, non-governmental organizations and civil society) and how they interact with each other;
- building internal and external partnerships by taking a multi-disciplinary approach and creating strategic networks and alliances;
- identifying stakeholders, building relations with internal and external stakeholders, engaging in dialogue and balancing competing demands;
- respecting diversity, understanding difference and adjusting one's approach to different situations;
- taking a strategic view of the business environment."

With new technological advancements, digitalization and digital transformations, companies need to also bear in mind that digitalization can be seen, as Knaut (2017, p. 255) emphasizes, a new holistic and hybrid driver of organizational business model. New models will change organizations and stress their decentralized and dynamic networks, where CR will have an active role in shaping these advancements and stress aspects related to sustainability and organizational responsibility (Gangloff, 2018).

New CR strategies can be explored as a result of digitization, especially related to new interactive communication channels (Spanos, 2018) that can help to communicate but also interact with even wider stakeholder groups. These communication channels, like social media, not only introduce dynamics into the communication process but also present effective monitoring tools, that can help to identify emerging issues related to responsibility (Camilleri, 2017). Social media contribute and strengthen corporate-society-citizenship relations (Spanos, 2018; Whelan, Moon & Grant, 2013) and instead of maintaining legitimacy through traditional stakeholder management, that is through more corporate- and stakeholder-centric frameworks (Whelan, Moon & Grant, 2013), organizations need to understand the polyphony of social media, and other forms used for legitimating process (Castelló, Morsing & Schultz, 2013).

Moreover, CR initiatives driven by new digital and technological tools, for example as the Internet of Things, as Roscoe (2015) argues, will also enhance organizations to embrace new technologies and tools that can be helpful to address constant economic, environmental and societal issues organizations confront daily. Spanos (2018) argues that benefits of modern tools can help organizations achieve significant benefits not only in greater communication with shareholders but also in an increase in available funding for various CR initiatives, increased public awareness, engagement, and marketing reach.

NON-FINANCIAL OR CORPORATE RESPONSIBILITY REPORTING

Non-financial or CR reporting presents a systematic communication between organization, its stakeholders and the general public about organizational social, environmental and other responsible activities, as well as corporate governance and ethical activities, which are not covered by financial performance indicators (Schreck, 2013). As it is the case with CR, it is also variously called social accounting, social disclosure, social auditing, social review, social reporting, triple-bottom-line or sustainability reporting (Douglas, Doris & Johnson, 2004). In 2017 KPMG member firm professionals reviewed CR and sustainability reporting from top 100 companies in 49 countries and regions (total 4,900 companies) where they discovered that non-financial is standard practice for large and mid-cap companies around the world where around three-quarters of the 4,900 companies issue non-financial reports (KPMG, 2017).

Increased demand for non-financial reporting is the result of many international initiatives to make reporting standardized with relevant business indicators. Although the financial statements are the dominant form of corporate reporting, the non-financial reporting is gaining ever-greater importance in corporate reporting, primarily to provide stakeholders with a complete picture of the company's overall performance. Companies have, for a long-time published a non-financial report, but there was no standardization between reports and lack of assurance that activities were genuine and actual. Another challenge is to link financial and non-financial into integrative reporting. Integrated reports have been created to establish linkage between non-financial and financial reports. The whole idea was to enhance accountability, stewardship, reputation, and trust among stakeholders, as well as to foster information flow and transparency of business activities. These remain fundamental challenges, although assurance of corporate responsibility data has more than doubled among the global top 250 companies in the last 12 years (now 67% of reports) (KPMG, 2017).

The primary objective of the non-financial report is thus to provide, to all interested stakeholders, public, reliable and complete information about the economic, social and environmental activities of the company, which can be compared and analyzed. It is a form of increased transparency that implies providing relevant and timely information to all interested users to optimize understanding and decision making and to ensure a more transparent and efficient business. Non-financial reports allow the dissemination of information related to the impact of organizational business practices and activities on the environment, social, and governance issues (Sethi, Martell & Denim, 2017). As such, non-financial reporting includes an analysis of three aspects of socially responsible business: economic, ecological, and social one. The economic aspect of social responsibility represents all the relevant indicators by which the organization achieves its financial profit, while the ecological and social environment includes all relevant environmental and social indicators that affect or may affect the economic aspect or financial performance of the business. The ecological aspect of non-financial reporting includes relevant indicators by which organization evaluates the impact of its business and the impact of its products and services on the environment. The social aspect of non-financial reporting relates to the relevant indicator by which organization evaluates the impact of its business and the impact of its products and services on society — for instance, its involvement in local community development, and human and labor issues.

From a corporate perspective, in his research Adams (2002) identified several benefits of non-financial reporting, including: (1) minimization of risks (e.g. of consumer boycotts, unforeseen issues); (2) better understanding of corporate activities reduces criticisms; (3) influence or delay legislation; (4) attract and retain the most talented people; (5) inclusion in ethical investment funds; (6) better internal systems and control leading to better decision making and cost savings; and (7) communicates the group's values and targets to all group companies. Furthermore, non-financial reporting is important to promote the efficiency and stability of an economy. Non-financial reporting discloses the information of society and is essential in the formulation of national policies (Khan Halabi & Samy, 2009).

An issue that arises with non-financial reports is that much of the reporting is voluntary, thus providing space for possible manipulation regarding data provided, leading to decrease in transparency and reliability of the information provided (Ringham & Miles, 2018). Therefore, several organizations have developed their guidelines for reporting, trying to certify the quality of reports. Among them, Global Reporting Initiative (GRI) presents the most widely adopted guidelines for non-financial reporting, as 93% of the world's largest 250 corporations report on their sustainability performance by using GRI (KPMG, 2017). GRI offer companies a framework and guidelines for triple-bottom-line reporting, where a third-party verification body accredits the reports and stakeholders can compare companies and their activities and efforts (Bjartmarz & Pedersen, 2015).

GRI was established in 1997 as an offshoot of the Coalition for Environmentally Responsible Economies (CERES) (Reynolds & Yuthas, 2008). Standardization of reporting has been achieved by setting the terminology of the framework for non-financial reporting, such as the current G4 guidelines. The GRI development of reporting guidelines is based primarily on cooperation with organizational stakeholders, followed by a strategic partnership with the United Nations Environment Program, the UNGC, the OECD and the International Organization for Standardization (ISO), and many other companies and non-governmental organizations around the world. The GRI reporting framework enables the organization to carry out a comprehensive business evaluation, reporting, and communication with all interested stakeholders, and due to its attractive and easy-to-format communications also to the general public. Content of the GRI report is detailed, and sections of the GRI set out: „objectives in relation to other initiatives, outlines of principles and qualitative characteristics, classification of performance reporting elements, ratio indicators and the disclosure of reporting practices." (Reynolds & Yuthas, 2008).

GRI guidelines also defined the principles of good reporting on CR: (1) the principle of balance, according to which the report contains positive and negative aspects of CR business; (2) the principle of comparability according to which the reports allow stakeholders a comparison (3) the principle of clarity according to which the report is easily understandable and accessible to the stakeholders, (4) the principle of accuracy of the data in the report, (5) the principle of timeliness according to which the reports are published at a given interval, and (6) the principle of reliability according to which the methods used to prepare the report can be verified by a third party who checks the quality of reporting (GRI).

CONNECTING NON-FINANCIAL REPORTING AND RESPONSIBLE STRATEGY - AN EMPIRICAL ANALYSIS

Sample and Procedure

In order to test the impact of the level of non-financial reporting on the type of socially responsible strategy implemented by the company, empirical research was designed and conducted. A total of 527 companies listed on the Fortune 500 largest companies or are registered in the GRI database were contacted from Austria, Belgium, Croatia, France, Germany, Italy, Netherlands, United Kingdom, Sweden, and Switzerland. In total, 157 organizations responded to our survey questionnaire, presenting a response rate of 29.22%. Table 1 provides the distribution of companies by headquarters countries.

Most companies are from the financial services industry (14.94%), followed by those in the food and beverages industry (11.69%). More than half of the sample (64,29%) are large organizations, while there is an approximately equal number of medium (18.83%) and small (16.88%) companies. By our survey questionnaire, we gathered data related to CR strategy, while information related to non-financial reports were retrieved from GRI Database or the official website of the company. Statistical analysis was done using Statistics software, ver. 12. For the non-financial reports, content analysis was done to assess relevant data and quality of non-financial reports.

Research Instrument

The level and quality of non-financial reporting are described by using the GRI framework, reporting, and the intent of reporting. Five levels of non-financial or corporate responsibility reporting are defined as presented in Table 2.

CR strategy is described by using three dimensions:

Table 1. Number of companies by their headquarters country

Country	Number of Companies	Structure [%]
Austria	6	3.90
Belgium	5	3.25
Croatia	84	54.54
France	4	2.60
Germany	12	7.79
Italy	13	8.44
Netherlands	5	3.25
United Kingdom	4	2.60
Sweden	9	5.84
Switzerland	12	7.79
Total	154	100.00

Table 2. The measure of the quality of non-financial reporting

Quality of Non-financial Reporting	Modalities	Variable Measurement
Level of application of the GRI framework for nonfinancial reporting	High Level Comprehensive	coded as 1
	Medium level Core;	coded as 2
	Low level Others	coded as 3
Have the intention	Companies that have not implemented corporate non-financial reporting, but they intend to (Low, Medium, High, Very high, Entirely)	coded as 4
No intention	Companies that don't intend to implement non-financial reporting (Not at all, slightly, No answer).	coded as 5

1. Companies' relationship towards the impact of non-financial reporting on their reputation.
2. Interactive connection with their stakeholders and
3. The theoretical framework, according to Lee (2011) + innovative strategy.

As regards to the third dimension and theoretical framework, according to Lee (2011) + innovative strategy, the questions used to measure strategies have been formulated based on Lee's description of individual corporate social strategies. As previously mentioned, Lee (2011) defined CR strategy according to the institutional and stakeholder pressures and differed among obstructionist, defence, accommodative, and proactive strategy as presented in Table 3. It is important to acknowledge and recognize that the sources of societal pressure are manifold and depend on a wide range of firm-industry, stakeholder-industry, and country-specific factors (Pedersen, 2015).

Based on our literature review, we can even extend Lee's framework, and state that an additional strategy can be defined, namely, innovative strategy. Their characteristics are as it follows (Lee, 2011, p. 286) with the authors' definition of innovative strategy:

- **Innovative Strategy**: involves building company reputation by using non-financial reporting, where stakeholders have a very big influence on non-financial reporting; non-financial reporting is a source of innovation for a company with an innovative strategy development of new products and services based on environmental protection and social inclusion; non-financial

Table 3. Strategy types according to institutional and stakeholder pressures (Lee, 2011)

	Institutional pressure	
Stakeholders pressure	Weak	Strong
Weak	Lack of external pressures (Obstructionist strategy)	Institutional pressure without stakeholder pressure (Defence strategy)
Strong	Stakeholder pressure without institutional pressure (Accommodative strategy)	Equal pressures (Proactive strategy)

report is a reputation-building instrument that represents the company's competitive advantage.

- **Proactive Strategy**: represents a full recognition and acceptance of corporate responsibility while the organization actively participates in social development and minimizes its negative impacts on the general well-being of its stakeholders. Companies build their reputation by using non-financial reporting to achieve a competitive advantage, with stakeholders having a big influence on non-financial reporting;
- **Accommodative Strategy**: includes acceptance of a certain degree of ethical responsibility, especially towards stakeholders. Organizations applying this strategy respect legal requirements and attach a certain importance to stakeholder. However, these companies mostly take a passive attitude and take minimal measures to respond to the demands of their stakeholders. Companies maintain their reputation by using non-financial reporting, with stakeholders having a moderate influence on non-financial reporting
- **Defensive Strategy**: includes rejection of the wider ethical responsibility of the organization. However, organizations that apply this strategy retain their interest by passively following legal requirements to remain legitimate. Companies protect their reputation by using non-financial reporting, with stakeholders having little influence on non-financial reporting
- **Obstructionist Strategy**: a complete rejection of any social and ethical responsibility of an organization that surpasses the economic interest of an organization. Organizations that apply this strategy ignore society demands for responsibility. Companies' reputation is not related to non-financial reporting, and stakeholders have very little influence on non-financial reporting.

Questions used to determine the type of CR strategy implemented by the company, according to Lee (2011) and the author's definition for innovative strategies are shown in Table 4.

Belonging of any company to a particular strategy has been calculated in the following way:

Step 1: Standardized values of individual variables have been calculated

Step 2: Average values of standardized variables for every strategy have been calculated

Step 3: The strategy with the highest rating has been attributed to a particular company as their dominant strategy

Overview of all variables used in this research is presented in Table 5.

Table 4. Specification of variables used to determine types of CR strategy

Strategy	Questions	Modalities of the variables
Innovative	Our company develops new features of products and services which are eco-friendly and socially responsible	1 - I strongly disagree 7 - I strongly agree
	Our company develops new products and services which are eco-friendly and socially responsible	1 - I strongly disagree 7 - I strongly agree
Proactive	Our company fully recognizes its corporate responsibility	1 - I strongly disagree 7 - I strongly agree
	Our company is actively engaged in minimizing negative impacts on the environment and stakeholders	1 - I strongly disagree 7 - I strongly agree
	Our company is actively engaged in improving the welfare of our stakeholders	1 - I strongly disagree 7 - I strongly agree
Accommodative	Our company maximizes its efforts in full-filling stakeholders demands	1 - I strongly disagree 7 - I strongly agree
	Our company rarely takes initiatives for further social goods voluntarily	1 - I strongly disagree 7 - I strongly agree
	Our company tries to pay attention to various stakeholders' voices	1 - I strongly disagree 7 - I strongly agree
Defensive	Our only concern is to comply with legal requirements and to remain legally legitimate	1 - I strongly disagree 7 - I strongly agree
	Our company believes that corporate responsibility initiatives create too much pressure for doing business	1 - I strongly disagree 7 - I strongly agree
Obstructionist	Our company fights against corporate responsibility initiatives	1 - I strongly disagree 7 - I strongly agree
	Our company believes that corporate responsibility initiatives lie outside our economic interest	1 - I strongly disagree 7 - I strongly agree

Research Results

Table 5. Variables used for empirical research

	Variable	Modalities of the variables	
Level of non-financial reporting from the aspect of the GRI framework, reporting outside the framework and the intent of reporting	Level of non-financial reporting	o Comprehensive o Core o Others o Have intention (Low, Medium, High, Very high, Entirely) o No intention (Not at all, slightly, No answer)	
The type of socially responsible strategy implemented by the company	Companies' relationship towards the impact of non-financial reporting on their reputation	Innovative	Building companies' reputation
		Proactive	Building companies' reputation for achieving competitive advantage
		Accommodative	Maintaining companies' reputation
		Defensive	Protecting companies' reputation
		Obstructionist	No influence on companies' reputation
	Interactive connection with their stakeholders	Innovative	Stakeholders have a very big influence on non-financial reports
		Proactive	Stakeholders have a big influence on non-financial reports
		Accommodative	Stakeholders have a moderate influence on non-financial reports
		Defensive	Stakeholders have little influence on non-financial reports
		Obstructionist	Stakeholders do not influence non-financial reports, nor is their inclusion planned in the future
	The theoretical framework according to Lee (2011) + innovative strategy	Innovative	The highest rating for questions that define an innovative strategy
		Proactive	The highest rating for questions that define a proactive strategy
		Accommodative	The highest rating for questions that define an accommodative strategy
		Defensive	The highest rating for questions that define a defensive strategy
		Obstructionist	The highest rating for questions that define an obstructionist strategy

Descriptive Statistics

Level of non-financial reporting is assessed through five levels, described by using the GRI framework, reporting, and the intent of reporting. As shown in Table 6 approximately the same number of companies are grouped within the following three levels of non-financial reporting: Comprehensive (16.88%), Core (17.53%) and Have intention (Low, Medium, High, Very high, Entirely) (16.23%). Within others, there is 23.38% of sampled companies. Most companies (25.97%) have no intention to report on CR matters.

As previously stated, CR strategy is described by using three dimensions related to reputation, stakeholders' interaction, and Lee's (2011) framework. Table 7 presents the distribution of companies regarding their CR strategy, and dimension used for measuring CR strategy.

According to the dimension related to the impact of reporting on companies' reputation, most sampled companies use Innovative CR strategy (27.9%). Approximately the same number of companies implements Proactive (23 companies) and Accommodative (21 companies) CR strategy. Only 15.4% of companies use 'Obstructionist' as their CR strategy of choice.

According to the dimension related to interactive communication with their stakeholders, the largest number of sampled companies uses 'Proactive' CR strategy (34.45%). Almost one-third of the companies use 'Accommodative' strategy (26.89%), followed by 'Obstructionist' (15.97%) and 'Defensive' (14.29%) CR strategy. Only 8.40% of companies use 'Innovative' as their CR strategy of choice when it comes to the criterion of interactive communication with their stakeholders.

Distribution of companies by dimension, according to Lee (2011) + innovative strategy framework reveals that the largest number of companies use innovative as their main CR strategy (27.73%). This is followed by 'Defensive' CR strategy (22.69%) 'Obstructionist' (18.49%) and 'Proactive' (17.65%). Only 13.45% of the companies, use 'Accommodative' as an acceptable CR strategy.

Table 6. Level of non-financial reporting among sampled companies

LEVEL	N	%
Comprehensive	26	16.88
Core	27	17.53
Others	36	23.38
Have intention (Low, Medium, High, Very high, Entirely)	25	16.23
No intention (Not at all, Slightly, No answer)	40	25.97
Total	154	100

Testing Research Propositions

To test a connection between the level of non-financial reporting and the type of CR strategy implemented by the company, the variables used for testing are treated as ranked variables with the following values:

- **Level of Non-Financial Reporting:** 1-Comprehensive, 2-Core; 3-Others; 4-Have intention; 5-No intention
- **CR Strategy Type:** 1-Innovative, 2-Proactive, 3-Accommodative, 4-Defensive, 5-Obstructionist. The higher rankings of CR strategies point to lower levels of development of the strategy and vice versa.

To examine whether there is a statistically significant difference between the strategies observed considering the level of non-financial reporting, a Kolmogorov-Smirnov test has been carried out. This test was chosen because the observed variables were treated as ranked variables. Table 8 shows the results of Kolmogorov-Smirnov's test of difference in the implementation of the CR strategies considering the level of non-financial reporting. The test proved that there is a difference between reporting levels at 1% (p-value = 0.000) for all three dimensions used to assess CR strategies (Companies' relationship towards the impact of non-financial reporting on their reputation; The level of interactive connection with stakeholders; Theoretical framework by Lee (2011) + innovative strategy).

The results of the Kolmogorov-Smirnov test have proven that there is a statistically significant difference in the implementation of the CR strategies considering the level of non-financial reporting.

Table 7. Distribution of companies regarding their CR strategy, and dimension used for measuring CR strategy

Type of CR strategy	Companies' relationship towards the impact of non-financial reporting on their reputation* (%)	The level of interactive connection with their stakeholders** (%)	Theoretical framework by Lee (2011) + innovative strategy** (%)
Innovative	37.72	8.40	27.73
Proactive	20.18	34.45	17.65
Accommodative	18.42	26.89	13.45
Defensive	14.91	14.29	22.69
Obstructionist	8.77	15.97	18.49

*N=114 (40 missing values)
**N=119 (35 missing values)

Table 8. Descriptive statistics of CR strategies according to the level of non-financial reporting

Level of non-financial reporting		Companies' relationship towards the impact of non-financial reporting on their reputation	The level of interactive connection with their stakeholders	Theoretical framework by Lee (2011) + innovative strategy
Comprehensive	Mean	2.520	2.577	2.423
	N	25	26	26
	Std. Dev	1.005	1.065	1.474
Core	Mean	2.000	2.407	2.259
	N	25	27	27
	Std. Dev	1.000	0.844	1.430
Others	Mean	2.667	2.500	2.889
	N	18	18	18
	Std. Dev	1.085	1.043	1.491
Have intention	Mean	3.250	3.440	3.240
	N	24	25	25
	Std. Dev	1.294	1.261	1.508
No intention	Mean	3.318	3.826	3.652
	N	22	23	23
	Std. Dev	1.129	1.193	1.229
Total	Mean	2.737	2.950	2.866
	N	114	119	119
	Std. Dev	1.198	1.213	1.501
Kolmogorov-Smirnov test	Chi-square (p-value)	18.221 (0.001***)	14.168 (0.007***)	25.191 (0.000***)

Note: *** Statistically significant with 1% probability

Furthermore, we tested the connection between the level of non-financial reporting and the type of CR strategy implemented by the company. Following research propositions (RP) about the connection between the level of non-financial reporting and the types of CR strategy implemented by the company have been proposed:

- **RP1:** Companies with a high degree of non-financial reporting use an innovative CR strategy
- **RP2:** Companies with a middle degree of non-financial reporting use a proactive CR strategy

- **RP3:** Companies with a low degree of non-financial reporting use an accommodative CR strategy
- **RP4:** Companies who do not report about CR, but plan to report, use a defensive CR strategy
- **RP5:** Companies who do not report about CR, use an obstructionist CR strategy

In order to examine these research propositions, a comparison of individual strategies (depending on three dimensions used) concerning the level of non-financial reporting was carried out.

Table 9 shows the distribution of companies concerning the CR strategy implemented by the company (according to dimension impact of non-financial reporting on companies' reputation) and the level of non-financial reporting.

Out of the companies that fall into the category Comprehensive, most of them use Innovative as their CR strategy of choice (30%). Companies that fall within Levels Core most often use Proactive as their chosen CR strategy (43%). Companies that fall within Levels Others mostly use Defensive as their CR strategy (24%). Out of all the companies that fall within categories: Have intention and No intention half of them (50%) in both categories use Obstructionist as their chosen CR strategy.

In order to determine whether the relationship between the CR strategy implemented by the company (dimension impact of non-financial reporting on companies' reputation) and the level of non-financial reporting is statistically significant, the Chi-square test was performed (Table 10). Test results showed that

Table 9. Distribution of companies concerning the CR strategy implemented by the company (dimension impact of non-financial reporting on companies' reputation) and the level of non-financial reporting

Reputation	Comprehensive	Core	Others	Have intention (Low, Medium, High, Very high, Entirely)	No intention (Not at all, Slightly, No answer)	Total
Innovative	30%	14%	19%	19%	19%	100%
Proactive	26%	43%	17%	13%	0%	100%
Accommodative	14%	33%	10%	14%	29%	100%
Defensive	18%	12%	24%	29%	18%	100%
Obstructionist	0%	0%	0%	50%	50%	100%
Total	26%	43%	17%	13%	0%	100%

the correlation is statistically significant with 1% probability (Chi-square = 33.90, p-value = 0.006).

Table 11 shows the distribution of companies concerning the CR strategy implemented by the company (dimension level of interactive connection with their stakeholders) and the level of non-financial reporting. Out of the companies that fall into the category Comprehensive, most of them use Proactive as their CR strategy of choice (42%). The same goes for the categories: Core and Others, with an even higher percentage of companies using Proactive as their CR strategy- up to 48% of companies for Core and 50% for Level Others. Companies that fall within categories Have intention and No intention mostly use Obstructionist as their chosen CR strategy, 28%, and 39% respectively.

Table 10. Results of Chi-square test (dimension impact of non-financial reporting on companies' reputation)

	Value	df	Asymp. Sig. (2-sided)
Pearson Chi-Square	33.90	16	0.006***
Likelihood Ratio	40.68	16	0.001
Linear-by-Linear Association	13.29	1	0.000
N of Valid Cases	119		

Note: *** Statistically significant with 1% probability

Table 11. Distribution of companies concerning the CR strategy implemented by the company (dimension level of interactive connection with their stakeholders) and the level of non-financial reporting

Stakeholders	Comprehensive	Core	Others	Have intention (Low, Medium, High, Very high, Entirely)	No intention (Not at all, Slightly, No answer)	Total
Innovative	12%	11%	11%	4%	4%	8%
Proactive	42%	48%	50%	24%	9%	34%
Accommodative	31%	30%	22%	24%	26%	27%
Defensive	8%	11%	11%	20%	22%	14%
Obstructionist	8%	0%	6%	28%	39%	16%
Total	100%	100%	100%	100%	100%	100%

Table 12. Results of Chi-square (dimension level of interactive connection with their stakeholders)

Chi-Square Tests	Value	df	Asymp. Sig. (2-sided)
Pearson Chi-Square	29.813	16	0.019***
Likelihood Ratio	33.963	16	0.005
Linear-by-Linear Association	20.586	1	0.000
N of Valid Cases	119		

Note: *** Statistically significant with 1% probability

In order to determine if this relationship is statistically significant, the Chi-square test was performed (Table 12). Results show that the correlation is statistically significant with 1% probability (chi-square = 29,813, p-value = 0.019).

Table 13 shows the distribution of companies concerning the CR strategy implemented by the company (dimension Lee (2011) + innovative strategy) and the level of non-financial reporting. Within Comprehensive and Core, most companies use Innovative as their CR strategy, 38% and 44% of companies respectively. Out of companies that fall within Others, most companies use Proactive as their CR strategy (39%). Within the categories Have intention and No intention, most companies use Defensive as their CR strategy, 40%, and 39% of companies respectively.

Results of Chi-square test showed that the correlation between the CR strategy implemented by the company (according to Lee (2011) + innovative strategy) and

Table 13. Distribution of companies concerning the CR strategy implemented by the company (dimension Lee (2011) + innovative strategy) and the level of non-financial reporting

Lee (2011) + innovative strategy	Comprehensive	Core	Others	Have intention (Low, Medium, High, Very high, Entirely)	No intention (Not at all, Slightly, No answer)	Total
Innovative	38%	44%	17%	24%	9%	28%
Proactive	19%	19%	39%	8%	9%	18%
Accommodative	19%	15%	6%	8%	17%	13%
Defensive	8%	11%	17%	40%	39%	23%
Obstructionist	15%	11%	22%	20%	26%	18%
Total	100%	100%	100%	100%	100%	100%

the level of non-financial reporting is statistically significant with 5% probability (chi-square = 29.392, p-value = 0.021), as shown in Table 14.

Table 15 shows the summary on the relationship between the level of non-financial reporting and type of CR strategy implemented.

DISCUSSION

Our intention, in this chapter, was to provide more understanding between CR strategy and non-financial reporting and investigate their connection. The empirical analysis was conducted on a sample of 154 companies from European countries using an

Table 14. Results of Chi-square test (dimension Lee (2011) + innovative strategy)

Chi-Square Tests	Value	df	Asymp. Sig. (2-sided)
Pearson Chi-Square	29.392	16	0.021**
Likelihood Ratio	29.967	16	0.018
Linear-by-Linear Association	12.936	1	0.000
N of Valid Cases	119		

Note: ** Statistically significant with 5% probability

Table 15. Summary on the relationship between the level of non-financial reporting and type of CR strategy implemented

Criteria for determining the CR strategy	Level of non-financial reporting				
	Comprehensive	Core	Others	Have intention (Low, Medium, High, Very high, Entirely)	No intention (Not at all, Slightly, No answer)
Stakeholders	Proactive	Proactive	Proactive	Obstructionist	Obstructionist
Reputation	Innovative	Proactive	Defensive	Obstructionist	Obstructionist
Lee (2011) + innovative	Innovative	Innovative	Proactive	Defensive	Defensive
Total	Innovative	Proactive	Proactive	Obstructionist	Obstructionist
Assumption	Innovative	Proactive	Accommodative	Defensive	Obstructionist
RP	RP1	RP2	RP3	RP4	RP5
Acceptance	Yes	Yes	No	No	Yes

online questionnaire and GRI database. In this study, we tried to examine whether there is a statistically significant difference between the CR strategies implemented by the companies and the level of non-financial reporting. Based on the results, the following can be concluded:

1. Companies with a high degree of non-financial reporting (Comprehensive) and concerning CR strategy (according to Lee (2011) + innovative strategy and the impact of non-financial reporting on companies' reputation) are implementing an innovative strategy. Considering that according to at least two criteria companies with a high degree of non-financial reporting are implementing an innovative strategy, we confirm our RP1
2. Companies with a middle degree of non-financial reporting use a proactive strategy (Core) and concerning CR strategy (according to the level of interactive connection with their stakeholders and the impact of non-financial reporting on companies' reputation) are implementing a proactive strategy. Considering that according to at least two criteria companies with a middle degree of non-financial reporting are implementing a proactive strategy, we confirm our RP2
3. Companies with a low degree of non-financial reporting (Others) and concerning CR strategy (according to the level of interactive connection with their stakeholders and according to Lee (2011) + innovative strategy) are implementing a proactive strategy. Considering that according to at least two criteria companies with a low degree of non-financial reporting are not implementing an Accommodative strategy, we do not confirm our RP3
4. Companies that currently do not report on CR but intend to in the future and concerning CR strategy (according to the level of interactive connection with their stakeholders and the impact of non-financial reporting on companies' reputation) are implementing an obstructionist strategy. Considering that according to at least two criteria companies that currently do not report on CR but intend to in the future, are not implementing a defensive strategy, we do not confirm our RP4
5. Companies that currently do not report on CR and have no intentions to start in the future and concerning CR strategy (according to the level of interactive connection with their stakeholders and the impact of non-financial reporting on companies' reputation) are implementing an obstructionist strategy. Considering that according to at least two criteria companies that currently do not report on CR and have no intentions to start in the future, are implementing an obstructionist strategy, we do confirm our RP5.

CONCLUSION

CR has become an increasingly popular concept among business, policymakers, media, and academics. One should bear on his/her mind that it is unlikely that CR alone can be the solution to problems caused by the lack of CR (Pedersen, 2015). Organizations are trying to achieve long term profitability and attract more and more new clients to be better than the competition and to survive in the global market. However, that is no longer sufficient, and they have to think about how to gain public trust, build a reputation, manage risks, and satisfy local community needs. By linking non-financial reporting to their strategy, companies are offering their stakeholders' channel through which they can address one another's needs and wants. As the research suggests, one of the main challenges for organizations is balancing between the need for profitability and contribution to the local community. In the global world characterized by digitization, where information is available 24/7, organizations try to create a culture of the non-financial report. Besides the implementation of CR into their strategic decisions, the organization should disseminate information and enable the local community to know and understand that they are doing their best to effect on economic, socio-political and environmental issues positively.

In this chapter, we have described connection among CR strategy and non-financial reporting, which are relevant for the successful business of organizations. Results of our empirical research showed that there is a statistically significant difference regarding the implementation of CR strategies and non-financial reporting. Regarding five proposed research questions, research results confirmed that companies with a high degree of non-financial reporting implement an innovative strategy, companies with a middle degree of non-financial reporting implement a proactive CR strategy and companies that currently do not report on CR and have no intentions to start in the future, implement an obstructionist strategy.

Still, our results have to be seen in the light of several research limitations. First one is related to the cross-sectional characteristics of our study as we analyzed data at one point in time. Consequently, data point to current state and level of non-financial reporting and CR strategy and thus present the main limitation of this study. Second is related to sample characteristics, as the sample encompasses organizations from a limited number of countries and countries that are quite similar in terms of legally and socially defined frameworks related to the application of CR. Therefore, to overcome these issues, future research might employ longitudinal study allowing for testing dynamics of relations, but also to capture additional mediation and moderation effects of different factors and trends that can affect the connection between CR strategy and reporting. Also, in future research sample needs to include a larger number of organizations from countries with more diverse cultural, political, and institutional frameworks.

Furthermore, while CR strategy and report present one of the influential factors for organizational sustainability, many additional factors (such as legal and governmental issues, non-governmental institutions, customer perception, etc.) influence organizational CR and should be considered and analyzed in future research. We should not also neglect the impact of digitalization on both CR strategy and non-financial reporting, but also to analyze various changes brought by the digital era that shape CR activities in organizations. By embracing digital changes and advancements, organizations can ensure new improvements in their CR strategies and develop unique non-financial reporting, allowing them to be more effective and efficient in the local community, which will positively reflect on their business model and in achieving higher sustainability levels.

REFERENCES

Adams, C. A. (2002). Internal organizational factors influencing corporate social and ethical reporting: Beyond current theorizing. *Accounting, Auditing & Accountability Journal, 15*(7), 223–250. doi:10.1108/09513570210418905

Adams, C. A. (2016). *CSR and Accountability in an Age of Digitization*. Retrieved from: https://drcaroladams.net/csr-and-accountability-in-an-age-of-digitization/

Aguinis, H. (2011). Organizational responsibility: Doing good and doing well. In S. Zedeck (Ed.), *APA handbook of industrial and organizational psychology* (pp. 855–879). Washington, DC: American Psychological Association. doi:10.1037/12171-024

Aguinis, H., & Glavas, A. (2012). What we know and don't know about corporate social responsibility: A review and research agenda. *Journal of Management, 38*(4), 932–968. doi:10.1177/0149206311436079

Bjartmarz, T., & Pedersen, E. R. G. (2015). CSR Standards and Social Accounting. In C. S. Responsibility (pp. 103–123). Croydon: SAGE.

Bonn, I., & Fisher, J. (2011). Sustainability: The missing ingredient in strategy. *The Journal of Business Strategy, 32*(1), 5–14. doi:10.1108/02756661111100274

Camilleri, M. A. (2017). *Corporate Sustainability, Social Responsibility and Environmental Management: An Introduction to Theory and Practice with Case Studies*. Heidelberg, Germany: Springer International. doi:10.1007/978-3-319-46849-5

Campbell, J. L. (2007). Why would corporations behave in socially responsible ways? An institutional theory of corporate social responsibility. *Academy of Management Review, 32*(3), 946–967. doi:10.5465/amr.2007.25275684

Castelló, I., Morsing, M., & Schultz, F. (2013). Communicative Dynamics and the Polyphony of Corporate Social Responsibility in the Network Society. *Journal of Business Ethics, 118*(4), 683–694. doi:10.100710551-013-1954-1

Dahlsrud, A. (2008). How corporate social responsibility is defined: An analysis of 37 definitions. *Corporate Social Responsibility and Environmental Management, 15*(1), 1–13. doi:10.1002/csr.132

Douglas, A., Doris, J., & Johnson, B. (2004). Corporate social reporting in Irish financial institutions. *The TQM Magazine, 16*(6), 387–395. doi:10.1108/09544780410563301

European Commission (EC). (2019). *Commission Staff Working Document Corporate Social Responsibility, Responsible Business Conduct, and Business & Human Rights: Overview of Progress.* Retrieved from: https://ec.europa.eu/docsroom/documents/34482

Fontaine, M. (2013). Corporate Social Responsibility and Sustainability: The New Bottom Line? *International Journal of Business and Social Science, 4*(4), 110–119.

Fung, B. (2014). The Demand and Need for Transparency and Disclosure in Corporate Governance. *Universal Journal of Management, 2*(2), 72–80.

Galbreath, J. (2009). Building corporate social responsibility into strategy. *European Business Review, 21*(2), 109–127. doi:10.1108/09555340910940123

Galpin, T., & Whittington, L. J. (2012). Sustainability leadership: From strategy to results. *The Journal of Business Strategy, 33*(4), 40–48. doi:10.1108/02756661211242690

Gangloff, T. (2018). *Corporate Social Responsibility and Digitization.* Retrieved from: https://novunex.com/corporate-social-responsibility-and-digitization/

GRI. (2019). *GRI Database.* Retrieved from: http://database.globalreporting.org

Heikkurinen, P. (2010). Image differentiation with corporate environmental responsibility. *Corporate Social Responsibility and Environmental Management, 17*(3), 142–152. doi:10.1002/csr.225

Husted, B. W., & Allen, D. B. (2011). *Corporate Social Strategy, Stakeholder Engagement and Competitive Advantage.* Cambridge, UK: Cambridge University Press.

Khan, H. U. Z., Halabi, A. K., & Samy, M. (2009). Corporate social responsibility (CSR) reporting: A study of selected banking companies in Bangladesh. *Social Responsibility Journal*, *5*(3), 344–357. doi:10.1108/17471110910977276

Knaut, A. (2017). How CSR Should Understand Digitalization. In T. Osburg & C. Lohrmann (Eds.), *Sustainability in a Digital World. CSR, Sustainability, Ethics & Governance* (pp. 249–256). Cham: Springer. doi:10.1007/978-3-319-54603-2_21

KPMG. (2017). *The road ahead, The KPMG Survey of Corporate Responsibility Reporting, 2017*. Retrieved from: http://www.kpmg.com/crreporting

Lee, K. H. (2009). Why and how to adopt green management into business organizations? The case study of Korean SMEs in manufacturing industry. *Management Decision*, *47*(7), 1101–1121. doi:10.1108/00251740910978322

Lee, M.-D. P. (2011). Configuration of external influences: The combined effects of institutions and stakeholders on corporate social responsibility strategies. *Journal of Business Ethics*, *102*(2), 281–298. doi:10.100710551-011-0814-0

Leon, R.-D. (2013). From the sustainable organization to sustainable knowledge-based organization. *Economic Insight – Trends and Challenges, 2*(2), 63-73.

Mikolajuk, Z., & Yeh Gar-On, A. (2002). Sustainable Development and Decision Support Systems. In G. E. Kersten, Z. Mikolajuk, & A. Gar-On Yeh (Eds.), *Decision Support Systems for Sustainable Development* (pp. 13–27). Boston: Springer. doi:10.1007/0-306-47542-1_2

Moneva, J. M., & Ortas, E. (2010). Corporate environmental and financial performance: A multivariate approach. *Industrial Management & Data Systems*, *110*(2), 193–210. doi:10.1108/02635571011020304

Okoye, A. (2009). Theorising Corporate Social Responsibility as an Essentially Contested Concept: Is a Definition Necessary? *Journal of Business Ethics*, *89*(4), 613–627. doi:10.100710551-008-0021-9

Pedersen, E. R. G. (2015). The Anatomy of CSR. In C. S. Responsibility. Croydon: SAGE.

Pejić-Bach, M., Omazić, M. A., Aleksić, A., & Zoroja, J. (2018). Massive Open Online Courses (MOOCs) as an education tool for corporate social responsibility. In A. Stachowicz-Stanusch & W. Amann (Eds.), *Management Education for Corporate Social Performance* (pp. 157–179). Charlotte, NC: IAP.

Porter, M. E., & Kramer, M. R. (2006). Strategy and society: The link between competitive advantage and corporate social responsibility. *Harvard Business Review, 84*(12), 78–92. PMID:17183795

Quinn, L., & Dalton, M. (2009). Leading for sustainability: implementing the tasks of leadership. Corporate Governance. *The International Journal of Business in Society, 9*(1), 21-38.

Reynolds, M., & Yuthas, K. (2008). Moral discourse and corporate social responsibility reporting. *Journal of Business Ethics, 78*(1-2), 47–64. doi:10.100710551-006-9316-x

Ringham, K., & Miles, S. (2018). The boundary of corporate social responsibility reporting: The case of the airline industry. *Journal of Sustainable Tourism, 26*(7), 1043–1062. doi:10.1080/09669582.2017.1423317

Roscoe, A. (2015). *The Digital Future of Corporate Responsibility*. Retrieved from: https://ssir.org/articles/entry/the_digital_future_of_corporate_responsibility#

Schreck, P. (2013). Disclosure (CSR Reporting). In *Encyclopedia of Corporate Social Responsibility* (pp. 801–810). Berlin: Springer-Verlag. doi:10.1007/978-3-642-28036-8_145

Sethi, S. P., Martell, T. F., & Demir, M. (2017). An evaluation of the quality of corporate social responsibility reports by some of the world's largest financial institutions. *Journal of Business Ethics, 140*(4), 787–805. doi:10.100710551-015-2878-8

Smith, P. A. C., & Sharicz, C. (2011). The shift needed for Sustainability. *The Learning Organization, 18*(1), 73–86. doi:10.1108/09696471111096019

Spanos, L. (2018). Complementarity and Interconnection Between CSR and Crowdfunding: A Case Study in Greece. In G. Grigore, A. Stancu, & D. McQueen (Eds.), *Corporate Responsibility and Digital Communities. Palgrave Studies in Governance, Leadership and Responsibility* (pp. 29–49). Cham: Palgrave Macmillan. doi:10.1007/978-3-319-63480-7_3

Suchman, M. C. (1995). Managing legitimacy: Strategic and institutional approaches. *Academy of Management Review, 20*(3), 571–610. doi:10.5465/amr.1995.9508080331

Wang, H., Tong, L., Takeuchi, R., & George, G. (2016). Corporate social responsibility: An overview and new research directions. *Academy of Management Journal, 59*(2), 534–544. doi:10.5465/amj.2016.5001

Wheeler, D., Colbert, B., & Freeman, R. E. (2003). Focusing on value: Reconciling corporate social responsibility, sustainability and a stakeholder approach in a network world. *Journal of General Management*, *28*(3), 1–27. doi:10.1177/030630700302800301

Whelan, G., Moon, J., & Grant, B. (2013). Corporations and Citizenship Arenas in the Age of Social Media. *Journal of Business Ethics*, *118*(4), 777–790. doi:10.100710551-013-1960-3

World Business Council for Sustainable Development (WBCSD). (2000). *Corporate Social Responsibility: Making Good Business Sense*. Geneva: World Business Council for Sustainable Development.

KEY TERMS AND DEFINITIONS

Corporate Responsibility: Also known as corporate social responsibility, represents the business concept that integrates socially and environmentally beneficial programs and practices into a distinctive business model. It is a form of corporate social and environmentalist self-regulation integrated into its everyday activities, or how companies manage the business processes to produce an overall positive impact on society.

Corporate Responsibility Strategy: Presents a framework for managing a company's current and future impacts on the economy, society, and environment.

Integrated Reports: Are linking non-financial reports to financial reports to provide transparent and reliable information on business activities and how corporation creates value on short, mid and long-term. It follows trend that companies should expand their reporting system beyond the stewardship of financial capital, and include all the resources they use as inputs to their business processes.

Non-Financial Reporting: It reflects a company's claim to portray, in printed or/and on the website, an account of its environmental, social and governmental performance and impact on economic performance, and to inform their stakeholders on company's role in society.

Chapter 4
Social Entrepreneurs as Servant Leaders:
Revealing the Implied Nature of Power in Servant Leadership

Devi Akella
Albany State University, USA

Niveen Eid
https://orcid.org/0000-0003-0284-8155
Birzeit University, Palestine

ABSTRACT

This chapter critically examines the servant leadership style of social entrepreneurs. Qualitative data from social enterprises in Palestine and Lukes third dimension power framework are used to explore the intricate forces of power, manipulation, and domination hidden within the service and follower-oriented model of servant leadership. Insights are provided on how the concept of 'service', the focal aspect in both social enterprises and servant leadership, could be another facet of soft and insidious power exercised by the social entrepreneurs over their followers. A political model of servant leadership is developed that demonstrates how social entrepreneurs could be imposing power and control over their followers under the guise of social mission, creation of social value, serving and empowering their followers and the community.

DOI: 10.4018/978-1-7998-1108-4.ch004

INTRODUCTION

In an effort, to move from the highly political and unethical business environments, ideologies pertaining to social enterprises and servant leadership have gained momentum. Social entrepreneurship is a phenomenon which combines the ideologies of both private organizations and non-profit organizations, i.e., social mission and a profit objective. Social enterprises have been described as a "radical innovation in the nonprofit sector" (Dart, 2004, p. 411), as "innovative approaches for dealing with complex needs" (Johnson, 2000, p. 1) and the "answer to worklessness, social isolation and inequality" (Organization for Economic Co-operation and Development, 2006 as cited in Bull, 2008, p. 268). These organizations primarily revolve around upliftment of the society, with a "tremendous hunger for doing things for others" (Saldinger, 2015, p. 1). Social enterprises undertake a strategic role in the economic development of the society and its youth. The leaders of social enterprises believe in service to the society and its citizens (Sabella & Eid, 2016). Social entrepreneurs thus are a special type of entrepreneurs, a "special breed of leaders" (Dees, 1998, p. 6) who are motivated and passionate about service to the community, creation of social value by addressing and resolving social problems and providing services to the poor (Seelos & Mair, 2005). Social entrepreneurs thus show an affinity towards servant leadership, where the leaders are also "affirmative builders of a better society" (Greenleaf & Spears, 2002, p. 24).

It was Greenleaf (1977) who first coined the oxymoron term of "Servant Leadership", to describe a leader whose chief motive was to serve as opposed to lead. A leader who placed the interests of others' needs, aspirations and interests above his/her own (Greenleaf & Spears, 2002). Servant leaders are essentially altruistic and moral people. These leaders have been described as stewards (Block, 1993; Senge, 1990), who regard their followers as their responsibility. The role of the leader is that of a role model, a risk taker, a servant who promotes others (The Greenleaf Center for Servant Leadership, 1997). Servant leaders are value and character-driven, performance and process oriented as well. Individuals who are driven by inner motivation, possess a shared vision, with the ability to develop and hone the talents of others. They strive to serve, add value and make a difference (Eva, Mulyadi, Sendjaya, van Direndonck & Liden, 2018; Page & Wong, 2002; van Dierendonck, 2010).

"Social entrepreneurs [thus] may have specific leadership attributes that allow classifying them as servant leaders" (Petrovskaya & Mirakyan, 2018, p. 755). In other words, social entrepreneurs or rather good and ethical entrepreneurs can be conceptualized as good leaders or servant leaders (Petrovskaya & Mirakyan, 2018).

However, recently questions have been raised about the problematic issues surrounding social enterprises (Akella & Eid, 2018; Bull, 2008; Dey & Steyaert,

2012). Social enterprises, it has been argued possess all ingredients of a "dangerous ideology" which requires reconciling two contrasting ideologies such as 'social' and 'enterprises', i.e., social mission with profit objectives (Bull, 2008). Social enterprises use market-based solutions to achieve social agendas and missions (Dart, 2004). This could result in social enterprises behaving as "businesses closer to the private system…" (Pearce 2003 as cited in Bull, 2008, p. 269) thereby raising questions about different aspects of social entrepreneurship and strongly advocating for critiques which would bring social enterprises, their objectives (Dey & Steyaert, 2012) and their leaders under empirical scrutiny (Klotz & Neubaum, 2016; Miller, 2015; Reid, Anglin, Baur, Short & Buckley, 2018).

This chapter will add towards the literature on leadership styles of social entrepreneurs by critically investigating the servant leadership style of social entrepreneurs. Servant leadership and its various facets are examined under the critical lens of power, domination and manipulation to understand how social entrepreneurs effectively manage their followers to fulfill the social mission and profit objectives of their organizations. Furthermore, there are very few critical studies which investigate sensitive but significant issues pertaining to the dark side of the servant leadership style of social entrepreneurs i.e., issues pertaining to the role of power, control, influence and domination (Klotz & Neubaum, 2016; Miller, 2015; Reid et al., 2018). For instance, it can be argued that the exterior façade of the spiritual side of servant leadership could be camouflaging another stronger invisible form of domination and control. Power can also be exerted while serving the employees, the organization and the community at large. Service need not imply an attitude of servility instead it could be another facet of power and its dimensions (van Dierendonck, 2010).

This chapter develops an empirically oriented power framework which facilitates an understanding of the power dynamics behind the social entrepreneur's goals of service and development of followers. The chapter will integrate Lukes (1974; 2005) and his arguments of third dimensional power to understand power and its intricate forces within this leadership model concerned with service and development of followers. Qualitative data gathered from two case studies, in the form of semi-structured and in-depth interviews with the owners and senior executives of two social enterprises in Palestine will be used to deconstruct the servant leadership style of social entrepreneurs.

BACKGROUND

Entrepreneurs are usually concerned with discovering business opportunities and exploiting them to earn profits (Santos, Caetano, Baron & Curral, 2015; Shane &

Venkataraman, 2000). In contrast social entrepreneurs, in addition to earning a profit, have a social mission involving solving social problems within the society and the creation of a social value (Gurvits, Nikitina-Kalame & Sidorova, 2015; Iyengar, 2014; Petrovskaya & Mirakyan, 2018). Social entrepreneurs respond to the market, institutional and state failures by establishing and operating commercial enterprises which contribute towards the upliftment of the society. More emphasis is laid on social values, social returns and social impacts instead of profitability (Iyengar, 2014). These entrepreneurs use business acumen to tackle a variety of social issues existing within the society such as poverty, hunger, lack of education and health care facilities and caring for the environment (Fowler, 2000). The focus of these social enterprises is on "creation of viable socio-economic structures, relations, institutions, organizations and practices that yield and sustain social benefits" (Fowler, 2000; p. 649). Social entrepreneurs thus are sensitive to the suffering experienced by people within a society (Petrovskaya & Mirakyan, 2018). They possess personality characteristics such as empathy (Mair & Noboa, 2003), compassion (Miller, Grimes, McMullen & Vogus, 2012) and altruism (Tan, Williams & Tan, 2005).

These entrepreneurs are focused on helping people in the society, a conceptual element which integrates them with the style of servant leadership. Servant Leadership describes a leadership style where the leader places emphasis on serving his/her followers. Greenleaf (1977) constructed this entire notion based on Herman Hesse's story, "Journey to the East". In this story there are a band of men who go on a mythical journey, where the central figure is Leo who accompanies the band in the capacity of a servant. Leo happens to have extraordinary presence whose value is appreciated when he disappears, and the entire group falls apart. Later the narrator of the story finds out that Leo was the titular head of the Order, a great and noble leader. Servant leadership is also rooted in the religion of Christianity and the teachings of Jesus Christ. It was Jesus Christ who taught his disciplines the concept of servant leadership. His words were "I am among you as one who serves" (The Holy Bible, Luke 22, p.27). Here the greatness of a leader was measured in terms of his/her ability to serve his/her employees or followers. A leadership philosophy where the leaders were interested in transforming their followers to become "healthier, freer, autonomous and more likely themselves to become servants" (Greenleaf, 1977, p. 13-14). The leader here is a servant first "...one [who] wants to serve...then conscious choice brings one to aspire to lead...the difference manifests itself in the care taken by the servant—first to make sure that other people's highest priority needs are being served" (Greenleaf, 1977, p. 13).

Servant leadership attributes include humility, trust, integrity, serving others needs before one's own, collaboration, unconditional love and learning (Buchan, 1998; DePree, 1992; Graham, 1991; Spears, 1998), modeling, pioneering, teaching and delegation, credibility, visibility, influence and appreciation of others (Russell,

2000; Russell & Stone, 2002). The new roles of leaders are that of coordinating, facilitating, coaching, supporting and nurturing the employees, ensuring open and effective communication and resolving intergroup conflict (Wong & Page, 2003). There is a clear emphasis on the service aspect, within the organization- service to the followers, outside the organization- service to the stakeholders, to the community and to the society (Barbuto & Wheeler, 2006). Servant leaders aim at creation of a better society by serving others (Barbuto & Wheeler, 2006).

Servant leadership in this chapter is approached from a contingency perspective or a skill-based ideology. Social entrepreneurs are conceptualized as servant leaders in terms of their behavior as a leader, which is dependent on the leader's values, followers' expectations from the leader as social beneficiaries and the failure of the society and government in solving social issues such as poverty and unemployment.

However, it can't be denied that the concept of servant leadership does overlap with other types of leadership styles such as transformational, ethical, virtuous and authentic in terms of the moral dimension and commitment to the followers. For instance, like transformational leadership, servant leadership is also follower centered and people oriented, encompassing concepts such as vision, trust, integrity, modeling and credibility (Stone, Russell & Patterson, 2004). However, while the focus of transformational leadership is on engaging followers to support the objectives of the organization. Servant leaders focus more on service, building an environment of trust for their followers and in their professional development (Greenleaf, 1977). Ethical leadership is a combination of transformational leadership with a strong ethical foundation where the leader demonstrates "normatively appropriate conduct through personal actions and interpersonal relationships and [promotes] such conduct to the followers through two-way communication, reinforcement and decision-making" (Brown, Trevino & Harrison, 2005, p. 120). For servant leaders as well, ethics is an important component in their dealings with their followers, stakeholders and with the society at large. But the essence of servant leadership is on the element of service in the leader-follower relationship which is not the basis of ethical leadership. In ethical leadership, in the context of leader-follower relationships, it is more about moral values and open dealings with one's followers (Reddy & Kamesh, 2016). Similarly, authentic leadership as a leadership construct incorporates morals, integrity and concern for others. Authentic leaders encompass attributes such as hope and trust, being self-aware and true to oneself (Brown & Trevino, 2006; Crawford, Dawkins, Martin & Lewis, 2019). The focus here is on the leader and his/her characteristics, being "deeply aware of how they think and behave" and are perceived by their followers in terms of moral values (Avolio, Luthans & Walumbwa, 2004, p. 4). Service to others is not the core concept here. Another style of leadership, gaining a major foothold within the theories of leadership is virtuous style of leadership (Newstead, Dawkins, Macklin & Martin,

2019), where virtues such as fairness, justice, care and integrity are fundamental to be considered a good leader. A virtuous leader is a leader who by integrating good values enables "eudemonia" or a prosperous and flourishing society (Newstead et al., 2019). Virtuous leadership concentrates on explaining how virtues instead of service to followers are relevant in the creation of a thriving society. Servant leadership's focus is on the attention given to the followers, in the interest shown by the leaders in nurturing and developing their followers to their full potential (Reid et al., 2018). The interests of the followers are given priority over the leader's own interests. The servant leader gives importance to helping others and on developing the resources of the community (Reid et al., 2018). Hence, the service aspect in servant leadership, makes it different from other moral forms of leadership styles such as ethical, virtuous, authentic and transformational (Reid et al., 2018).

Servant leadership's concern for social service and welfare and its stewardship dimension (Van Dierendonck & Nuijten, 2011) binds it to social entrepreneurship. Social entrepreneurship discusses the role of the leader at the external environmental level, social impact in terms of opportunity exploitation and establishment of business ventures with a social purpose (Reid et al, 2018). While servant leadership deals with the social agenda at the internal organizational level, in terms of management of followers, relationships with stakeholders and the society. In other words, both social entrepreneurship and servant leadership are driven by social mission, while the former is at the entrepreneurial level, the later is at the leadership level (Petrovskaya & Mirakyan, 2018). There is thus a strong likelihood that "…social entrepreneurs may be an appropriate population where servant leaders may be located" Petrovskaya & Mirakyan, 2018, p. 762). It can therefore be surmised that social entrepreneurs and social enterprises with their agendas of social mission and economic development lend themselves to the study of servant leadership.

However, in spite of social entrepreneurship's affiliation to being "good" entrepreneurs and of servant leadership to being "good" leaders, who are concerned about their followers, community and the society (Petrovskaya & Mirakyan, 2018). The concept of social entrepreneurship is "poorly defined" trying to merge two irreconcilable terms such as "social" and "enterprise" (Bull, 2008; Mair & Marti, 2006). The fact, cannot be ruled out or ignored that social entrepreneurs despite having the term 'social' in front of 'entrepreneurship', are still entrepreneurs who operate businesses where the element of profit is intact (Peredo & McLean, 2006; Zahra, Gedajlovic, Neubaum & Shulman, 2009). In other words, social entrepreneurs can be motivated by concepts similar to traditional entrepreneurs (Peredo & McLean, 2006; Zahra, et. al, 2009) like generation of profits and control of employees. This loosely juxtaposition of social and profit agendas within social entrepreneurship has strongly advocated the need for research projects especially empirically driven, which critically analyze the concept of social entrepreneurship, features of social enterprises, their leaders and their leadership styles (i.e., servant leadership) (Akella & Eid, 2018; Bull, 2008; Dey & Steyaert, 2012).

But there has been a singularly lack of debate in the form of critical studies on social enterprises (except to mention a few such as Akella & Eid, 2018; Bull, 2008; Dey & Steyaert, 2012) and their leaders and their leadership style, in this context servant leadership (excluding a few like Wong & Page, 2003; Christman, 2007; Minnis & Callahan, 2010). In fact, issues pertaining to power and domination have largely remained sidelined in empirical studies pertaining to social enterprises and leadership styles (Reid et al., 2018). But ironically even social entrepreneurs need to use power over their followers, even if it is an altruistic style of leadership like servant leadership. Of course, leaders do not have to always use strict and exploitative techniques to influence subordinates and one's followers. But leadership can be tough, power can still be exercised, control still be maintained, and regime of domination still be sustained under the domain of servant leadership. For instance, servant leadership covers two areas: servanthood and leadership. Servanthood aspect covers responsibilities pertaining to developing people and in helping build the organizations. The emphasis is on visioning, teambuilding, training and motivating. While leadership revolves around establishment of the business, developing people, valuing people, building community, displaying authenticity and integrity and sharing leadership and decision making (Wong & Page, 2003). The servant leader builds a community which believes in shared vision, s/he relinquishes control and legitimate authority (Peck, 1997). Leaders are no "longer bosses, commanders, controllers or big shots. Instead they are supporters, partners and providers" (Kouzes & Posner, 1993, p. 7). A servant leader just needs greater leadership skills, in the area of relationships, to seek employee obedience. The exterior façade of spiritual power could be camouflaging another stronger invisible form of domination and control. As aptly remarked "a leader is best when people barely know that s/he exits. Not so good when people obey and acclaim him. Worse when they despise him [sic]. If you fail to honor people, they [will] fail to honor you. But a good leader, who talks little, when his work is done, his [sic] aim fulfilled they will all say, we did this ourselves" (Lao Tzu cited in Wong & Page, 2003, p.11). Power can be exerted through service to the employees, the organization and to the community at large. Service need not imply an attitude of servility instead it could be another facet of power and its dimensions (van Dierendonck, 2010).

The objective of this chapter is not to measure the traits nor the behavior of the social entrepreneurs, but the types of power dimensions used by them to accomplish their goals and objectives and to ensure the cooperation of their followers in totality. The next section explores different types of power and which facet of power is appropriate for social entrepreneurs in fulfilling their leadership responsibilities.

MAIN FOCUS OF THE CHAPTER

This chapter develops a political model of servant leadership which demonstrates a positive exercise of invisible and insidious power used by social entrepreneurs using Lukes (1974; 2005) third dimension power and empirical analysis of qualitative data gathered from two social enterprises located in the West Bank region of Palestine.

To explore the power dimensions of servant leadership, it is essential to study "the less visible dimensions of power, operating collective forces and social arrangements to suppress potential issues and avert conflict by helping to shape men's [sic] beliefs and preferences and acting against their real interests" (Lukes, 1974, p. 10). Lukes (1974) argues that both one dimensional (Dahl, 1961) and two-dimensional power (Bachrach & Baratz, 1962) frameworks are too committed to the study of overt, actual behavioral conflicts, and tend to ignore how power can be used to prevent conflicts from arising in the first place. However, power can also be a process whereby it is possible to shape the cognitions, preferences and perceptions of individuals to promote the interests of one group over the other (Lukes, 1974; Tilly, 1991). Lukes (1974, 2005), Nye (1990); Gaventa (2006) and Gramsci (1971), Foucault (1977) have conceptualized and discussed such forms of latent and invisible power mechanisms.

These power philosophies question "how can willing compliance to domination be secured?" (Lukes, 2005, p. 10). If domination and influence hurt, why do individuals comply? Why don't they rebel or resist it? Lukes (2005) argues that power is more than saying "A exercises power over B when A affects B in a manner contrary to B's interests. Power is a capacity not the exercise of that capacity (it may never be, and never need to be, exercised), and you can be powerful by satisfying and advancing others' interests" (Lukes, 2005, p. 12). This is just one type of power, the domination form. In this exercise of power, the domination force, the followers' beliefs and values are shaped, so as to believe that this is the way of doing things, and no other way exists. Followers therefore consent to this domination and influence even when coercion is lacking. For instance, John Stuart Mill's (1869) account of the subjection of Victorian women and Pierre Bourdieu study where all followers see their subjugated condition as natural and fail to recognize their true desires and beliefs. The followers have forgotten their autonomous state of mind, and willingly accept their leader's vision as destined and natural. Again, the example of caste system in Hinduism, where a genuine consensus was reached on different values. With a complete acceptance of the societal hierarchical levels by the lower castes (Srinivas, 1962).

This would constitute the third dimension of power in working which misleads people's judgement and works against their free judgements. Power here conceals the real interests of individuals and in a subtle fashion guides them on the path of their leader's judgement and beliefs. "Men make their own history, but they do not make it just as they please, they do not make it under circumstances chosen by themselves but under circumstances directly encountered, given and transmitted from the past" (Marx, 1976, p. 15). People here don't have any problems or grievances yet there are interests that have been harmed. Power here "is an exercise [which] prevents people, to whatever degree, from having grievances by shaping their perceptions, cognitions and preferences in such a way that they accept their role in the existing order of things, either because they can see or imagine no alternative to it, or because they see it as natural and unchangeable or because they value it as divinely ordained and beneficial" (Lukes, 2005, p. 28; Gaventa, 1980; 2006).

This is similar to Nye's (1990) ideology of soft power, an alternative to hard power, the power of attraction—"the ability to shape what others want" (Bilgin & Elis, 2008, p. 11), "...can rest on the...ability to manipulate the agenda of political choices in a manner that makes others fail to express some preferences because they seem to be too unrealistic" (Nye, 1990, p. 7). Similarly, Gramsci (1971) defined hegemony as a process where "the intellectual, moral and philosophical leadership provided by the class or alliance of class and class fractions which is ruling, successfully achieves its objectives of providing the fundamental outlook for the whole society" (Bocock, 1986, p. 63).

Hegemony was effective in enforcing the beliefs and values of the dominating group but with the active consent of the entire organization (Clegg, 1989). Examples of hegemonic control can be seen in schools, unions, mass media and the church. In these situations, leaders just don't respond to their subjects but also shape their preferences as well. For instance, in schools "where the entire adult population has been subjected to some degree of indoctrination" (Dahl, 1961, p. 317). Similar examples of power exercises can be seen in teaching, mentoring, parenting, therapy, and apprenticeship relationships.

Power thus as domination could restrain people and their choices, secure their compliance, dictate over their own nature and judgement. Power could constrain the choices of other individuals and secure their compliance. To control, dominate, subjugate and conform them, seek their acquiescence and compliance.

METHODOLOGY AND EMPIRICAL ANALYSIS

Qualitative methodology encompassing the case study design was used to investigate power dimensions within servant leadership style of social entrepreneurs. Case study method is appropriate when trying to understand the issues pertaining to "why" and "how" (Merriman, 1998; Stake, 1998; Yin, 2003). It allows the researchers to gain perceptions about servant leadership from the social entrepreneurs themselves. And provides insights on the opinions and explanations of the social entrepreneurs regarding their roles, functions and responsibilities and whether their style of leadership imposes power and control over the employees. Further within the case study method it is possible to use multiple sources of evidences like interviews, physical artifacts and archival records (Yin, 1989). Six in-depth, semi-structured interviews were conducted. The two enterprises were contacted for in-person interviews with their owners and other senior executives by the secondary author. The two cases studies used in this paper were identified from a group of four cases through a snowball technique. The two case studies could be considered as a pilot sample which could pave the way for further analysis and investigation on other leaders of social enterprises in Palestine and in other countries. Both the enterprises are situated in the West Bank villages of Taybeh and Birzeit in the district of Ramallah. Both the social enterprises have been operating in Palestine for a period of more than seven years and are primarily concerned with resolving social issues within the region. These organizations will be referred to as SE1 and SE2. SE1 was an organization concerned with revitalizing an entire town, rebuilding its infrastructure, developing the commercial, cultural and economic structure of the region and increasing tourism within that area. While SE2, the other social enterprise was interested in increasing non-violence and disseminating peace in that geographical area and creating job opportunities for the youth.

All interviews were undertaken by the secondary author, were tape-recorded and lasted for sixty to ninety minutes. The interviews were in Arabic language and later translated into English during the time of transcription. The secondary author also reviewed newsletters and other forms of documentation published by and on the two social enterprises to get an overview of their leaders. She emailed all transcripts and her observations to the first author who analyzed the entire empirical data. All transcripts were thematically analyzed.

The data analysis started by reviewing the interview transcripts. The interview transcripts were read and then reread, and several codes were extracted. Examples of codes which emerged throughout the empirical analysis included: '*visioning*', '*building community*', '*developing people*', '*initiative*', '*teamwork*', '*partnerships*', '*social gaps/problems*', '*strategize*', '*resources*', '*entrepreneurship*', '*leader*'. Next commonalities and differences were identified, the themes were revised and then categorized under three broad emerging themes. Servant leadership encompasses leadership qualities of possessing a shared vision, being consistent and honest, practicing open and clear communication with a personal conviction in one's beliefs and ideas, qualities which overlap with other leadership styles such as authentic and ethical leadership styles. However, servant leaders in addition need to possess traits which enable service to one's followers, such as listening skills, empathy, awareness, persuasion ability, foresight, commitment to the community and to the growth of others and stewardship (Greenleaf, 1977). The entire concept of servant leadership can be understood using three themes which encompass a servant leader's responsibilities, the service aspect unique to this style of leadership and strategies used to influence and control one's followers i.e., three themes of Leadership, Servanthood and Third Dimension Power which provide insights on the basic issue being investigated within this chapter. These three themes will be discussed next.

Leadership

The leadership process for social entrepreneurs started with an objective to resolve social gaps left unresolved by the government within the economy. As stated during the interview:

Solving social problems, it is a crucial priority under this poor and unstable political, economic and social context. Our organization tries to solve bits and pieces of those social gaps that are underemphasized by the formal parties such as the governments and the other NGOs. (SE1)

All business endeavors of SE1 and SE2 revolved around serving people and the society and in building strong communities.

We feel that we are socially responsible for helping our community prosper and survive. Because people in this country live in extremely stressful and unsafe conditions due to the occupation and the poor economic downturns. We are here as positive social agents this is the core value where our work stands for and is created from. (SE1)

The social entrepreneurs had visions for the society, towards its upliftment, helping its citizens and generating methods to transform the economy. For instance, the owner of SE2, was a priest who resided in Taybeh, a small Christian village in middle of Palestine. He truly believed in the philosophy of non-violence and peace and this became his sole inspiration in starting social enterprises.

I served for over 13 years in Taybeh village a small Christian village in the middle of Palestine and that has religious history. I created the enterprise because I believed that I have a philosophy of life that should be communicated to my chapel and the youth citizens, this philosophy is the philosophy of non-violence and disseminating peace…. (SE2)

This dream and vision motivated him to start new projects, at least one each year. So that he could provide employment to Palestinian youth and eliminate immigration and brain drain from that region. He was referred to as "*businessman of hope*" and he liked this name because he wanted to serve humanity.

I dream that every year I should start at least one project so that I can employ people, generate profits.... My aim is that the projects should succeed to eliminate unemployment and immigration... People call me "the business man of hope" I liked this nick name and accepted my role as a human to serve others through these projects... I spread hope through this small social enterprise... (SE2)

The social entrepreneurs in the identified case studies achieved their social mission and agendas through shared leadership and decision making, collaboration and cooperation, participation and consensus with an emphasis on service to others. The social entrepreneur sought the support of his/her employees by inculcating strong beliefs and values of openness, empathy and a willingness to share one's experiences. The leader here encouraged equal participation, took up the roles of a coach and mentor who shared his/her knowledge and provided constructive feedback to his/her followers (Senge, 1990; Spears, 1995; 1998). The leader respected his/her followers and they in return trusted him/her.

People at SE1 have shared strong beliefs and values of sharing, giving, empathy and openness... We encourage participation and knowledge sharing, we are cooperative not competitive, we help weak people and respect individuals as human beings. (SE1)

Servant leadership tapped into the spiritual side of the social entrepreneurs which emphasized philanthropy and service allowing them to earn the trust and respect of their followers. The social entrepreneurs emerged as genuine and authentic in wanting to serve and help others. As the owner of SE2 told the authors:

... they think I don't find time to sleep. ... I tell them I sleep endlessly because I am having fun in what I am doing. The secret of success is to love what you do.... (SE2)

The leaders took the initiative, invested funds, tapped into their social networks and worked hard to transform the society. The owner of SE2, the priest published articles about the current situation in Palestine. He arranged for visitors to visit Taybeh to increase tourism within the region. His words were:

I also published articles in a national newspaper about the people, history, political and economic and social situation in the country.... I therefore, created outreach through the publications via the website and the other channels. Using my networks,

I was able to arrange for tour rounds to come and visit Taybeh and its chapel and its historical place.

In the above interview accounts, the social entrepreneurs respected the values and principles of fair trade and were committed to serving the disadvantaged groups. They possessed high levels of integrity, were honest and socially responsible. They followed transparent business policies like "triple bottom line" which generated high levels of goodwill, reputation and popularity with their followers and community.

We have created a triple, bottom line with our community, we started our work and initiated the organization from the bottom levels of the community's hierarchy, and we engaged everyone.

However, it cannot be denied that in all these interview quotes the social entrepreneurs also emerged as businessmen, who identified social gaps and converted them into business opportunities. They developed strategic initiatives, action plans, missions and visions, marketing and networking strategies. They were corporate leaders who were empathetic, participatory decision makers, systemic thinkers and mentors. Corporate leaders who were aware and interested in profits and further expansion of their business ventures just like traditional entrepreneurs. Ethics and surplus were now placed next to each other. Power and service were juxtaposed next to each other (Akella & Eid, 2018).

Servanthood

The focus of servant leadership is on serving the followers and developing their potential (Greenleaf, 1997; Kouzes & Posner, 1993; Sendjaya & Sarros, 2002; van Dierendonck, 2010). Similarly, with the leaders in the empirical data, they were interested in social and community development, empowerment of their followers, and helping poor and marginalized people. As the owner of SE1 explained in his interview:

We create awareness about social and community development. We empower people.... We initiate change even when no resources are available..., seek help from poor and marginalized people and engage them in the first place not to serve us but to let us serve them. We head towards social development....

The leaders achieved the above-mentioned social objectives through personal qualities such as self-sacrifice, service to others, sharing, perseverance, self-motivation and systemic thinking. They were good listeners, empathetic, persuasive and sincere (Kim, 2016). In the words of a leader at SE1:

1. *"Yes, they should have a strong sense of mission.*
2. *They need to sacrifice and give up things they really loved ... instead share these with others.*
3. *This is a tiring work and therefore they need to have perseverance and stay committed. No one from outside will give them the support they expect, they should be self-motivated and have a strong belief in their competence.*
4. *It is also important to stay focused and look at the overall picture and not the details or you'll lose vision.*
5. *They should be good listeners as opportunities are told sometimes... the engagement of others requires wisdom and critical and attentive listening.*
6. *They are expected to train themselves on respecting others' needs and let go certain things to receive other things in return.*
7. *They should influence the hearts, minds and behaviors of others.*
8. *The top of all of these is that they should be sincere and convinced down beneath their souls that they intend to benefit their societies."*

The leaders of the two case studies served their employees, other stakeholders and the community at large. They explored business opportunities, provided community service, helped the underprivileged families and motivated and developed their employees. They were stewards who shaped the entire organization according to the needs of the society (Block, 1993; Senge, 1990; Spears, 1998). As a social entrepreneur explained how their social enterprise improved the community, by providing educational opportunities to the youth, investing in rural villages, exploring hidden business opportunities in various tourism, catering and hospitality related areas.

We have balanced learning, community service and development, as education is utilized to improve the community...increased the development and investment in rural villages. Encouraged the underprivileged families to explore business opportunities in farm related production, tourism, catering, housing, hospitality and other business areas. We have explored and discovered hidden and neglected historical areas and sites such as those found in near villages including the Byzantine churches, crusaders sites and temples. (SE1)

Efforts were made by these social entrepreneurs to integrate people, their skills and abilities at all levels in all capacities. All interviewees who were occupying senior level positions in these social enterprises were actively engaged in developing and training their employees to become empowered and motivated. These leaders supported their employees, listened to them and guided them to become confident and independent. The followers were trained, coached and mentored by these leaders. The authors came across terms such as "*empowered*", "*develop*", "*connected*" and "*opportunities*". As commented:

We create... but we do not tell them how to exploit these opportunities, they are to depend on themselves and to develop their conditions. We are the center of this work. We created social network of businesses, individuals, places and houses that are connected together in an organized and novel fashion as a cluster of tourism place that offer tourists an unforgettable and a humble experience. (SE1)

And again:

We don't do the work and offer them money and philanthropy, we instead help them create their opportunities, exploit these opportunities, and let them share their experiences and problems with us when they seek help. (SE2)

For instance, in the interview account below, the leader trained local farmers about farming techniques, installed water networks for them and marketed their agricultural produce to local small businesses.

In an agricultural village called Ein Ariek ...I noticed that poor farmers needed basic network of infrastructure that allowed them to plant their lands and reach the distant areas. So, with the help of the Caritas organization, I helped those farmers by installing water networks and training farmers in planting and farming. I promised them that once their fruits, vegetables and other organic products are harvested, I will come and buy all their products and sell these products to local small businesses.... (SE1)

Similarly, another social entrepreneur offered coaching and mentoring classes to local chemical producers on how to sell, market and export cosmetics. As explained by him:

I also offered coaching and mentoring on how to market and sell/export locally produced products for a chemical producer of dead sea cosmetics.... (SE2)

The social entrepreneurs by serving their followers, stakeholders and the society were able to make a positive impact upon the economy and on the lives of the citizens. The interviewees proudly mentioned the number of jobs created, the number of families who were involved in their projects and were beneficiaries of their projects.

We have created jobs for over 50 persons and to more than 20 families of those who worked closely with us on construction tasks.... (SE1)

Third Dimension Power

The West Bank is the geographical area of Palestine and Palestine is under Israeli occupation. It is faced with numerous political, social and economic issues. Unemployment, political instability and poverty are serious issues which Palestine is currently facing. Palestinian youth prefer to immigrate abroad, rather than face the violence and political instability back home. This has meant brain drain, lack of qualified and educated youth who could contribute towards the economic development within that region (Akella & Eid, 2018). The social entrepreneurs in Palestine have therefore concentrated their efforts on developing the culture and history of Palestine, to inculcate in youth a pride in their identity, origin and create employment opportunities. As the owner of SE2 explained in detail:

The business is based on fundamental principles that can be summarized as:

1. Christians in this Holy Land are not needy; they should work in order to earn money for living, they work and maintain their dignity. Some NGOs make people dependent and useless because they feed them money and do projects that can help them without engaging them to endure the success and challenge of managing the projects and businesses by themselves. It is better to teach them how to catch a fish better than having them eat a fish.

2. In the West Bank and Gaza strip we had over 50,000 citizens, if we are under occupation and the government is not helping us find jobs should we leave the country or conquer our problems by creating jobs for ourselves. I prefer to teach others the latter behavior.

To live in this country, it is simply the fact that a young citizen should find a job/ create a job for himself/herself- buy a house/build it- get married and so on. So, if we have social enterprises that allow people to live by creating jobs, we will solve these social gaps.... (SE2)

To further clarify, one of the social entrepreneurs interviewed, described his project on dairy farming:

One of my most admired beliefs is that: 'I do believe that people should work to earn living'. I could have distributed the funds collected to each family as a gift, but I decided not to let them feel dependent or to think as needy people. I wanted them to conquer their situation and help themselves by encouraging them through buying a cow for each family and asking them to produce dairy products that can be sold later in farm and organic food markets. (SE2)

The social entrepreneur organized all families into a collective group and distributed cows to them. He then used the milk to manufacture dairy products and sold the products in the organic food market. The leader here did not ask the people if they wanted to engage in dairy farming or not. He just imposed his will on his parish. This is similar to the ideologies of third-dimension power (Lukes, 1974; 2005); soft power (Nye, 1990) and hegemonic control system where the perceptions and opinions of the other individuals are shaped and manipulated.

All social enterprises followed a similar pattern i.e., social problems were identified and then these problems were turned into valuable business opportunities. People were trained, coached and mentored to make these projects a financial success. The entrepreneur's business objective was to assist the society and its citizens, commercialize their efforts and market their artifacts and products i.e., vegetables and fruits, handiwork, books etc. As explained by a leader working in SE1:

We discover social problems from this continuous analysis of the community's anatomy and try to turn these problems into valuable opportunities and value generation projects. For example, we found that over 50% of the population in Birzeit is composed of graduates of young people under 30 years old and therefore, we wanted some of those who are interested to enroll in becoming a tourist guide... they were directed to study hospitality and tourism programs in collaboration with Bethlehem University and upon graduation they will work for SE1 not as volunteers but as employees. (SE1)

In the above interview account, graduates below the age of 30 years, were encouraged to study hospitality and tourism with their cost of education covered by SE1, under a contractual understanding that these individuals would work for SE1 in the future. This example shows signs of subtle and insidious form of manipulation and domination.

Another example relates to the publishing project managed by the owner of SE2. The owner advised the authors on what to write, used his contacts and then published their books in the foreign markets for a profit.

I noticed that some young citizens were good at writing and I had some networks with publishers, so I created a fourth project aimed at helping authors. Five textbooks were published and sold in foreign markets. (SE2)

In the above interview accounts, the social entrepreneurs showed characteristics of servant leaders who were serving the society. They were actively involved in community building, developing and motivating employees and transforming the societies. They were empathetic, good at listening, sharing, collaborative with participatory skills and sincere with high levels of integrity.

But to comment that they did not use power would be wrong. Power was used, it had shades of insidious, soft and third dimension elements where the values and preferences of others were shaped and influenced to reflect the needs and desires of the leader.

To summarize, the above empirical analysis suggests that social entrepreneurs in Palestine identified social problems (i.e., unemployment and immigration) prevailing within the community which they converted into viable business opportunities such as tourism, dairy farming and publishing. The social entrepreneurs then recruited Palestinian youth, educated, trained and mentored them to participate in these social ventures which improved the economic conditions of the economy. Here the philosophy and beliefs of the social entrepreneurs was shaping the preferences of the citizens and the outlook of the entire Palestinian society. The youth and community were being manipulated to integrate the beliefs and values of their leaders, yet they did not portray any external forms of grievances or conflicts. These arguments can become the three founding pillars of a political model of servant leadership [see Figure 1] where service to followers was achieved through persuasive domination.

THEORETICAL AND PRACTICAL IMPLICATIONS

This chapter contributes theoretically towards the literature on servant leadership, social enterprises, power and control and critical management studies. Critical empirical insights were provided on the dark side of social enterprises, in specific the ability of social entrepreneurs to dominate and influence their employees to push their own business agendas and objectives forward. At a practical level this chapter offers critical insights to social entrepreneurs, corporate leaders, managers and organizational consultants about the third-dimension power quotient existing in servant leadership. It focuses on the exploitative and manipulative component of servant leadership and its negative outcomes if it is not wielded properly. This study demands the formulation of political laws and regulations at the institutional level which would regulate social enterprises especially in developing and conflicted regions like Palestine to ensure these organizations don't deviate from their social mission agendas. And their leaders don't dominate their followers under the name of social welfare and upliftment of the society.

Figure 1.

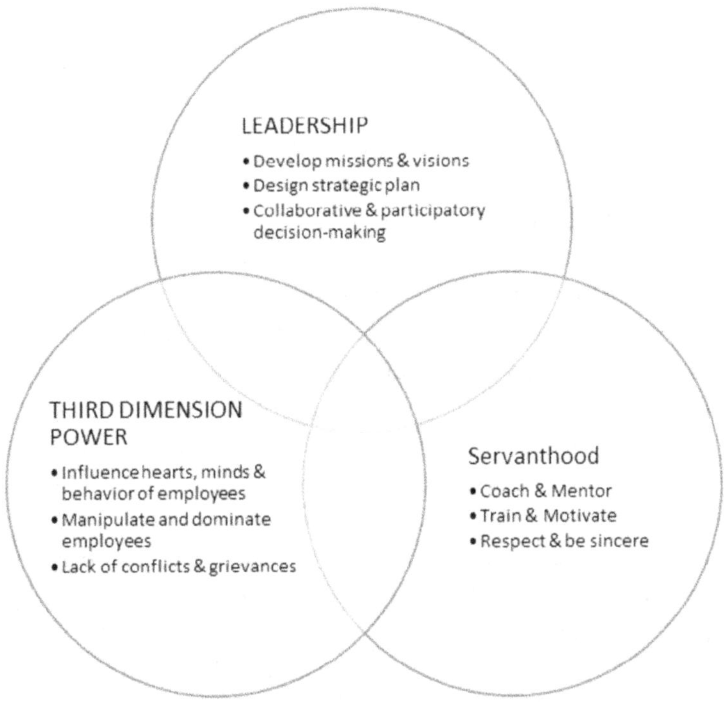

CONCLUSION AND DIRECTIONS FOR FURTHER RESEARCH

This chapter explored the power facets integrated within social entrepreneurship and servant leadership by developing an empirically driven political model of servant leadership. The empirical data collected from the two social enterprises in Palestine suggested that social entrepreneurs can be classified as servant leaders. Social entrepreneurs are leaders of organizations which have a social mission and agenda, are primarily concerned with upliftment of the society, with a focus on service and community development and in helping the poor and marginalized sections of the society. However social entrepreneurs as servant leaders in the above case studies showed traces of subtle and insidious hegemonic type of domination and manipulation (Lukes, 1974; 2005; Gramsci, 1971; Nye, 1990). This chapter argues that social entrepreneurs might be using soft or third dimension power, an insidious and invisible form of domination, where the leaders shape and influence the values and beliefs of their followers without the existence of any forms of overt or covert conflicts or grievances. For instance, the social entrepreneurs in Palestine had clear opinions about the political and economic conditions prevailing in that geographical region and how to improve these conditions. All their actions and decisions were influenced by these preexisting values and beliefs. Their employees were guided, motivated and trained according to these preexisting values and beliefs. The followers and their values and interests never entered the picture. This thus is a supreme exercise of power, where all followers were manipulated and dominated according to the interests and values of their leaders.

This chapter did not consider the views and opinions of the employees who were hired by the social entrepreneurs. Without the followers, employees or subordinates' perspectives it is not possible to comment whether the third-dimension power being exerted was exploitative or discriminatory, whether the outcomes of this domination had any negative intonations. Future research studies could also examine social entrepreneurs in different geographical locations, to understand the influence of cultural attributes on leaders and their values and interests. Also, whether these cultural attributes differ in terms of servant leadership concepts for social entrepreneurs in different geographical locations.

Moreover, it would also be interesting to explore whether servant leadership attributes favor the steward-principal relationship as opposed to the agent-principal relationship (Davies, Schoorman & Donaldson, 1997) and how this might impact the dual objectives of social enterprises—social mission and profit motive. Further studies could also investigate the impact of servant leadership style on the financial performance of an organization. What is the relation between the financial performance of an organization in terms of business profitability and growth and value creation when leaders of that corporation practice servant leadership? Does the level of employee performance increase when leaders practice servant leadership? What is the relationship between servant leadership and employee motivation levels? Are some questions which can be explored to understand the outcome of servant leadership in terms of business performance and efficiency (Miloloza, 2018).

Studies in future could also explore the role of third dimension power in servant leaders, other than social entrepreneurs, and how it could turn into an exploitative force. And what restraining forces need to be in place to hinder it from escalating beyond a certain limit. Servant leadership could also be examined within commercial organizations, which are being operated solely for-profit motives. This would provide insights on whether servant leaders could ever overcome the divisions created by the structural frameworks of the society.

REFERENCES

Akella, D., & Eid, N. (2018). Social enterprises in Palestine: A critical analysis. *Journal of Enterprising Communities: People and Places in Global Economy*, *12*(4), 510–544. doi:10.1108/JEC-01-2017-0010

Avolio, B., Luthans, F., & Walumbwa, F. O. (2004). *Authentic leadership: Theory building for veritable sustained performance*. Working paper. Lincoln, NE: Gallup Leadership Institute, University of Nebraska.

Bachrach, P., & Baratz, M. S. (1962). Two faces of power. *The American Political Science Review*, *56*(4), 947–952. doi:10.2307/1952796

Barbuto, J. E. Jr, & Wheeler, D. W. (2006). Scale development and construct clarification of servant leadership. *Group & Organization Management*, *31*(3), 300–326. doi:10.1177/1059601106287091

Bilgin, P., & Elis, B. (2008). Hard power, soft power: Toward a more realistic power analysis. *Insight Turkey*, *10*(2), 5–20.

Block, P. (1993). *Stewardship: Choosing service over self-interest*. San Francisco, CA: Berrett-Koehler.

Bocock, G. B. J. (1986). *Hegemony*. London: Tavistock.

Brown, M. E., & Trevino, L. K. (2006). Ethical leadership: A review and future directions. *The Leadership Quarterly*, *17*(6), 595–616. doi:10.1016/j.leaqua.2006.10.004

Brown, M. E., Trevino, L. K., & Harrison, D. (2005). Ethical leadership: A social learning perspective for construct development and testing. *Organizational Behavior and Human Decision Processes*, *97*(2), 117–134. doi:10.1016/j.obhdp.2005.03.002

Buchen, I. H. (1998). Servant leadership: A model for future faculty and future institutions. *The Journal of Leadership Studies*, *5*(1), 125–128. doi:10.1177/107179199800500111

Bull, M. (2008). Challenging tensions: Critical, theoretical and empirical perspectives on social enterprise. *International Journal of Entrepreneurial Behaviour & Research*, *14*(5), 268–275. doi:10.1108/13552550810897641

Christman, R. (2007). Servant-leadership and power in positional-led organizations. In *Servant Leadership Research Roundtable presentations*. School of Global Leadership and Entrepreneurship, Regent University. Retrieved from http://www.regant.edu/acad/global/publications/sl-proceedings/2007/christman.pdf

Clegg, S. (1989). *Frameworks of power*. Newbury Park, CA: Sage. doi:10.4135/9781446279267

Crawford, J., Dawkins, S., Martin, A., & Lewis, G. (2019). Putting the leader back into authentic leadership: Reconceptualizing and rethinking leaders. *Australian Journal of Management, 10*. doi:10.1177/0312896219836460

Dahl, R. A. (1961). *Who governs? Democracy and power in an American city*. London: Yale University Press.

Dart, R. (2004). The legitimacy of social enterprise. *Nonprofit Management & Leadership, 14*(4), 411–424. doi:10.1002/nml.43

Davis, J. H., Schoorman, F. D., & Donaldson, L. (1997). Towards a stewardship theory of management. *Academy of Management Review, 22*(1), 20–47. doi:10.5465/amr.1997.9707180258

Dees, J. G. (1998). *The meaning of social entrepreneurship*. Durham, NC: Center for the Advancement of Social Entrepreneurship, Fuqua School of Business, Duke University. Available at https://centers.fuqua.duke.edu/case/wp-content/uploads/sites/7/2015/03/Article_Dees_MeaningsofSocialEntrepreneurship_2001.pdf

DePree, M. (1992). *Leadership jazz*. New York, NY: Dell.

Dey, P., & Steyaert, C. (2012). Social entrepreneurship: Critique and the radical enactment of the social. *Social Enterprise Journal, 8*(2), 90–107. doi:10.1108/17508611211252828

Eva, N., Mulyadi, R., Sandjaya, S., Direndonck, D., & Liden, R. (2018). Servant leadership: A systematic review and call for future research. *The Leadership Quarterly*. doi:10.1016/j.leaqua.2018.07.004

Foucault, M. (1977). *Discipline and punish*. London: Penguin.

Fowler, A. (2000). NGDOs as a moment in history: Beyond aid to social entrepreneurship or civic innovation? *Third World Quarterly, 21*(4), 637–654. doi:10.1080/713701063

Gaventa, J. (1980). *Power and powerlessness: Quiescence and rebellion in an Appalachian Valley*. University of Illinois Press.

Gaventa, J. (2006). Finding the spaces for change: A power analysis. *IDS Bulletin, 37*(6), 23–29. doi:10.1111/j.1759-5436.2006.tb00320.x

Graham, J. W. (1991). Servant-leadership in organizations: Inspirational and moral. *The Leadership Quarterly, 5*(1), 105–119. doi:10.1016/1048-9843(91)90025-W

Gramsci, A. (1971). *Selections from the prison notebooks of Antonio Gramsci*. London: Lawrence and Wishant.

Greenleaf, R. (1977). *Servant leadership: A journey into the nature of legitimate power and greatness*. New York: Paulist Press.

Greenleaf, R., & Spears, L. (2002). *Servant leadership: A journey into the nature of legitimate power and greatness, 25th Anniversary Edition*. New York, NY: Paulist Press.

Gurtvis, N., Nikitina-Kalame, M., & Sidorova, I. (2015). Social enterprise in Estonia: Present situation and the perspectives of future development, survey of Estonian opinion. *Procedia: Social and Behavioral Sciences, 213*, 497–502. doi:10.1016/j.sbspro.2015.11.440

Iyengar, V. (2014). For goodness, we will change-An ultimate goal of the entrepreneurs for effecting social enterprise. *Procedia Economics and Finance, 11*, 767–774. doi:10.1016/S2212-5671(14)00240-8

Johnson, S. (2000). *Literature review on social entrepreneurship*. Canadian Centre for Social Entrepreneurship.

Kim, N. (2016). *Servant leadership style: What is it and what are the benefits and contrasts to other leadership styles?* (Unpublished Masters Thesis). University of North Carolina at Chapel Hill.

Klotz, A. C., & Neubaum, D. O. (2016). Research on the dark side of personality traits in entrepreneurship: Observations from an organizational behavior perspective. *Entrepreneurship Theory and Practice, 40*(1), 7–17. doi:10.1111/etap.12214

Kouzes, J. M., & Posner, B. Z. (1993). *Credibility: How leaders gain and lose it, why people demand it*. San Francisco: Jossey-Bass.

Lukes, S. (1974). *Power: A radical view*. London: Macmillan. doi:10.1007/978-1-349-02248-9

Lukes, S. (2005). *Power: A radical view*. New York: Palgrave Macmillan. doi:10.1007/978-0-230-80257-5

Mair, J., & Marti, I. (2006). Social entrepreneurship research: A source of explanation, prediction and delight. *Journal of World Business, 41*(1), 36–44. doi:10.1016/j.jwb.2005.09.002

Mair, J., & Noboa, E. (2003). *Social entrepreneurship: How intentions to create a social enterprise get formed.* IESE Working Paper No. D/521. Available at SSRN: https://ssrn.com/abstract=462283

Marx, K. (1976). Capital: Vol. 1. *Introduced by E. Mandel* (B. Fowkes, Trans.). Harmondsworth, UK: Penguin, New York.

Merriman, S. B. (1998). *Qualitative research and case study applications in education.* San Francisco, CA: Jossey-Bass.

Miller, D. (2015). A downside to the entrepreneurial personality? *Entrepreneurship Theory and Practice*, *39*(1), 1–8. doi:10.1111/etap.12130

Miller, T. L., Grimes, M. G., McMullen, J., & Vogus, T. (2012). Venturing for others with heart and head: How compassion encourages social entrepreneurship. *Academy of Management Review*, *37*(4), 616–640. doi:10.5465/amr.2010.0456

Mills, J. D. (1869). *The subjection of women.* London: Longmans, Green, Reader & Dyer. doi:10.1037/12288-000

Miloloza, I. (2018). Impact of leadership style to financial performance of enterprises. *Business Systems Research*, *9*(1), 93–106. doi:10.2478/bsrj-2018-0008

Minnis, S., & Callahan, J. (2010). *Servant Leadership in question: A critical review of power within servant leadership.* In The 17th International UFHRD Conference, Manchester, UK.

Newstead, T., Dawkins, S., Macklin, R., & Martin, A. (2019). The virtues project: An approach to developing good leaders. *Journal of Business Ethics*, *15*, 1–18.

Nye, J. S. (1990). Soft power. *Foreign Policy*, *80*(80), 153–171. doi:10.2307/1148580

Page, D., & Wong, P. (2002). A conceptual framework for measuring servant leadership. In S. Adjibolosoo (Ed.), *The human factor in shaping the course of history and development.* Boston: University Press of America.

Peck, S. (1997). *The different drum: Community making and peace.* New York: Simon and Shuster.

Peredo, A. M., & McLean, M. (2006). Social entrepreneurship: A critical review of the concept. *Journal of World Business*, *41*(10), 56–65. doi:10.1016/j.jwb.2005.10.007

Petrovskaya, I., & Mirakyan, A. (2018). A mission of service: Social entrepreneur as a servant leader. *International Journal of Entrepreneurial Behaviour & Research*, *24*(1), 755–767. doi:10.1108/IJEBR-02-2016-0057

Pomerantz, M. (2003). The business of social entrepreneurship in a down economy. In Business, 25(2), 25-28.

Reddy, A. V., & Kamesh, A. V. S. (2016). Integrating servant leadership and ethical leadership. In M. Chatterji & L. Zsolnai (Eds.), *Ethical leadership*. London: Palgrave Macmillan. doi:10.1057/978-1-137-60194-0_7

Reid, S. W., Anglin, A. H., Baur, J. E., Short, J. C., & Buckley, M. R. (2018). Blazing new trails or opportunity lost? Evaluating research at the intersection of leadership and entrepreneurship. *The Leadership Quarterly*, *29*(1), 150–164. doi:10.1016/j.leaqua.2017.11.005

Russell, R. F. (2000). *Exploring the values and attributes of servant leaders*. Doctoral dissertation, 5359435.

Russell, R. F., & Stone, A. G. (2002). A review of servant leadership attributes: Developing a practical model. *Leadership and Organization Development Journal*, *23*(3), 145–157. doi:10.1108/01437730210424

Sabella, A. R., & Eid, N. L. (2016). A strategic perspective of social enterprise sustainability. *Journal of General Management*, *41*(4), 71–89. doi:10.1177/030630701604100405

Saldinger, A. (2015). Can redefining humanity be Muhammad Yunus legacy? an interview with Muhammad Yunus. *Yunus Centre*. Available at http://muhammadyunus.org/index.php/news-media/in-the-media/1451-can-redefining-humanity-be-muhammad-yunus-legacy

Santos, S. C., Caetano, A., Baron, R., & Curral, L. (2015). Prototype models of opportunity recognition and the decision to launch a new venture: Identifying the basic dimensions. *International Journal of Entrepreneurial Behaviour & Research*, *21*(4), 510–538. doi:10.1108/IJEBR-04-2014-0058

Seelos, C., & Mair, J. (2005). Social entrepreneurship: Creating new business models to serve the poor. *Business Horizons*, *48*(3), 241–246. doi:10.1016/j.bushor.2004.11.006

Sendjaya, S., & Sarros, J. (2002). Servant leadership: Its origin, development and application in organizations. *Journal of Leadership & Organizational Studies*, *9*(2), 57–64. doi:10.1177/107179190200900205

Senge, P. (1990). The leader's new work: Building learning organizations. *Sloan Management Review*, *32*(1), 7–24.

Shane, S., & Venkataraman, S. (2000). The promise of entrepreneurship as a field of research. *Academy of Management Review*, *25*(1), 217–226.

Spears, L. (1995). *Reflections on leadership: How Robert K. Greenleaf's theory of servant leadership influenced today's top management thinkers*. New York: John Wiley.

Spears, L. C. (1998). *Insights on leadership: Service, stewardship, spirit and servant-leadership*. New York: John Wiley and Sons.

Srinivas, M. N. (1962). *Caste in modern India and other essays*. London: Asia Publishing House.

Stake, R. E. (1995). *The art of case study research*. Thousand Oaks, CA: Sage.

Stone, A. G., Russell, R. F., & Patterson, K. (2004). Transformational versus servant leadership: A difference in leader focus. *Leadership and Organization Development Journal, 25*(4), 349–361. doi:10.1108/01437730410538671

Tan, W. L., Williams, J., & Tan, T. M. (2005). Defining the 'social' in 'social entrepreneurship': Altruism and Entrepreneurship. *The International Entrepreneurship and Management Journal, 1*(3), 353–365. doi:10.100711365-005-2600-x

The Greenleaf Center for Servant Leadership. (1997, Fall). *The Servant Leader*.

The Holy Bible (New International Version). (1984). Grand Rapid, MI: Zondervan Publishing House.

Tilly, C. (1991). Domination, resistance, compliance...discourse. *Sociological Forum, 6*(3), 593–602. doi:10.1007/BF01114480

van Dierendonck, D. (2010). Servant leadership: A review and synthesis. *Journal of Management, 10*, 1–34.

Van Dierendonck, D., & Nuijten, I. (2011). The servant leadership survey: Development and validation of multidimensional measure. *Journal of Business and Psychology, 26*(3), 249–267. doi:10.100710869-010-9194-1 PMID:21949466

Wong, P., & Page, D. (2003). *Servant leadership: An opponent-process and the revised servant leadership profile*. Virginia Beach, VA: Servant Leadership Research.

Yin, R. K. (1989). *Case study research: Design and method*. Sage Publications.

Yin, R. K. (2003). *Case study research: Design and methods*. Thousand Oaks, CA: Sage Publications.

Zahra, S. A., Gedajlovic, E., Neubaum, D. O., & Shulman, J. M. (2009). A typology of social entrepreneurs: Motives, search processes and ethical challenges. *Journal of Business Venturing*, *24*(5), 519–532. doi:10.1016/j.jbusvent.2008.04.007

ADDITIONAL READING

Alvesson, M., & Willmott, H. C. (1992). *Critical management studies*. London: Sage Publications.

Bowman, M. A. (1997). Popular approaches to leadership. In P. G. Northouse (Ed.), *Leadership: Theory and practice*. Thousand Oaks, CA: Sage Publications.

Buchen, I. H. (1998). Servant leadership: A model for future faculty and future institutions. *The Journal of Leadership Studies*, *5*(1), 125–128. doi:10.1177/107179199800500111

Conger, J. A. (2000). Motivate performance through empowerment. In E. A. Locke (Ed.), *The Blackwell handbook of principles of organizational behavior* (pp. 137–149). Oxford: Blackwell.

Dowding, K. (2006). Three-dimensional power: A discussion of Steven Lukes' Power: A radical view. *Political Studies Review*, *4*(2), 136–145. doi:10.1111/j.1478-9299.2006.000100.x

Garfield, Z. H., von Rueden, C., & Hagen, E. H. (2018). The evolutionary anthropology of political leadership. *The Leadership Quarterly*, *30*(1), 59–80. doi:10.1016/j.leaqua.2018.09.001

Goldman, A. (1972). Towards a theory of social power. *Philosophical Studies*, *23*(4), 221–268. doi:10.1007/BF00356228

Graham, J. W. (1991). Servant-leadership in organizations: Inspirational and moral. *The Leadership Quarterly*, *5*(1), 105–119. doi:10.1016/1048-9843(91)90025-W

Greene, R. (1998). *The 48 laws of power*. NY: Penguin Books.

Hathaway, T. (2016). Lukes reloaded: An actor-centred three-dimensional power framework. *Politics*, *36*(2), 118–130. doi:10.1111/1467-9256.12099

Latiff, A., Majid, I. A., & Mohamad, M. (2017). Exploring servant leadership for social enterprise. *European Journal of Multidisciplinary Studies*, *2*(3), 7–23. doi:10.26417/ejms.v4i3.p7-23

Lorenzi, M. (2006). Power: A radical view by Steven Lukes. *Crossroads*, *6*(2), 87–95.

McGee-Cooper, A., & Looper, G. (2001). *The essentials of servant-leadership: Principles in practice*. Waltham, MA: Pegasus Communication.

Mills, C. W. (1959). *The power elite*. New York: Oxford University Press.

Newman, A., Neesham, C., Manville, G., & Tse, H. (2018). Examining the influence of servant and entrepreneurial leadership on the work outcomes of employees in social enterprises. *International Journal of Human Resource Management, 29*(20), 2905–2926. doi:10.1080/09585192.2017.1359792

Rose, N. (1990). *Governing the soul: The shaping of private self.* London: Routledge.

Yukl, I. G. (2006). *Leadership in organization*. New Jersey: Prentice Hall.

KEY TERMS AND DEFINITIONS

Domination: Is a process where the cognitions, preferences and perceptions of the followers are insidiously manipulated by the leader.

Entrepreneurship: Means to discover business opportunities and then to start a business venture to earn profits.

Hegemony: Means to enforce the beliefs and values of the powerful over the entire group but with their active consent.

Servant Leadership: Describes a leadership style where the leader is primarily interested in serving his/her followers. The leader here places the interests of his/her followers needs, interests, and aspirations before his/her own.

Social Entrepreneurship: Is a phenomenon which combines the concepts of social mission and business objectives to resolve social issues prevailing within the economy.

Soft Power: An approach usually found in international relations where power is exercised over others through persuasion, or through use of economic or cultural influence.

Third Dimension Power: Is a type of power exercise where followers have no complaints or grievances against their leader and yet their interests have been harmed. Power here shapes the perceptions and preferences of the followers to such degree that they accept the current situation as the natural order of things because they feel this is the predestined and unchangeable.

Chapter 5
Autocratic, Transactional, and Servant Leadership in Japan, France, and Mexico:
A Cross-Culture Theoretical Analysis

Cynthia Maria Montaudon-Tomas
https://orcid.org/0000-0002-2595-6960
UPAEP Universidad, Mexico

Ivonne M. Montaudon-Tomas
https://orcid.org/0000-0001-5794-7762
UPAEP Universidad, Mexico

Yvonne Lomas-Montaudon
https://orcid.org/0000-0002-6793-8330
Universidad Iberoamericana, Mexico

ABSTRACT

A theoretical study was conducted based on three different leadership styles: autocratic, transactional, and servant. The most relevant characteristics of each leadership style were summarized. A cross-culture study was proposed considering three countries from diverse clusters according to the Global Leadership and Organizational Behavior Effectiveness Project (GLOBE). Each culture was analyzed separately, and relevant statistics were presented as elements for comparison. Different models and tools to evaluate cultural differences were used to create a multiple perspectives overview. Information from the three leadership styles was used to further determine whether those styles fit the cultural descriptions in order to establish the most frequent and suitable leadership styles in the selected countries and to understand how leadership styles can vary from region to region.

DOI: 10.4018/978-1-7998-1108-4.ch005

INTRODUCTION

Cultural studies have become relevant in different areas of knowledge, including leadership. Studies that analyze cultural differences abound, as they are key factors to understand both leadership efficiency, and the ways in which the same situation can be considered either positive or negative in different cultures. National cultural values are powerful and ever present in individual and organizational lives (Rubio-Sánchez, Bosco, & Melchar, 2013).

There is an increasing interest in identifying commonalities on leadership styles around the world (Mittal & Dorfman, 2012), and a need to create models and theories that can provide insights on leader characteristics that are universal, and those that are culture-dependent. There are universal leadership attributes, such as leaders' supportiveness and charisma, and, at the same time, there are culture-specific attributes that include participation and directedness (Dickson, Castaño, Magomaeva, & Den Hartog, 2012).

Leaders can be found in every walk of life and business: leadership studies are relevant because they are strong predictors of organizational cultures (Nazari, Saud, Emami, Regagholiyan & Soltani, 2012), and cultural social and psychological factors can even predict entrepreneurial intentions (Pruett, Schinnar, Torey, Llopis, & Fox, 2009).

This chapter analyzes autocratic, transactional and servant leadership as three distinctive styles in the leadership continuum, in three particular countries, using a variety of tools for cultural analysis. The selection was based on creating possibilities for contrast by selecting countries from three different continents of the world, with distinct cultural patterns, and more or less equidistant. It has been suggested that contextual factors may influence the way leadership styles are effective in inducing desired behaviors in followers (Crossan,Vera & Najad, 2008). The three leadership styles were selected based on their main characteristics. The objective is to identify the main similarities and differences between cultures and leadership traits through a theoretical analysis. Cross cultural variances influence global innovations capabilities (Mc Carthy, 2019), affect innovation adoption processes, and help explain motivational perspectives to measure entrepreneurial intentions (Liñan & Chen, 2009).

LEADERSHIP AND SOCIAL ENTREPRENEURSHIP

There seems to be a leadership-entrepreneurship gap world-wide, especially in relation to the way leadership styles based on specific cultures can either enhance or hinder entrepreneurship activity. Social entrepreneurship differences can be explained in part by considering cultural leadership ideals at country-level (Lee & Kelly, 2019).

Studies on social entrepreneurial behavior suggest that comportment like commercial entrepreneurial behavior is influenced by leadership skills (Stephan & Drencheva, 2017).

Social entrepreneurship is an inspiring phenomenon (Petrovskaya & Miraykan, 2018); it refers to the development of innovative projects whose main aim is not individual profit but the transformation of society (Rivera, Santos, Martin-Fernandez, Requero, & Cancela, 2018). It involves the recognition, evaluation, and exploitation of opportunities that result in addressing the basic and long-standing needs of societies or, in other words, in the creation and establishment of social values (Austin, Stevenson & Wei Skillern, 2006). Its effectiveness is based on the reflection of the leader in their employees' intrinsic motivation, satisfaction, and commitment, and in the way their leadership style affects the attitudes of followers.

Petrovskaya & Miraykan (2018) have suggested that social entrepreneurship is based on a mission for service, and they established conceptual links between social entrepreneurship and servant leadership in facilitating social transformation because it leads to significant changes in the social, political and economic context for poor and marginalized groups (Letts, Brown, Alvord, 2010). Social entrepreneurship can also be identified as a style between transactional and transformational leadership; between the leaders and the inspirer. Transactional leadership is underdeveloped in social entrepreneurship. It is important to provide followers with guidance, and to manage the organization in an effective way. Transactional leadership creates a positive association with the intention of staying in the organization, and it gives positive feedback, which is related to effectiveness (WEF, 2014)

Position is obtained by analyzing the following variables: Government support, making a living, attracting skilled staff, gaining momentum, public understanding of social entrepreneurship, and access to investment (Thompson Reuters Foundation, 2016).

Table 1. Leadership styles of social entrepreneurs; Schwab Foundation survey results

Range	Leadership style		Social entrepreneurship
High	Ethical/ servant		6.0
Low	Transactional	Positive feedback Negative feedback Material reward	5.5 4.9 4.4
Medium	Autocratic		2.8

Source: World Economic Forum & Schwab Foundation, 2014.

Table 2. Social entrepreneurship ranking by selected country

Country	Position out of 44 countries
France	10
Mexico	28
Japan	40

Source: Thompson Reuters Foundation Social Entrepreneur Poll 2016.

SELECTED LEADERSHIP STYLES

Leadership theories, models and styles abound; some are being transformed and updated, while others are being constantly developed. Interest in particular leadership styles has shifted as there is a renewed interest in more participative styles. As an example of the earlier, a bibliometric analysis from the past 20 years regarding autocratic, transactional, and servant leadership is presented (Table 3).

Transactional leadership has attracted more attention of researchers, while autocratic leadership is the style that has the least number of publications and citing articles. Theories and methods regarding leadership are diverse, and their effects differ. Positive leadership situations in a country might be considered as negative in another as they can operate in certain conditions or fail in others. Furthermore, frameworks tend to overlap, so it is important to clearly differentiate the leadership styles that were selected for the analysis.

Autocratic Leadership

At one time, the study of autocratic leaders was one of the most researched topics (Meloen, 1993), but interest has decreased due to the emergence of additional leadership styles, and to the fact that it is often considered as negative to morale and performance because of a detachment from employees (Milgrun, 1991), which

Table 3. Bibliometric analysis on autocratic, servant and transactional leadership 1999-2018

Statistics	Autocratic	Transactional	Servant
Total publications	198	1631	517
Average citations per item	20.37	40.89	13.5
Sum of time cited	4,033	66,687	6,981
Citing articles	3,620	27,671	4,609

Web of Science, Clarivate analytics, 2018.

makes them appear as belligerent (Franz, Kendall-Taylor & Ezfow, 2014). By definition, autocratic leadership refers to the centralization of power, and is usually characterized by centralized decision making, power concentration, and control (de Hoog, Greer & den Hartog, 2015). It is a monitoring style where followers are treated as non-valuable contributors to goals, but are meant to follow orders (Amah, 2017).

General qualities include making all the important decisions; concern with task accomplishment rather than satisfaction; maintaining considerable social distance from followers, and motivating by the use of punishment or threat (Rast, Hogg & Geissner, 2013). According to Farh and Cheng (2000), and Wu, Hsu & Cheng (2003), autocratic leadership is based on top–down communication, underestimated subordinate competence, and disregard of subordinates' suggestions and contributions. Autocratic leaders create an image or *persona*, demanding achievement and specific conducts from followers. Leaders do not see the value in developing people and do not believe that they are able to make independent and valuable contributions (Amah, 2017).

Nevertheless, it is an appropriate style when crisis emerge or there is a time limit for decision making; when workers are uninformed and unaware regarding job descriptions (Amah, 2017); when it is required to deal with untrained employees, or with those who do not respond to their leadership styles. Studies have also suggested that it can yield positive effects when followers accept hierarchy (de Hoog, Greer & Den Hartog, 2015) because it allows rapid initiatives by leaders and agreement by followers (Mohiuddin, 2017; Adair, 2003), since the person in charge never has his actions or decisions cross-examined. However, autocratic leadership is only effective in decision making situations when the task has a clear structure and when follower commitment and support is high (Vroom and Yetton, 1973).

Independence of the group is weakened by the authority of the leader, creating defiance; and rivalry among group members tends to increase, leading to violence in extreme cases. Relevant characteristics of autocratic leadership have been summarized (Table 4).

Servant Leadership

Servant leadership became famous with the writings of Greenleaf (1979), but its origins go far back in religion and philosophy. It is perceived as important for effective leadership (Mittal & Dorfman, 2012) because there is a renewed interest in more caring and responsible leadership styles (Peterson, Galvin & Lange, 2012). This particular style has been considered as a key predictor of attitudes and behaviors of social entrepreneurship (Rivera, Santos, Martín-Fernández, Requero, Cancela, 2018).

Current models are anchored in the human drive to bond with others and to contribute to the betterment of society (Mittal & Dorfman, 2012); therefore, it

Table 4. Characteristics of autocratic leadership

Characteristics	Authors
Authority and control	Farh & Cheng, 2000, Adair, 2003
Bullying	Hoel, Glaso, Hetland, Cooper, Einarsen, 2010; Vartia, 1996
Centralized decision making	De Hoog, Greer & Den Hartog; Steer, Sanchez-Runde & Nardon, 2012
Centralized power	De Hoog, Greer & Den Hartog
Demands obedience	Pellegrini & Scandura; Chen, Chou, Wu, Huang & Farh, 2004
Despotic	Bass & Bass, 2008
Detachment	Milgrun, 1991
Disregard of others contributions	Farh & Cheng, 2000
Domination	Amah, 2017
Efficient	Amanchukwu, Stanley, & Ololube 2015
Hierarchical	Steer Sanchez Runde & Nardon, 2012
Monitoring	Amah, 2017
Manipulative	Thoroughghood, Padilla, Hunter & Tate, 2012; Warrick, 1991
Orders and rules	Amah, 2017
Power abuse	Warrick, 1991
Punishment and rewards	Chen, Chou, Wu, Huang & Farh, 2004
Quick response	Mohiuddin, 2017
Self interest	Amah, 2017; Collins, 2009
Social distance	Rast, Hogg & Geissner, 2013; Zhang, Huai & Xie, 2015; Zhang, Tsui & Wang, 2011
Stressful	Hoel & Salin, 2003
Task oriented	Rast, Hogg, Geissner, 2013; Einarsen, Aasland & Skogstad; Nystedt, 1997
Top down communication	Farh & Cheng, 2000
Use of force	Hoel & Salin, 2003

might seem that the construct overlaps transformational leadership, as it resembles idealized influence and intellectual stimulation, since servant leaders set an example for followers and inspire them with enthusiasm (Liden, Wayne, Zhao & Henderson, 2008).

Servant leaders set aside their self-interest for the advancement of their followers and organizations (Boyatzis & McKee, 2005). Altruism, humility, empathy, trust in others and integrity are considered as its essential attributes (Petrovskaya & Miraykan, 2018). Servant leaders are ready to serve first, sacrificially (Manz & Sims, 1987). Leaders strive selflessly to assist others before themselves and encourage others to do the same (Greenleaf, 1977), positively influencing overall performance (Van Knippenberg & van Knippenberg, 2005), something that is highly appreciated (Pekerti & Sendjaya, 2010) and results in increased commitment (De Cremer and Van Knippenberg, 2004). It emphasizes the benefits for followers and the community at large; it is follower-centric, altruistic, and it promotes moral, ethical and spiritual values (Sendjaya & Sarros, 2002). Followers develop a proactive and self-confident attitude that facilitates personal growth. Trust is an essential ingredient in servant leadership since followers are trusted to act in the best interest of the organization (Liden, Wayne, Zhao & Henderson, 2008).

Servant leadership can be learned and applied by those who have the will and the intent to change, grow and improve (Schwantes, 2015); the learning process and the influence are multidirectional, serving others from the bottom of the pyramid. Followers tend to respond well to this leadership style because leaders prove themselves as trustworthy servants. This leadership style is most effective in cultures with low power distances, low to moderate individualism, and low to moderate masculinity, low uncertainty avoidance and moderate to high long-term orientation (Hannay, 2000, p.9). The downsides include that it is difficult for the leader to always be empathetic and perseverant. It places the leader in a sometimes vulnerable position in which he needs to decide how to deal with risk sharing (Wilson, 1998), since the emphasis is on the long run.

Servant leadership is based on the idea that beneath cultural differences, there is a common basis, namely, being human. It is value-based, team-oriented, participative, self-protective and human-oriented. It moves from linear to cyclical thinking, from one direction to multiple, choosing between two opposite values to combine them; from one dimensional to a holistic perspective, from top down to bottom up, from analysis to synthesis (Coker, Flight & Valle, 2017).

The most recent studies on servant leadership by Eva, Robin, Sendjaya, van Dierendonck & Liden (2019), have suggested that despite the academic interest in servant leadership, research is fragmented across disciplines and there is a lack of coherence and clarity on servant leadership when compared to other leadership styles. Their 20-year systematic review revealed that servant leadership is similar to authentic leadership in that it acknowledges the importance of being true to one's actions with others, because people are driven either by their inner conviction or a higher calling, to make a positive difference. They proposed that servant leadership holds some similarities with ethical leadership because it incorporates the idea of stewardship, and that it resembles transformational leadership in that both styles focus on followers' needs.

After an extensive theoretical review, Eva et al, (2019: 114) proposed the following definition for servant leadership: Servant leadership is an other-oriented approach to leadership, manifested through one-on-one, prioritizing follower individual needs and interests, and outward reorientation of their concern for self towards concern for others within the organization and the larger community. Its basic motto is: *The more you serve the more you lead* (Tompenaars & Voerman, 2010).

The most relevant characteristics of servant leadership are presented as follows (Table 5).

Transactional Leadership

Transactional leadership was first called contingent reinforcement, and it was considered a core component of effective leadership (Bass, Avolio, Jung & Berson, 2003) and an independent and distinct construct underpinned by separate ethical foundations (Groves and La Rocca, 2011). Theory is founded on the idea that leader-follower relations are based on a series of exchanges (den Hartog, Van Muijen & Koopman, 1997) and its success is contingent on the leaders' abilities to meet and respond to the reactions and changing expectations of their followers (Kellerman, 1984). It evolved from the marketplace of fast simple transactions in search for gratification (McCleskey, 2014), allowing the followers to fulfill their own self-interest (Burns, 1978), and achieve organizational objectives while reducing anxiety (Mc Cleskey, 2014).

The purpose is an exchange of something valued (Kuhnert & Lewis, 1987; Yukl, 1999). It can also be defined as a bargaining process with subordinates to motivate behavior (Vito, Higgings and Denney, 2014). Leaders produce sufficient confidence in followers and support them in completing their tasks. They recognize follower's needs and desires and also clarify the way in which those needs will be fulfilled if followers show the expected performance (Winker, 2010). Leaders and followers must agree, accept and comply with the terms and provisions of the exchange of rewards for the work performed, because it is the foundation of leader-follower relationship (KK & Kumar, 2004; Howell & Hall-Merenda, 1999). Requirements and conditions of the task are explicit and leaders provide rewards for fulfilling those requirements (Kirschner-Bromley, 2007). It relies on a reinforcement strategy (Waldman, Bass and Einstein, 1989) in which leaders approach followers with an eye towards exchanging something (Burns, 1978) that can be material or symbolic, immediate or delayed, partial or whole (Bass, 1997). In transactional leadership, the prototype reward is a pay check, with other factors such as social recognition and pride in a job well done; transactional leadership capitalizes on these and other motivators to help teams work better and feel more satisfied with the job. It focuses on the here and now rather than on the future.

Potential rewards include an increase in salary, promotions, and other benefits, while penalties may include pay cuts, demotions, and terminations (Fernández, & Awamleh, 2004). Monetary rewards and recognitions are said to encourage knowledge

Table 5. Characteristics of servant leadership

Characteristics	Authors
Altruism	Patterson, 2003; Reed, Vidaver-Cohen & Colwell, 2011; Yukl, 2010
Authenticity	Van Dierendonck & Heeren, 2006; Van Dierendonck & Nuitjen, 2011; Wong & Page, 2003; Sendjaya, 2003; Laub, 1999
Awareness	Spears 1996; Barbuto & Wheeler, 2006; Liden, Wayne, Zhao & Henderson, 2008
Community building (value)	Barbuto & Wheeler, 2002, 2006; Spears, 1995, 1996; Laub, 1999; Linen,Wayne, Zhao & Henderson, 2008; Reed, Vidaver-Cohen and Colwell, 2011; Mittal & Dorfman, 2012
Credibility	Farling, Stone & Wilson, 1999; Russell, 2001; Russel & Stone, 2002
Egalitarianism	Mittal & Dorfman, 2012
Empathy	Spears, 2004; Barbuto & Wheeler; Aarum -Andersen, 2009; Ciulla, 2010; George, 2000; Holt and Marques, 2012; Mill-Chalmers, 2010
Empowerment	Patterson, 2003; Wong & Page, 2003; Dennis, 2004; Vann Dierendonck & Heeren; Van Dierendock & Niutjen; Russell, 2001; Liden, Wayne, Zhao & Henderson, 2008
Ethical	Sandjaya & Sarros, 2002; Liden, Wayne, Zhao & Henderson, 2008; Liden, Wayne, Zhao & Henderson, 2008
Healing	Spears, 1996; Barbuto & Wheeler, 2006; Liden, Wayne, Zhao & Henderson, 2008; Mittal & Dorfman, 2012
Humility	Patterson, 2003; Van Dierendonck & Nuitjen, 2011; Dennis, 2004; Van Dierendonck & Heeren, 2006; Chan et al. (2011), Dennis and Bocarnea, 2005, Owens and Hekman 2012, Van Dierendonck, 2011; Mittal & Dorfman, 2012
Influence	Russel & Stone, 2002; Schwarz et al, 2015
Inspiring	Spears, 1996; Graham, 1991
Integrity	Wong & Page, 2003; Van Dierendonck & Heeren, 2006; Reed, Vidaver-Cohen & Colwell, 2011; Russel & Stone, 2002
Listening	Spears 2003; ; Russel & Stone, 2002; Barbuto & Wheeler, 2002, 2006
People development	Spears, 1996; Liden, Wayne, Zhao & Henderson, 2008; reed, Vidaver-Cohen & Colwell, 2011; Hale and Fields, 2007; Laub, 1999
Relationships/ appreciation of others	Liden, Wayne, Zhao & Henderson, 2008; Wong & Page, 2003; Buchen, 1998; Russell, 2001; Sendjaya, 2003; Eva et al., 2019
Self-sacrifice	Whittington, Fran, May & Goodwin, 2006; Whittington, Fran, May & Goodwin, 2006; Choi & Mai-Dalton, 1998; Barbuto & Wheeler, 2006
Servanthood	Liden, Wayne, Zhao & Henderson, 2008
Shared leadership	Laub, 1999
Stewardship	Russel & Stone, 2002; Spears, 1999; Van Dierendonck & Heeren, 2006
Trust	Patterson, 2003; Dennis, 2003; Farling, Stine & Wilson, 1999; Russell, 2001; Russel & Stone, 2002;
Vision/foresight	Patterson, 2003; Wong & Page, 2003; Dennis, 2004; Farling, Stone & Wilson, 1999; Russell, 2001; Goodwin, 2006; Greenleaf, 2002

sharing in organizations (Hussain, Abbas, Lei, Jamal, Haider & Akram, 2017). Since the benefits are attractive, followers end up agreeing (Loya, 2006). Along with the followers, leaders agree, explicitly or implicitly, that the desired follower behaviors will be rewarded, while undesirable behaviors will be punished. When members are rewarded for their accomplishments, it is more likely that they will repeat the efforts in the future (Corrigan & Garman, 1999). They function best when they are adequate for the person receiving them and when they are not abused (Loya, 2006).

Effectiveness of transactional leadership depends on the acceptance of differences in organizational hierarchies and structures, and the collective work towards mutual exchange of benefits (Cheung, 2012; Bass, Avolio, Jung & Berson, 2003); success is positively correlated to followers' commitment, satisfaction and performance (Bycio, Hackett & Allen, 1995). Goals are stated in positive ways. According to Bass (1990), organizations whose leaders are transactional are less effective than those whose leaders are transformational, particularly if it is passive management-by-exception. Downsides include leading followers to short term relationships of exchange with the leader that can create resentments between participants (McCleskey, 2014). It can also create a habit of being compensated frequently, and employees stop working as they should unless they are compensated (Loya, 2006). Still, transactional leadership is a middle ground for social entrepreneurship linking employee performance to predictable financial reward.

A summary of relevant characteristics of transactional leadership is presented (Table 6).

A brief summary of leadership styles was developed to ease comparisons (Table 7).

The most relevant differences relate to individualistic or community practices, interests, decision making, the relation to others, the nature of ethics and the time span for leadership actions.

COUNTRIES SELECTED

Three countries were selected from the clusters proposed in the GLOBE study (House, Javidan, Hanges, & Dorfman, 2002): Mexico was chosen from the Latin American cluster; France from the Latin European; and Japan from the Confucian Asia cluster. While being completely different cultures, all three countries share some commonalities (Table 8).

Mexico has the lowest per capita Gross Domestic Product (GDP), human capital and development rankings. Surprisingly, its global skill index is identical to that of France.

Among the selected countries, Japan has the highest GDP per capita, the highest human capital and human development ranking (United Nations Development

Table 6. Characteristics of transactional leadership

Characteristics	Authors
Confidence	Winkler, 2010
Respond and react	Kellerman, 1984
Rewards and sanctions	Fernández & Awmaleh, 2004; Bass, 1990
Incentivize	Loya, 2006
Recognition based	Corrigan & Garman, 1999; Bass, Avolio, Jung & Berson, 2003
Positive reinforcement	Corrigan & Garman, 1999
Influence	Kanungo, 2001
Formal authority	Kanungo, 2001
Monitoring	Kanungo, 2001
Control	Bass & Steidlemeir, 1999; Kanungo, 2001
Concentrate on methods	Burns, 1978
Utilitarian/mutual altruism	Kanungo & Conger, 1993; Kanungo, 2001
Contractual agreements	Mc Clesky, 2014
Minimize anxiety	Mc Clesky, 2014
Reciprocity	Mc Clesky, 2014
Short term relationships	Burns, 1978, Mc Clesky, 2014
Shallow relationships	Mc Clesky, 2014
Contingent reward	Bass, 1990
Management by exception	Bass, 1990

Table 7. Leadership styles comparison

	Autocratic	Transactional	Servant
Motive/intent	Self interest	Mutual altruism	Altruistic
Norms and rules	Directive	Reciprocity norm internalized	Collective
Decision making	Individualistic, leader centered	Collective based on the exchange	Collective, follower centered
Relation to others	Distant	Independent	Network
Nature of goals	Personal goals	Pragmatic goals	Community goals
Behavioral strategy to influence others	Control, force	social contract exchange of resources	Personal development
Nature of ethics	Situational, interest based	Teleological and situational ethics.	Collective based
Time span	Here and now	Short term	Long term

Table 8. General statistics on selected countries

Variables	Mexico	Japan	France
GDP per capita	8,209	38,883	38,578
Competitive ranking	51	9	22
Human capital ranking	65	4	17
Human development Index	77	17	21
Global skills index	5.5	5.7	5.5

Sources: Hays Global Skills Index, 2008; United Nations Development Program, 2017.

Program, 2017). The uniqueness of Japan might derive from three factors: isolation, being an island country or archipelago; crowded conditions, and language itself, which is only spoken in their territory (Lewis, 2006).

France has a GDP per capita almost as high as that of Japan. The Human Development Index is slightly behind Japan, and the same level in the global skills index as Mexico. Lewis (2006) suggests that French people live in their own world and go by the *French manner,* which makes them unique.

A cross culture analysis on the selected countries was developed based on Hofstede's model (Hofstede's Insights, 2018). This model is based on measuring culture values. The dimensions are based on correlations considering two different variables, and the coefficient of correlation was measured to determine the strongest relationships of predominant values. A questionnaire was constructed which evolved over time, being first used at IBM and later in numerous companies around the world, creating country patterns. The development of the questionnaire required being cautions not to create stereotypes (Hofstede, Hofstede & Minkov, 2010). Results include the score in each dimension (Table 9).

Mexico is a hierarchical society. The ideal leadership/boss form is benevolent autocrat; it is a collectivistic society, very masculine with high uncertainty avoidance, long term orientation, respect for traditions and very high indulgence. France is a highly hierarchical and individualistic society, with a somewhat feminine culture, with high uncertainty avoidance, long term orientation, and considers indulgence versus restraint. Japan is the most masculine society of the three, with very high uncertainty avoidance, very high long-term orientation, guided by values, and is a culture of restraint.

Lewis' Culture Model was used to provide a different perspective (Table 10).

According to Lewis Culture Model (2006), Mexico is a multi-active culture; France is both multi-active and linear active, while Japan is reactive.

Hall & Hall (1990) cultural differences analysis provided additional insights by comparing space, context, time, and information flow in the three countries (Table 11).

Table 9. Cross-culture analysis based on Hofstede's dimensions

Dimensions	Mexico	France	Japan
Power distance	81 Hierarchical society shows inherent inequalities. Centralization. Ideal boss is a benevolent autocrat	68 Hierarchical society Fairly high Power Distance. Power is centralized	54 Borderline hierarchical society. Slow decision-making process Meritocratic society.
Individualism	30 Collectivistic society. Loyalty is paramount. Offence leads to shame.	71 Individualist society. Unique combination of a high Power Distance and high Individualism.	46 Collectivistic and Paternalistic society. Strong sense of shame and loosing face.
Masculinity	69 Masculine society. Managers are expected to be decisive and assertive. Conflicts are resolved by fighting them out.	43 Somewhat feminine culture. The upper class scores Feminine while the working class scores Masculine.	95 One of the most Masculine societies, combined with mild collectivism. Severe competition between groups.
Uncertainty avoidance	82 High uncertainty avoidance. Rigid codes of belief and behavior. Intolerant of unorthodox behavior and ideas. Emotional need for rules.	86 High Uncertainty Avoidance. Structure and planning are required as well as laws, rules and regulations. Need for emotional safety valves.	92 Very high uncertainty avoidance. Everything is prescribed for maximum predictability.
Long term orientation	24 Low long-term orientation. Culture is normative. Great respect for traditions Focus on achieving quick results.	63 High on long term orientation Pragmatic culture. Truth depends on situation, context and time. Adapt to change. Frugality and perseverance.	88 Very high long-term orientation. Life is guided by virtues and practical good examples.
Indulgence	97 Very high indulgence. Willingness to realize their impulses and desires with regard to enjoying life and having fun. Positive attitude. Tendency towards optimism.	48 Indulgence versus Restraint. Low score on happiness indices.	42 Culture of Restraint. Tendency to cynicism and pessimism. Not much emphasis on leisure time or indulgence.

Source: Hofstede Insights, 2018.

Table 10. Analysis on Lewis' (2006) culture model

Country	Mexico	France	Japan
Type of culture	Multi-active	Multi-active/linear active	Reactive
Traits	Emotional and impulsive. Attach great importance to family, relationships and feelings. Multitaskers. Problems with keeping up with time and agendas. Warm, emotional and impulsive.	Includes multi-active, mixed with being task oriented, highly organized planners who complete action chains by doing one thing at a time, preferably according to a linear agenda. A mix between warm, emotional / cool, factual.	Good listeners. Rarely initiate actions or discussions. Listen to others first. Establish others' positions, and then form their own opinion. Courteous, amiable, accommodating, compromisers.

Source: Lewis, 2006.

According to Hall & Hall (1990), Mexico, France and Japan share numerous cultural similarities. All of them are low territorial, high context and high information flow countries. The only difference is time consideration. While Mexico and France are considered polychronic, Japan is not. In low territorial cultures people have less ownership of space and boundaries are less important. High context cultures use more nonverbal communication cues that others, which can be confusing, and people share stronger bonds and a sense of affiliation. Because Japan is a monochronic culture, people are used to doing one thing at a time, carefully managing time, while France and Mexico manage their own time, and interactions are valued over time. Because of the high information flow, people in the three countries are expected to read other people's thoughts and be intuitive.

Trompenaars & Hampden-Turner proposed the Seven Dimension model of culture (Mind Tools, 2014), which was also used to compare the selected countries (Table 12). It is important to establish that the model does not consider personal experiences or differences between subcultures.

Table 11. Hall & Hall (1990) cultural differences analysis

Dimension	Mexico	France	Japan
Space	Low territorial	Low territorial	Low territorial
Context	High context	High context	High context
Time	Polychronic	Polychronic	Monochronic
Information Flow	High	High	High

Source: Hall & Hall, 1990.

An interesting fact is that while Hofstede considers Mexican culture to be collective, Trompenaars & Hampden-Turner defined it as individualistic. The analysis shows that Mexico is a particularistic and individualistic culture; personal involvement is diffuse and inhabitants are highly emotional. The culture is described as achieved; it uses synchronous time management and tends to follow internal directions. France is ranked in a level between universal and communitarian with a specific and low emotional culture. It is ascribed and follows both sequential and synchronous time, as well as internal directions. Japan is between universal and particular and is highly communitarian. It is defined as diffuse and neutral. It is ascribed, and it follows synchronous time and external directions.

According to Trompenaars & Hampden-Turner, (1998), the universal/particular dimensions explain whether people place high importance on laws, rules, values and obligations, versus people believing that their circumstances and relationships dictate the rules they live by.

The individual/ communitarian dimension determines whether people believe in personal freedom and achievement or if the group is more important than the person.

The specific/diffuse dimension refers to whether people keep work and personal life separated or not.

The neutral/emotional dimensions explain the way in which people manage their emotions: whether they make an effort to control their emotions and not reveal their feelings, or express them frequently and spontaneously.

The achieved/ ascribed status dimension explains whether people are valued because of their performance, or if they are valued for whom they are.

The sequential/synchronous time analyses whether activities follow a logical order, based on planning and scheduling, or if the past, present and future are interwoven periods.

Table 12. Trompenaars & Hampden-Turner (1998) seven dimensions model analysis

	Mexico	France	Japan
Universal/particular	Particularism	Middle Universalism	Middle Universalism
Individualism/communitarianism	Individualism	Communitarianism	High Communitarianism
Specific/diffuse	Diffuse	Specific	Diffuse
Neutral/emotional	High emotional	Low emotional	Extreme Neutral
Achieved/ascribed status	Achieved	Medium Ascribed	Ascribed
Sequential/synchronous time	Synchronous	Sequential/ Synchronous	Synchronous
Internal/external direction	Internal	Internal	External

Source: Trompenaars & Hampden-Turner, 1998.

Finally, the internal/external dimension explains whether there is a belief that people can control their environment, or if they consider that the environment controls them.

The only common dimension among the three selected countries in Trompenaars & Hampden-Turner's (1998) model is time. Although France has a mixed form, they all share the use of synchronous time.

Value analysis was performed using data from the 2014 GLOBE study (Table 13).

Meyer (2014) proposed a Culture Map based on eight dimensions that helps explain how people work. Dimensions are evaluated based on two opposing constructs. A comparison between Mexico, Japan and France was developed (Table 14), using the Culture Map Tool (2018).

According to Meyer (2014), Mexico has medium to high context communication; uses medium indirect negative feedback; has hierarchical leadership; decisions are made top-down; trust is relationship based; they mildly avoid confrontation; work on flexible time, and persuasion is slightly oriented towards principles first. France has medium to high context communication; evaluations are done through direct negative feedback; leadership leans towards being hierarchical; decisions are normally made top-down; trust is slightly more relationship based; when disagreeing they use confrontation. The French plan on medium to high flexible time and put principles first for persuasion. Japan has the distinction of being the highest context country in the world (Meyer, 2014); it provides completely indirect negative feedback; has a highly hierarchical leadership; decision making is consensual; trust is relationship based; they strongly avoid confrontation; use linear time, and persuade using applications first.

Table 13. Cultural practices and values by country, 2014 GLOBE study

Variables	Mexico	France	Japan
Performance orientation	4.2	4.1	4.1
Assertiveness	4.4	4.1	3.6
Future orientation	3.8	3.5	4.2
Humane orientation	3.9	3.3	4.2
Institutional collectivism	4.1	3.9	5.1
In-group collectivism	5.7	4.4	4.6
Gender egalitarianism	3.6	3.6	3.2
Power distance	5.2	5.2	5.1
Uncertainty avoidance	4.2	4.4	4.1

Source; GLOBE 2014*Approximate values

**Very low (1); low (2); relatively low (3); medium (4); relatively high (5); high (6); very high (7)

Table 14. Meyer's (2014) culture map analysis

Dimension		1	2	3	4	5	6	7	8	9	10	
Communication	Low context communication						MX	FR			JP	High context communication
Evaluation	Direct negative feedback		FR					MX			JP	Indirect negative feedback
Leading	Egalitarian							FR	MX		JP	Hierarchical
Deciding	Consensual	JP						FR	MX			Top-down
Trusting	Task based						FR	MX	JP			Relationship based
Disagreeing	Confrontational	FR						MX			JP	Avoids confrontation
Scheduling	Linear time		JP				FR		MX			Flexible time
Persuading	Principles first		FR			MX				JP		Applications first

Source: Culture Map, 2018.

The GLOBE project was designed to examine the relationships between culture, leadership and management patters. The study was conducted in 62 countries around the world (House, Javidan, Hanges, & Dorfman, 2002), leading to the selection of six distinctive dimensions that help establish differences and similarities between countries. The comparison between Mexico, France and Japan is presented as follows (Table 15).

The GLOBE Project is based on clusters and on individual cultures. Latin America is defined as highly charismatic, and team oriented, with a medium level of both participative and humane orientation, low level of autonomy and medium to high self-protectiveness. Latin Europe is characterized by medium to high charisma, medium team orientation and participation; low humane orientation and autonomy,

Table 15. GLOBE Study CLT dimensions

GLOBE CLUSTERS and countries	Latin America	Latin Europe	Confucian Asia
Dimensions and levels			
Charismatic/value based	High	Medium/High	Medium
Team oriented	High	Medium	Medium/High
Participative	Medium	Medium	Low
Humane oriented	Medium	Low	Medium/High
Autonomous	Low	Low	Medium
Self-protective	Medium/High	Medium	High

Source: GLOBE, 2004.

and medium self-protectiveness. Confucian Asia possess a medium level of charisma and values, medium high team orientation, low participative styles, medium high humane orientation, medium level of autonomy and high self-protectiveness.

Discussion

The country comparison, as well as the cultural differences that were reviewed using different tools and models helps explain each country's leadership style under a different light.

In Mexico, hierarchical power is the most effective form of leadership (Rodríguez-Rubio & Kaiser, 2013). Mexicans prefer leaders who stand above the crowd and command respect, and business is not predictable (Crane & Rizowy, 2005). Leadership is very autocratic, particularly in business circles. The Mexican leader may have acquired his position initially by birthright or nepotism but he will not be successful unless he develops a network of family, friends, partners and officials around him (Lewis, 2006). Power distance is a hindrance to servant leadership in Mexico. It is an obstacle because in high power distance societies, people do not necessarily see anything wrong with power being held by a central leader. Mexican face-saving and obsession with respect and status, resembles that of the Japanese (Lewis, 2006). Senior managers would normally command unquestioned obedience and respect from subordinates, but as in the case of Japan, they are obligated to reward them with loyalty, courtesy and protection (Lewis 2006). Directive leadership was related to organizational commitment. There is widespread distrust and fear of power. The dislike of privilege prevails (Steers, Sánchez-Runde & Nardon, 2012).

In France, the tight culture has implicit expectations so people behave exactly as specified by the leader (Mazdar, 2005). French management style is autocratic, although it is not made evident (Lewis, 2006). The highly organic nature of French enterprises implies interdependence, tolerance and teamwork (Lewis, 2006). Historically, France has been considered a country of revolutionary leadership (Burns, 1978). Leaders are more approachable and less intimidating (Steers, Sánchez-Runde & Nardon, 2012). Leadership is individualistic; power is placed in a single person, and it is highly visible (Steers, Sanchez-Runde & Nardon, 2012). Participatory methods are not common. Leadership is based on wisdom; the French use instinct rather than analysis. They are outgoing, energetic, intentional and emotional, and they want to create value through supportive relationships (Taleghani, Salmani, Taatian 2010). Work is an intellectual challenge; power is based on mental and intellectual superiority; structures are centralized; there is an emphasis on hierarchy, and respect for authority, it resists flexible structures, commanding leaders, and a formal respect for superiors. The French expect their leaders to be cultivated and highly educated

(Steers, Sanchez-Runde & Nardon, 2011). The best image of French leadership is the military general (Taleghani, Salmani, Taatian 2010).

In Japan, top executives have great power in conformity with consigned hierarchy, but they have little involvement in the everyday affairs of the company. They initiate policies that are conveyed to middle managers (Lewis, 2006). The Japanese prefer leaders who are much less visible. Major characteristics include self-discipline, continuous work, cooperation and coordination, close relationships with colleagues, support from senior leaders, sympathy with followers, less power of control than other cultures, fidelity, loyalty, decreased conflicts, oppositions, tensions and anxieties (Taleghani, Salmani & Taatian, 2010). Since the Japanese are long term planners, synchronic notion of time has been translated into industrial practices such as Just in Time. There is growing awareness that shared leadership can be beneficial for complex organizations in turbulent environments.

Japanese managers use more written memos for directions and to channel negative feedback through peers (Yukl, 2013). Leaders are not expected to be autocratic or authoritarian but sensitive to the feelings of others (Reischauer, 1989). Ideas often originate down to top; suggestions, ideas and inventions climb the hierarchical ladder by a process of signature collection, in which many people are involved (Lewis, 2006). Transactional leadership is not very effective in this country because transactional leaders are sometimes seen as being too mercenary and manipulative. They prefer something called gate-keeping leadership where they work to reduce barriers to achieve successful performance alongside their subordinates. Japan leaders are publicly willing to accept both the responsibility and the consequences of corporate failure. Leaders promote social activities outside of work. The preferred style is protective leadership and network leadership. The best way to represent Japanese leadership is the consensus builder (Taleghani, Salmani, Taatian 2010); servant leadership is deeply rooted in Japan; leaders give their followers more than they could ever repay, and because workers feel indebted to their leaders, they try to fulfill their wishes. A Japanese leader is the perfect example of the integration of master as servant (Trompenaars & Voerman 2010 p. 51). Japanese and French leaders produce paternalistic attitudes, as they concern themselves with the personal and private problems of the staff (Lewis, 2006).

CONCLUSIONS AND FURTHER RESEARCH

This chapter presents a comparative analysis on different leadership styles and distinct cultural patterns. It shows different tools that are useful in evaluating cultural differences, including Hofstede's Dimensions, Lewis' Culture Model, Hall & Hall's Cultural Differences Analysis, Trompenaars & Hampden Turner's Seven Dimensions

Model Analysis, Cultural Practices and Values according to the 2014 GLOBE study, Meyer's Culture Map, and the CLT Dimensions GLOBE study.

General statistics on the selected countries were included to provide a general background that could help explain cultural differences; still, it is difficult to generate universal consensus about predominant leadership styles in specific countries and cultures because leadership is always changing (Rowley, Ishikawa & Oh, 2019). Additionally, since leadership is a continuum more often than not it is possible to identify crossovers between leadership styles. Servant leadership, for instance is often connected to ethical and moral leadership. Moral leaders communicate messages of ethics and values by visibly and intentionally role-modeling ethical behavior, and by using the reward system essential in transactional leadership to hold followers accountable for ethical conduct (Brown and Treviño, 2006).

Overlapping of leadership styles is also visible at the cultural level. Leadership practices in Japan often differ from styles in the West in that seniority and prestige comes with age (Takahashi, Ishikawa & Kanai, 2012). In Japan the leader is the consensus builder. Leadership can be identified as transformational and gatekeeping (Ishikawa, 2012), as well as paternalistic (Rowley, Ishikawa & Oh, 2019). Japanese lead in a way in which they become an example from the service perspective (Trompenaars & Voerman, 2010), which demonstrate service leadership traits as well. Japan is transactional in that it understands employee needs providing for those needs. Transactions are based on trust; leadership is practical, direct and manages details, but it can induce stress (Bass, 1999). There is a respect for harmony, collective responsibility, and participative management; both authoritarian and democratic, disciplined consensus and feedback concur (Dogan, Schroevers & Ventr, 2016).

In France, the leader is the General (Taleghani, Salmani & Taatian, 2010). France is among the most autocratic directive, paternalistic-authoritarian (Trompenaars & Voerman, 2010). Decision making is authoritarian using an unapologetic, direct language. Leaders are autocratic, tough minded and dominant heads who assume full responsibility, and force opinions and values onto others.

In Mexico, the leader is the one that holds the power position. The country has a patriarchal- paternalistic, egocentric approach (Martínez, 2005; Dorfman & Howell, 1997), tradition prevails, it is slightly directive, charismatic, and autocratic.

Among the three selected leadership styles, Transactional leadership has been the most studied in terms of research in the area and published works, while autocratic has been the least analyzed of the three. The different tools that were used to compare cultures provided different results, as they rated variables differently. Hall & Hall's cultural model analysis presents the three cultures as very similar, while others, such as Trompenaars & Hampden-Turner, reveal them as completely different from each other. This shows it is best to use a collection of different tools and techniques when performing cultural analysis in order to attain a more general perspective.

Additional studies can be conducted based on country statistics and leadership styles to determine whether certain variables, such as wealth, education level, population density, crime rate, and others, are related to specific leadership styles. It would be interesting to select additional countries and leadership styles, as well as dominant cultural traits to create a comprehensive study about cultural differences and their relation with leadership.

REFERENCES

Aarum Andersen, J. (2009). When a servant-leader comes knocking. *Leadership and Organization Development Journal*, *30*(1), 4–15. doi:10.1108/01437730910927070

Adair, J. (2003). *Effective Leadership. How to Develop Skill, How to Book*. London: Macmillan.

Alvord, S. H., Brown, L. D., & Letts, C. W. (2004). Social Entrepreneurship: Leadership that Facilitates Societal Transformation-An Exploratory Study. *The Journal of Applied Behavioral Science*, *40*(3), 260–282. doi:10.1177/0021886304266847

Amah, O. E. (2017). Leadership styles & relational energy in high quality mentoring relationship. *Indian Journal of Industrial Relations*, *53*(1), 59.

Amanchukwu, R. N., Stanley, G. J., & Ololube, N. P. (2015). A Review of Leadership Theories, Principles and Styles and Their Relevance to Educational Management. *Management*, *5*(1), 6–14.

Austin, J., Stevenson, H., & Wei-Skillern, J. (2006). Social and Commercial Entrepreneurship: Same, Different or Both? *Entrepreneurial Theory and Practice*, *30*(1), 1–22. doi:10.1111/j.1540-6520.2006.00107.x

Barbuto, J. E. Jr, & Wheeler, D. W. (2006). Scale development and construct clarification of servant leadership. *Group & Organization Management*, *31*(3), 300–326. doi:10.1177/1059601106287091

Bass, B., & Bass, R. (2008). *The bass handbook of leadership: Theory, research, and managerial applications* (4th ed.). New York: The Free Press.

Bass, B. M. (1985). *Leadership and performance beyond expectations*. New York, NY: Free Press.

Bass, B. M. (1990). From transactional to transformational leadership: Learning to share the vision. *Organizational Dynamics*, *18*(3), 19–31. doi:10.1016/0090-2616(90)90061-S

Bass, B. M. (1997). Does the transactional-transformational leadership paradigm transcend organizational and national boundaries? *The American Psychologist*, *52*(2), 130–139. doi:10.1037/0003-066X.52.2.130

Bass, B. M. (1999). Two decades of research and development in transformational leadership. *European Journal of Work and Organizational Psychology*, *8*(1), 9–32. doi:10.1080/135943299398410

Bass, B. M., Avolio, B. J., Jung, D. I., & Berson, Y. (2003). Predicting unit performance by assessing transformational and transactional leadership. *The Journal of Applied Psychology*, *88*(2), 207–218. doi:10.1037/0021-9010.88.2.207 PMID:12731705

Bass, B. M., & Steidlmeier, P. (1999). Ethics, character, and authentic transformational leadership behavior. *The Leadership Quarterly*, *10*(2), 181–217. doi:10.1016/S1048-9843(99)00016-8

Boyatzis, R., & McKee, A. (2006). Inspiring others through resonant leadership. *Business Strategy Review*, *17*(2), 15–19. doi:10.1111/j.0955-6419.2006.00394.x

Brown, M. E., & Treviño, L. K. (2006). Ethical leadership: A review and future directions. *The Leadership Quarterly*, *17*(6), 595–616. doi:10.1016/j.leaqua.2006.10.004

Burns, J. M. (1978). *Leadership*. New York: Harper & Row.

Bycio, P., Hackett, R. D., & Allen, J. S. (1995). Further assessments of bass's (1985) conceptualization of transactional and transformational leadership. *The Journal of Applied Psychology*, *80*(4), 468–478. doi:10.1037/0021-9010.80.4.468

Cheng, B. S., Chou, L. F., Wu, T. Y., Huang, M. P., & Farh, J. L. (2004). Paternalistic leadership and subordinates responses: Establishing a leadership model in Chinese organizations. *Asian Journal of Social Psychology*, *7*(1), 89–117. doi:10.1111/j.1467-839X.2004.00137.x

Chhokar, J. S., Brodbeck, F. C., & House, R. J. (Eds.). (2013). *Culture and leadership across the world: The GLOBE book of in-depth studies of 25 societies*. Routledge.

Ciulla, J. B. (2010). Being there: Why leaders should not "fiddle" while Rome burns. *Presidential Studies Quarterly*, *40*(1), 38–56. doi:10.1111/j.1741-5705.2009.03753.x

Coker, K. K., Flight, R. L., & Valle, K. N. (2017). Social entrepreneurship: The role of national leadership culture. *Journal of Research in Marketing and Entrepreneurship*, *19*(2), 125–139. doi:10.1108/JRME-10-2016-0040

Corrigan, P. W., & Garman, A. W. (1999). Transformational and transactional leadership skills for mental health teams. *Community Mental Health Journal, 35*(4), 301–312. doi:10.1023/A:1018757706316 PMID:10452698

Crawford, J., Dawkins, S., Martin, A., & Lewis, G. (2017). Understanding the organizational climate of unethical leadership in the Australian Football League. *The Journal of Leadership Studies, 11*(2), 52–54. doi:10.1002/jls.21525

Crawford, J., Dawkins, S., Martin, A., & Lewis, G. (2018). Conceptualising authentic followers and developing a future research agenda. In Authentic Leadership and Followership (pp. 271-293). Palgrave Macmillan. doi:10.1007/978-3-319-65307-5_11

Crossan, M., Vera, D., & Nanjad, L. (2008). Transcendent leadership: Strategic leadership in dynamic environments. *The Leadership Quarterly, 19*(5), 569–581. doi:10.1016/j.leaqua.2008.07.008

Culture Map Tool. (2018). *Country comparison: Mexico, France and Japan*. Available on line: www.erinmeyer.com/tools/

Daghfous, N., Petrof, J. V., & Pons, F. (1999). Values and adoption of innovations: A cross-cultural study. *Journal of Consumer Marketing, 16*(4), 314–331. doi:10.1108/07363769910277102

De Cremer, D. (2006). Affective and motivational consequences of leader self-sacrifice: The moderating effect of autocratic leadership. *The Leadership Quarterly, 17*(1), 79–93. doi:10.1016/j.leaqua.2005.10.005

De Cremer, D., & Knippenberg, D. V. (2004). Leader self-sacrifice and leadership effectiveness: The moderating role of leader self-confidence. *Organizational Behavior and Human Decision Processes, 95*(2), 140–155. doi:10.1016/j.obhdp.2004.04.002

De Hoogh, A. H. B., Greer, L. L., & Den Hartog, D. N. (2015). Diabolical dictators or capable commanders? An investigation of the differential effects of autocratic leadership on team performance. *The Leadership Quarterly, 26*(5), 687–701. doi:10.1016/j.leaqua.2015.01.001

Den Hartog, D. N., Van Muijen, J. J., & Koopman, P. K. (1997). Transactional versus transformational leadership: An analysis of the MLQ. *Journal of Occupational and Organizational Psychology, 70*(1), 19–34. doi:10.1111/j.2044-8325.1997.tb00628.x

Dennis, R. S., & Bocarnea, M. (2005). Development of a servant leadership assessment instrument. *Leadership and Organization Development Journal, 26*(8), 600–615. doi:10.1108/01437730510633692

Dickson, M. W., Castaño, N., Magomaeva, A., & Den Hartog, D. N. (2012). Conceptualizing leadership across cultures. *Journal of World Business*, *47*(4), 483–492. doi:10.1016/j.jwb.2012.01.002

Doğan, A., Schroevers, S., & Venter, I. (2016). *The Global Leadership-Lab: A comparative analysis of leadership styles across nations*. CCBS Press.

Dorfman, P. W., & Howell, J. P. (1997). Managerial leadership in the United States and Mexico: Distant neighbors or close cousins? In C. S. Granrose & S. Oskamp (Eds.), Cross-cultural workgroups (pp. 234-264). Thousand Oaks, CA: Sage Publications.

Einarsen, S., Aasland, M. S., & Skogstad, A. (2007). Destructive leadership behavior: A definition and conceptual model. *The Leadership Quarterly*, *18*(3), 207–216. doi:10.1016/j.leaqua.2007.03.002

Eva, N., Robin, N., Dendjaya, J., Van Dierendonck, D., & Liden, R. C. (2018). Servant Leadership: A systematic review and call for future research. *The Leadership Quarterly*, *30*(1), 111–132. doi:10.1016/j.leaqua.2018.07.004

Farh, J. L., & Cheng, B. S. (2000). A cultural analysis of paternalistic leadership in Chinese organizations. In J. Li, A. Tsui, & E. Weldon (Eds.), *Management and Organizations in the Chinese Context* (pp. 84–127). London: Macmillan. doi:10.1057/9780230511590_5

Farling, M. L., Stone, A. G., & Winston, B. E. (1999). Servant leadership: Setting the stage for empirical research. *The Journal of Leadership Studies*, *6*(1-2), 49–72. doi:10.1177/107179199900600104

Fernandez, C., & Awamleh, R. (2004). The impact of transformational and transactional leadership styles on employee's satisfaction and performance: An empirical test in a multicultural environment. *International Business and Economics Research*, *3*(8), 65–76.

Frantz, E., Kendall-Taylor, A., & Ezrow, N. (2014). Autocratic fate: How leaders' post-tenure expectations influence the behavior of dictatorships. *Seton Hall Journal of Diplomacy and International Relations*, *15*(1), 39.

George, J. M. (2000). Emotions and leadership: The role of emotional intelligence. *Human Relations*, *53*(8), 1027–1055. doi:10.1177/0018726700538001

Giacomin, O., Janssen, F., Pruett, M., Shinnar, R. S., Llopis, F., & Toney, B. (2011). Entrepreneurial intentions, motivations and barriers: Differences among American, Asian and European students. *The International Entrepreneurship and Management Journal*, *7*(2), 219–238. doi:10.100711365-010-0155-y

GLOBE. (2014). *Globe CEO study 2014*. Available online: http://globeproject.com/study_2014

Goodwin, N. (2006). *Leadership in Health Care: A European Perspective*. Oxfordshire, UK: Routledge.

Graham, J. W. (1991). Servant-leadership in organizations: Inspirational and moral. *The Leadership Quarterly*, *2*(2), 105–119. doi:10.1016/1048-9843(91)90025-W

Greenleaf, R. (1977). *Servant Leadership: A Journey into the Nature of Legitimate Power and Greatness*. Mahwah, NJ: Paulist Press.

Greenleaf Center for Servant Leadership. (2010). *What Is Servant Leadership?* Retrieved from http://www.greenleaf.org/whatissl

Groves, K. S., & LaRocca, M. A. (2011). Responsible leadership outcomes via stakeholder CSR values: Testing a values-centered model of transformational leadership. *Journal of Business Ethics*, *98*(S1), 37–55. doi:10.100710551-011-1019-2

Hoel, H., Glaso, L., Hetland, J., Cooper, C. L., & Einarsen, S. (2009). 2010). Leadership styles as predictors of self-reported and observed workplace bullying. *British Journal of Management*, *21*(2), 453.

Hoel, H., & Salin, D. (2002). Organisational antecedents of workplace bullying. In Bullying and emotional abuse in the workplace (pp. 221-236). CRC Press.

Hofstede Insights. (2018). *Country comparison tool*. Available on line: www.hofstede-insights.com/product/compare-countries/

Hoftede, G., Hofstede, G. J., & Minkov, M. (2010). *Cultures and organizations: software of the mind: intercultural cooperation and its importance for survival*. New York: McGraw-Hill.

Holt, S., & Marques, J. (2012). Empathy in leadership: Appropriate or misplaced? An empirical study on a topic that is asking for attention. *Journal of Business Ethics*, *105*(1), 105. doi:10.100710551-011-0951-5

House, R., Javidan, M., Hanges, P., & Dorfman, P. (2002). Understanding cultures and implicit leadership theories across the globe: An introduction to project GLOBE. *Journal of World Business*, *37*(1), 3–10. doi:10.1016/S1090-9516(01)00069-4

Howell, J. M., & Hall-Merenda, K. E. (1999). The ties that bind: The impact of leader-member exchange, transformational and transactional leadership, and distance on predicting follower performance. *The Journal of Applied Psychology*, *84*(5), 680–694. doi:10.1037/0021-9010.84.5.680

Hussain, S. T., Abbas, J., Lei, S., Jamal Haider, M., & Akram, T. (2017). Transactional leadership and organizational creativity: Examining the mediating role of knowledge sharing behavior. *Cogent Business & Management*, *4*(1), 1–11. doi:10.1080/23311975.2017.1361663

Ishikawa, J. (2012). Transformational leadership and gatekeeping leadership: The rules of norm for maintaining consensus and shared leadership in team performance. *Asia Pacific Journal of Management*, *29*(2), 265–283. doi:10.100710490-012-9282-z

Kanungo, R. N. (2001). Ethical values of transactional and transformational leaders. *Canadian Journal of Administrative Sciences*, *18*(4), 257–265. doi:10.1111/j.1936-4490.2001.tb00261.x

Kanungo, R. N., & Conger, J. A. (1993). Promoting altruism as a corporate goal. *The Academy of Management Executive*, *7*(3), 37–48. doi:10.5465/ame.1993.9411302345

Kellerman, B. (1984). *Leadership: Multidisciplinary perspectives*. Englewood Cliffs, NJ: Prentice-Hall.

KK, V. D., & Kumar, M. P. (2004). Augmenting subordinates' competencies: Leaders matter. *Journal of Management Research*, *4*(3), 164.

Kluckhohn, F. R., & Strodtbeck, F. L. (1961). *Variations in value orientations*. Evanston, IL: Row, Peterson.

Kuhnert, K. W., & Lewis, P. (1987). Transactional and Transformational Leadership: A Constructive/Developmental Analysis. *Academy of Management Review*, *12*(4), 648–657. doi:10.5465/amr.1987.4306717

Laub, J. A. (1999). Assessing the Servant Organization: Development of the Servant Organizational Leadership Assessment (SOLA) Instrument. *Dissertation Abstracts International*, *60*(2), 921–922.

Lee, B., & Kelly, L. (2019). Cultural leadership ideals and social entrepreneurship: An international study. *Journal of Social Entrepreneurship*, *10*(1), 108–128. doi:10.1080/19420676.2018.1541005

Lewis, R. D. (2006). *When cultures collide*. Boston: Nicholas Brealey.

Liden, R. C., Wayne, S. J., Zhao, H., & Henderson, D. (2008). Servant leadership: Development of a multidimensional measure and multilevel assessment. *The Leadership Quarterly*, *19*(2), 161–177. doi:10.1016/j.leaqua.2008.01.006

Liñán, F., & Chen, Y. W. (2009). Development and cross–cultural application of a specific instrument to measure entrepreneurial intentions. *Entrepreneurship Theory and Practice, 33*(3), 593–617. doi:10.1111/j.1540-6520.2009.00318.x

Loya, S. (2016). *Liderazgo en el comportamiento organizacional [Leadership and Organizational Behavior]*. Trillas.

Manz, C. C., & Sims, H. P. (1987). Leading workers to lead themselves: The external leadership of self-managing work teams. *Administrative Science Quarterly, 32*(1), 106–129. doi:10.2307/2392745

Martinez, P. G. (2005). Paternalism as a positive form of leadership in the Latin American context: Leader benevolence, decision-making control and human resource management practices. In M. Elvira & A. Davila (Eds.), Managing human resources in Latin America: An agenda for international leaders (pp. 75-93). Oxford, UK: Routledge.

Mazdar, S. (2005). Subordinates information inquiry in undertain times: A cross cultural consideration of leadership style effect. *International Journal of Cross Cultural Management, 5*(3), 255–275. doi:10.1177/1470595805058412

Mc Cleskey, J. A. (2014). Situation, transformational and transactional leadership and leadership development. *Journal of Business Studies Quarterly, 5*(4), 117–130.

McCarthy, M. (2019). Cross-cultural differences in creativity: A process-based view through a prism of cognition, motivation and attribution. *Thinking Skills and Creativity, 32*, 82–91. doi:10.1016/j.tsc.2019.04.002

Meloen, J. D. (1999). Authoritarianism in the Netherlands: Mission completed? Downward trends in authoritarianism in the Netherlands 1970-1992 with an international comparison of world data. *Social Thought & Research, 22*(1/2), 45–93.

Meyer, E. (2014). *The Culture Map*. Available online: https://www.erinmeyer.com/tools/culture-map-premium/

Milgron, P., & Holmstrom, B. (1991). *Incentive contracts, asset ownership and job design*. London: Prentice Hall.

Mittal, R., & Dorfman, P. W. (2012). Servant leadership across cultures. *Journal of World Business, 47*(4), 555–570. doi:10.1016/j.jwb.2012.01.009

Mohiuddin, Z. A. (2017). Influence of leadership style on employee's performance: Evidence from literatures. *Journal of Marketing Management, 8*(1), 18.

Nazari, K., Saud, N., Emami, M., Regagholiyan, Z., & Asi Soltain, H. (2012). Analysis of leadership styles in different countries. *Journal of Basic and Applied Scientific Research*, *2*(8), 7921–7129.

Nystedt, L. (1997). Who should rule? Does personality matter? *European Journal of Personality*, *11*(1), 1–14. doi:10.1002/(SICI)1099-0984(199703)11:1<1::AID-PER275>3.0.CO;2-H

Page, D., & Wong, T. P. (2000). A conceptual framework for measuring servant leadership. In S. Adjibolosoo (Ed.), *The Human Factor in Shaping the Course of History and Development*. Lanham, MD: University Press of America.

Patterson, K. A. (2003). *Servant leadership: a theoretical model* (Doctoral dissertation). Regent University.

Pekerti, A. A., & Sendjaya, S. (2010). Exploring servant leadership across cultures: A comparative study in Australia and Indonesia. *International Journal of Human Resource Management*, *21*(5), 754–780. doi:10.1080/09585191003658920

Pellegrini, E. K., & Scandura, T. A. (2005). Construct equivalence across groups: An unexplored issue in mentoring research. *Educational and Psychological Measurement*, *65*(2), 323–335. doi:10.1177/0013164404268665

Peterson, S. J., Galvin, B. M., & Lange, D. (2012). CEO servant Leadership: Exploring executive characteristics and firm performance. *Personnel Psychology*, *65*(3), 656–596. doi:10.1111/j.1744-6570.2012.01253.x

Petrovskaya, I., & Mirakyan, A. (2018). A mission of service: Social entrepreneur as a servant leader. *International Journal of Entrepreneurial Behaviour & Research*, *24*(3), 755–767. doi:10.1108/IJEBR-02-2016-0057

Pruett, M., Shinnar, R., Toney, B., Llopis, F., & Fox, J. (2009). Explaining entrepreneurial intentions of university students: A cross-cultural study. *International Journal of Entrepreneurial Behaviour & Research*, *15*(6), 571–594. doi:10.1108/13552550910995443

Rast, D. E., Hogg, M. A., & Geisser, S. R. (2013). Self-uncertainty and support for autocratic leadership. *Self and Identity*, *12*(6), 635–649. doi:10.1080/15298868.2012.718864

Reed, L. L., Cohen, D. V., & Colwell, S. R. (2011). A new scale to measure executive servant leadership: Development, analysis, and implications for research. *Journal of Business Ethics*, *101*(3), 41–434.

Reischauer, E. O. (1989). *Japan: The story of a nation* (4th ed.). New York: Knopf.

Rivera, R. G., Santos, D., Martín-Fernández, M., Requero, B., & Cancela, A. (2018). Predicting attitudes and behavioural intentions towards social entrepreneurship: The role of servant leadership in young people/Predicción de las actitudes y las intenciones conductuales hacia el emprendimiento social: el papel del liderazgo de servicio en los jóvenes. *Revista de Psicología Social, 33*(3), 650–681. doi:10.1080/02134748.2018.1482057

Rodríguez-Rubio, A., & Kiser, A. I. T. (2013). An examination of servant leadership in the United States and Mexico. Do age and gender make a difference? *The Global Studies Journal, 5*(2), 127–149. doi:10.18848/1835-4432/CGP/v05i02/40848

Rowley, C., Ishikawa, J., & Oh, I. (2019). Changing facets of leadership in East Asia: Globalization, innovation and performance in Japan, South Korea and China. *Asia Pacific Business Review, 25*(2), 161–172. doi:10.1080/13602381.2018.1557425

Rowley, C., Oh, I., & Jang, W. (2019). New perspectives on East Asian leadership in the age of globalization: Local grounding and historical comparisons in the Asia Pacific region. *Asia Pacific Business Review, 25*(2), 307–315. doi:10.1080/13602381.2018.1557424

Rubio-Sánchez, A., Bosco, S. M., & Melchar, D. E. (2013). Servant leadership and world values. *The Global Studies Journal, 5*(3), 19–30. doi:10.18848/1835-4432/CGP/v05i03/40803

Russell, R. F. (2001). The role of values in servant leadership. *Leadership and Organization Development Journal, 22*(2), 76–84. doi:10.1108/01437730110382631

Russell, R. F., & Stone, A. G. (2002). A review of servant leadership attributes: Developing a practical model. *Leadership and Organization Development Journal, 23*(3), 145–157. doi:10.1108/01437730210424

Schwantes, M. (2015). Servant leadership. The case for the best leadership philosophy. *Leadership Excellence Essentials, 32*(5), 30–31.

Sendjaya, S., Sarros, J. C., & Santora, J. C. (2008). Defining and measuring servant leadership behaviour in organizations. *Journal of Management Studies, 45*(2), 402–424. doi:10.1111/j.1467-6486.2007.00761.x

Spears, L. C. (1996). Reflections on Robert K Greenleaf and servant leadership. *Leadership and Organization Development Journal, 17*(7), 33–35. doi:10.1108/01437739610148367

Spears, L. C. (2004). Practicing servant-leadership. *Leader to Leader, 34*(34), 7–11. doi:10.1002/ltl.94

Steers, R. M., Sánchez-Runde, C., & Nardon, L. (2012). Leadership in a global context: New directions in research and theory development. *Journal of World Business*, *47*(4), 479–489. doi:10.1016/j.jwb.2012.01.001

Stephan, U., & Drencheva, A. (2017). The person in social entrepreneurship: A systematic review of research on the social entrepreneurial personality. In G. Ahmetoglu, T. Chamorro-Premuzic, B. Klinger, & T. Karcisky (Eds.), *The Wiley Handbook of Entrepreneurship* (pp. 205–229). Hoboken, NJ: John Wiley & Sons. doi:10.1002/9781118970812.ch10

Takahashi, K., Ishikawa, J., & Kanai, T. (2012). Qualitative and quantitative leadership studies in multinational settings: Meta analytics and cross cultural reviews. *Journal of World Business*, *47*(4), 530–538. doi:10.1016/j.jwb.2012.01.006

Taleghani, G., Salmani, D., & Taatian, A. (2010). Survey of leadership styles in different countries. *Iranian Journal of Management Studies*, *3*(3), 91–111.

Thoroughgood, C. N., Padilla, A., Hunter, S. T., & Tate, B. W. (2012). The susceptible circle: A taxonomy of followers associated with destructive leadership. *The Leadership Quarterly*, *23*(5), 897–917. doi:10.1016/j.leaqua.2012.05.007

Trompenaars, A., & Voerman, E. (2010). *Servant-leadership across cultures: Harnessing the strength of the world's most powerful management philosophy*. Oxford, UK: Infinite Limited Ideas.

United Nations Development Program. (2017). *Country profiles: Mexico, Japan, and France*. Available online: http://hdr.undp.org/en/countries

Van de Vliert, E. (2006). Autocratic leadership around the globe - do climate and wealth drive leadership culture? *Journal of Cross-Cultural Psychology*, *37*(1), 42–59. doi:10.1177/0022022105282294

Van Dierendonck, D. (2011). Servant leadership: A review and synthesis. *Journal of Management*, *37*(4), 1228–1261. doi:10.1177/0149206310380462

Van Dierendonck, D., Nuijten, I., & Heeren, I. (2009). Servant leadership, key to follower well-being. In D. Tjosvold & B. Wisse (Eds.), *Power and interdependence in organizations*. Cambridge, UK: Cambridge University Press. doi:10.1017/CBO9780511626562.020

Van Dierendonck, D., & Nuitjen, I. (2011). The servant leadership survey: Development and validation of a multi-dimensional measure. *Journal of Business and Psychology*, *26*(3), 249–267. doi:10.100710869-010-9194-1 PMID:21949466

Van Dierendonck, D., Sousa, M., Gunnarsdóttir, S., Bobbio, A., Hakanen, J., Armin, P. V., & Rodriguez-Carvajal, R. (2017). The cross-cultural invariance of the servant leadership survey: A comparative study across eight countries. *Administrative Sciences*, *7*(2), 8. doi:10.3390/admsci7020008

Van Knippenberg, B., & Van Knippenberg, D. (2005). Leader self-sacrifice and leadership effectiveness: The moderating role of leader prototypicality. *The Journal of Applied Psychology*, *90*(1), 25–37. doi:10.1037/0021-9010.90.1.25 PMID:15641888

Vartia, M. (1996). The sources of bullying: Psychological work environment and organizational climate. *European Journal of Work and Organizational Psychology*, *5*(2), 203–214. doi:10.1080/13594329608414855

Vito, F., Higgins, G. E., & Denney, A. (2014). Transactional and transformational leadership: An examination of the leadership challenge model. *Policing: An International Journal of Police Strategies & Management*, *37*(4), 809–822. doi:10.1108/PIJPSM-01-2014-0008

Vroom, V. H., & Yetton, P. W. (1973). *Leadership and decision making*. Pittsburgh, PA: University of Pittsburgh Press. doi:10.2307/j.ctt6wrc8r

Waldman, D. A., Bass, B. M., & Einstein, W. O. (1987). Leadership and outcomes of performance appraisal processes. *Journal of Occupational Psychology*, *60*(3), 177–186. doi:10.1111/j.2044-8325.1987.tb00251.x

Warrick, D. (1981). Leadership styles and their consequences. *Journal of Experiential Learning and Simulation*, *3-4*, 155–172.

Waterman, H. (2011). Principles of 'servant leadership' and how they can enhance practice. *Nursing Management*, *17*(9), 24–26. doi:10.7748/nm2011.02.17.9.24.c8299 PMID:21473215

Web of Science. (2018). *Clarivate analytics, sear on servant, transactional and autocratic leadership.* Available online: clarivate.com/products/web-of-science/

Wilson, R. T. (1998). Servant leadership. *Physician Executive*, *24*(5), 6. PMID:10185648

Winker, I. (2010). *Contemporary leadership theories. Enhancing the understanding of the complexity, subjectivity and dynamic of leadership*. Springer Science and Business Media.

World Economic Forum and Schwab Foundation for Social Entrepreneurship. (2014). *Leadership in social enterprise. How to manage yourself and the team.* Available on line: www3.weforum.or/docs/WEF-leadership-in-social-entreprise-2014.pdf

Wu, T. Y., Hsu, W. L., & Cheng, B. S. (2003). Expressing anger? Or suppressing anger? Subordinates' anger responses to supervisors' authoritarian leadership behaviors in Chinese business. *Indigenous Psychological Research in Chinese Societies*, *18*, 3–49.

Wulleman, M., & Hudon, M. (2016). Models of social entrepreneurship: Empirical evidence from Mexico. *Journal of Social Entrepreneurship*, *7*(2), 162–188. doi:10.1080/19420676.2015.1057207

Yukl, G. (1999). An evaluation of conceptual weaknesses in transformational and charismatic leadership theories. *The Leadership Quarterly*, *10*(2), 285–305. doi:10.1016/S1048-9843(99)00013-2

Yukl, G. (2006). *Leadership in organizations*. Upper Saddle River, NJ: Prentice-Hall.

Zhang, Y., Huai, M., & Xie, Y. (2015). Paternalistic leadership and employee voice in china: A dual process model. *The Leadership Quarterly*, *26*(1), 25–36. doi:10.1016/j.leaqua.2014.01.002

ADDITIONAL READING

Banutu-Gomez, M. B. (2004). Great leaders teach exemplary followership and serve as servant leaders. *The Journal of American Academy of Business, Cambridge*, *4*(1/2), 143–151.

Den Hartog, D. N., & Belschak, F. D. (2012). When does transformational leadership enhance employee proactive behavior? The role of autonomy and role breadth self-efficacy. *The Journal of Applied Psychology*, *97*(1), 194–202. doi:10.1037/a0024903 PMID:21842977

Dickson, M. W., Den Hartog, D. N., & Mitchelson, J. (2003). Research on leadership in a cross-cultural context: Making progress, and raising new questions. *The Leadership Quarterly*, *14*(6), 729–768. doi:10.1016/j.leaqua.2003.09.002

Frick, D. M. (1994). Robert K. Greenleaf A Life of Servant Leadership. In *Berrett-Koehler Publishers, Hamel, Gary & Prahalad, CK. Competing for Future*. Boston, MA: Harvard Business School.

Graen, G. B., Hui, C. (1999). Transcultural global leadership in the twenty-first century: challenges and implications for development. *Advances in global leadership, 1*, 9-26.

Harvey, M. (2001). The hidden force: A critique of normative approaches to business leadership. *S.A.M. Advanced Management Journal, 66*(4), 36–48.

Wang, Y., & Poutziouris, P. (2010). Leadership styles, management systems and growth: Empirical evidence from UK owner-managed SMEs. *Journal of Enterprising Culture, 18*(3), 331–354. doi:10.1142/S0218495810000604

Winston, B. E. (2003). *Extending Paterson's Model: Explaining How Leaders and Followers Interact in a Circular Model. Servant Leadership Roundtable– August 2003*. Regent University.

KEY TERMS AND DEFINITIONS

Autocratic Leadership: A leadership style in which one person controls all the decisions and takes very little input from other group members.

Cultural Differences: The various beliefs, behaviors, languages, practices and expressions considered unique to members of a specific ethnicity, race, or national origin.

Culture: The collective customs, arts, social institutions, or a particular group.

Leadership: The art of motivating people towards accomplishing a goal. The action of leading a group of people or an organization. It includes guidance, direction, management and control.

Servant Leadership: A leadership style that involves demonstrating the characteristics of empathy, listening, stewardship, and commitment to personal growth toward others.

Transactional Leadership: A leadership style that sets clear objectives and goals for the followers and uses either punishments or rewards in order to encourage compliance with these goals.

Chapter 6
Servant Leadership and Organizational Discipline:
A Qualitative Study

Ravinder Jit
Maharaja Agarsen Institute of Technology, India

Anju Bharti
Maharaja Agarsen Institute of Technology, India

Urvashi Sharma
Department of Commerce, Delhi School of Economics, University of Delhi, India

P. V. Rajeev
Government of India, India

ABSTRACT

The chapter examines various strategies adopted by servant leaders to ensure discipline in the organization. The study suggests that with their inherent pro-follower orientation, servant leaders' first preference is to create an organizational environment where employees do not feel the need to indulge in indiscipline. For this, they take certain proactive initiatives such as open door policy; leading by example; clarifying goals and roles; ensuring objectivity, fairness, and transparency; etc. In case some employees still indulge in undisciplined behavior, the servant leaders tend to adopt certain retroactive measures. They listen patiently and actively to transgressors' perspective in order to (1) diagnose the situation, (2) provide employees an opportunity to give catharsis to their feelings, and (3) persuade them to adopt disciplined behavior. Employees' failure to respond to these initiatives may force the servant leaders to take recourse to punishment as a last resort. But in servant leaders' punishment also there is a trace of care, concern, and compassion.

DOI: 10.4018/978-1-7998-1108-4.ch006

INTRODUCTION

Discipline is considered to be one of the most important prerequisites for achieving organizational effectiveness. Various researchers have tried to identify factors that can determine regulation of employee conduct and maintenance of organizational discipline. Amongst these factors the nature and style of leadership has attracted significant attention (Grojean, Christian, Marcus & Smith, 2004; Stein Bauer, Renn, Taylor & Njoroge, 2014).

Researchers have reported a variety of methods and approaches adopted by different styles of leadership to ensure discipline in the organization. Ferch (2003), for example, suggested that the industrial/new modernist model of leadership, relying heavily on a hierarchal structure and a designated chain of command, uses dominance and control to regulate employee conduct and increase organizational effectiveness. Control based leadership demonstrates decisiveness where each activity, process, and resource-use have to be monitored to check negligence that could harm the organization (Johnson & Klee, 2007; Martin, Liao & Campbell-Bush, 2012). Bolin and Heatherly (2001), and Dunlop and Lee (2004) suggested that an organization failing to control deviant behaviour and ensure adherence to its work ethics could experience failure in overall organizational performance.

But regulation of employee behaviour only through control and dominance has its own repercussions. Ouellette, Lazaer, and Chambers (1999); Agbolla and Salawu (2011) reported that control oriented leadership leads to hostility and undesirable behaviour among followers. Ferch (2003) suggested that such leadership becomes blind to the dignity of the led and tends to take action without accountability to the feelings and emotions of the subordinates. This leads not only to the moral decline of the relational environment in the organization but also a pervasive malaise that afflicts the psyche of the contemporary employees. Day and Hamblin (1964) found that close supervision tended to be connected with reduction in group harmony and productivity.

The failure of control and punitive leadership to ensure positive intra-and inter personal employee attitudes and behaviours while achieving organizational goals, has led to increased interest in studying the impact of flexible and non-punitive leadership on individual and organizational effectiveness. Efforts are also being made to understand the ways in which the perspectives and actions of these leaders intersect with the organization's specific policies, rules, code of conduct etc. in regulating and shaping the attitudes, behaviours and conduct of employees.

In this context, flexible and non-punitive leadership constructs like servant leadership, ethical leadership, relational leadership, spiritual leadership and transformational leadership, etc. have attracted the attention of researchers. In the present study, servant leadership(SL) model has been selected over other non-punitive leadership models as SL model not only manifests various dimensions of these leadership models but also goes past them by manifesting various pro-follower characteristics such as motivation to serve, humility, ensuring followers' development, care and concern for followers, listening, persuasive approach, patience, compassion, honesty and integrity etc. It is assumed that with these characteristics, servant leadership is likely to enforce discipline in a more peaceful, persuasive, humane, compassionate and transparent manner. In other words, the above characteristics of servant leadership can intersect with its administrative role - when it tries to formulate and implement policies, procedures, rules and regulations and code of conduct etc. for effective functioning of the organization.

The critics proclaim that due to above mentioned characteristics servant leaders may be rendered incapable of ensuring compliance to organization's code of conduct, rules and regulations. The inability of the servant leader to enforce discipline or ensure compliance to rules could expose the organization to various challenges on legal and financial fronts and ultimately lead to its failure. These critics argue that servant leadership may be good for the subordinates, but the basic idea of 'people first' clashes with the concept of organizational efficiency (Andersen, 2009; Anderson, 2008; Giampetro-Mayer, Timothy-Brown, Browne & Kubasek, 1998; Yukl, 2010). Many of the decision makers, especially within for-profit and mission driven institutions, react cynically to the concept of servant leadership (Andersen, 2009; Brumback, 1999; Di Stefano, 1995; Johnson, 2001; Whetstone, 2002). Many managers and executives also resist Greenleaf's idea of servant leadership and reject it as impractical (Di Stefano, 1995).

While the debate regarding appropriateness of servant leadership for mission driven and for-profit organizations continues, an increasing number of organizations are moving away from industrial/early modernist approach to leadership with its hierarchical, top down, patriarchal approach (Crippen, 2005; Magoni, 2003; Nwogu, 2004). Today quite a few of Fortune magazine's 100 best companies to work for in USA, profess servant leadership as a core company value (Levering & Maskowitz, 2000; Ruschman, 2002). A growing number of researchers are also espousing servant leadership as an appropriate approach for meeting today's complex organizational needs (Chin & Smith, 2006). With their increasing focus on innovation and digitalisation, organizations are trying to provide more employee centric organizational climate (Kelley & Littman,2001; Martin,2009). For example, IDEO, one of the Fortune Magazine's 100 best companies to work for, fosters an atmosphere which is conducive to freely expressing ideas, breaking the rules and designing one's own work. These concepts work only if leadership imparts trust in their employees and exercises forgiveness when experiments do not go according to plan. The focus is thus shifting toward a trustworthy, team-serving, and self-actualising leadership that wants to make a difference (Davis, Schoorman & Donaldson, 1997).

Servant leadership fits this conceptualization based on its 'other-oriented approach' rather than self-oriented approach (Eva et al.,2019), focus on the aspirations and needs of followers (Patterson, 2003; Van Dierendonck, 2011), service to all stake holders (Hamilton & Bean, 2005), moral attitudes and behaviour (De Pree, 1992; Kouzes & Posner, 1995; Schein, 2010) interpersonal trust (Reinke, 2003; Sendjaya & Pekerti, 2010), and a number of other characteristics and competencies that have been listed above.

In this whole debate regarding appropriateness and inappropriateness of pro-follower and non-punitive servant leadership in the present day competitive and challenging business environment, the basic point of criticism seems to relate to servant leader's predominant orientation towards addressing mental, emotional and psychological needs of the followers at the cost of maintaining discipline or exercising authority (Bowie, 2000; Johnson, 2001) for achievement of organizational goals. The proponents of servant leadership, on the other hand, argue that by showing sensitivity to follower's needs, facilitating their growth and well-being, having ethical orientation (Mayer, Bardes & Piccolo, 2008), and having characteristic of love (Peck, 2004), servant leader creates a sense of fairness, self-discipline, self-respect and self-control among employees. He/she also delegates responsibility, empowers the employees and encourages participative decision making (Neuschel, 1998). His/her emphasis on the needs of the employees does not exclude his/her focus on the goals of the organization and societal good (McGee-Cooper & Looper, 2001). Covey (2004) also suggested that caring leaders can be as passionately committed to organizational excellence as any other leadership model.

In a climate where followers perceive greater safety and trust, they tend to reciprocate in trust (Sendjaya & Pekerti, 2010). Hence, they may voluntarily follow the leader who endeavours to create the right balance between autonomy and direction (Van Dierendonck, 2011).

OBJECTIVES OF THE STUDY

An analysis of the arguments of critics questioning the appropriateness of servant leader's pro-follower and non-punitive approach for organizational effectiveness and the fact that some of the 'high profit making companies' are preferring servant leadership approach over new modernist/industrial leadership approach, leads us to an assumption that along with being sensitive to the needs and aspirations of followers and supporting them in their growth and development, servant leaders must be ensuring that the organizational discipline, a prerequisite for organizational effectiveness, is not compromised.

The main objective of the present study was to investigate the approach/strategies adopted by servant leaders to ensure maintenance of discipline in the organization. We tried to find out whether the servant leaders adopt proactive measures, retroactive measures, or a combination of both for this purpose. Another objective of this study was to find out whether while ensuring discipline servant leaders focus on a non-punitive approach (as expected of servant leaders), a punitive approach or a balance of both approaches.

Insights provided by this study into the servant leaders' approach to maintain organizational discipline will guide the working managers to understand and use an appropriate balance of non-punitive and punitive as well as proactive and retroactive measures for establishing discipline.

The precedence of 'employee-first approach' over 'discipline-first approach' could help leaders to develop a culture of commitment, trust and self-discipline amongst employees which could lead to substantial reduction in the need for establishing external mechanisms of organizational discipline.

METHOD

Qualitative research often uses stories, conversations, interviews, life experiences, field notes, journals, autobiographies, photos etc. as sources of data. In the present study, semi-structured interviews were used to discover the deep values and perspectives of servant leaders as they interact with their subordinates for maintenance of discipline.

The aim of the researchers was to "capture immediate pre-reflective consciousness of the participant as his/her self-given awareness that belongs to and is processed by the participant rather than as something which is based on his/her perception and thus existing apart from self" (D'Cruz & Noronha, 2013).

We conducted fifteen semi-structured interviews with leaders in Education Sector, Corporate Sector and Public (Government) Sector. The respondents with an orientation for servant leadership were identified based on the feedback from present and past subordinates and colleagues of leaders. The feedback was derived from interviews of leaders' present and past subordinates and colleagues. From 8 to 10 subordinates and colleagues of each leader were interviewed. Interview questions were based on five key dimensions of servant leadership as identified by Reed, Cohen and Colwell (2011), namely Interpersonal Support; Building Community; Altruism; Egalitarianism; and Moral Integrity. These dimensions are comprehensive, exhaustive and robust, both conceptually as well as empirically.

Initially, references of these leaders were obtained from the researchers' network of colleagues, friends and acquaintances in the field of education and in public and private sector. Based on feedback from leaders' subordinates and colleagues, we generated a list of 20 servant leaders. Out of 20 shortlisted leaders, fifteen showed their willingness to participate in the study and completed the interviews.

Three of the respondents were from the Education sector, 4 from the Corporate Sector and the remaining 8 were from the Public sector. In all, 10 male and 5 female leaders, in the age range of 45 to 65, were interviewed. The respondents had 20 to 35 years of service experience. All the respondents belonged to upper/ top management level.

The researchers interviewed the respondents in person over a period of six months. Before starting the interview, the following three things were kept in mind:

1. The purpose of the study was thoroughly explained to the participants. The respondents were given an opportunity to ask questions, if any, from the researchers before the start of the interview.
2. Participants were given assurance about confidentiality.
3. The interviewer spent about 5-10 minutes for informal interaction with each participant to establish rapport.

Interviews were audio-recorded. The recordings of the interviews were transcribed verbatim and reviewed for accuracy by the researcher who conducted the interviews.

DATA ANALYSIS

Statements of the respondents were analysed using the model of Plowman, Baker, Beck, Kulkarni, Solansky and Travis, (2007), which espouses the use of a predetermined framework for carrying out analysis of data. In our study servant leadership has been taken as a guiding overall model. We have used the situations that call for maintenance of discipline to find out how servant leaders perceive; experience and manage various situations and whether typical characteristics and values of servant leaders emerge while handling these situations. Key themes were developed based on data obtained from interviews with the participants.

The primary analysis of the transcripts was conducted by two researchers independently. After initial coding of transcripts, two more researchers were involved for reviewing the initial analysis and conducting the final analysis along with the first two investigators. The key themes were discussed and reviewed by all the investigators for ensuring that saturation point regarding key themes had been reached.

In the present study, the researchers adopted two of the three approaches suggested by Van Manen (1998) to identify the themes. Initially, following the sententious/ holistic approach, each transcript was read as a whole so that core/essential meaning of respondents' experiences could be captured. In the second step researchers followed selective/highlighting approach through which they identified the sub-themes/categories that contributed to the core theme. In other words, at this stage significant statements illustrating the various dimensions of the core theme, were identified and demarcated.

Data analysis followed the under mentioned pattern:

1. Transcription of interviews
2. Preparation of analytical and reflective memos
3. Coding by individual researchers
4. Systematic inter-researcher discussions to resolve discrepancies regarding coding and thematic categories
5. Selection of central theme
6. Development of narrative themes

In step 5 and 6 the data was assembled in new ways. The team of researchers tried to identify the central theme, causal and intervening conditions and the specific context in which a particular characteristic or behaviour of a leader emerged. In this step, an effort is also made to see the relationship of various themes. For example, in our study, failure of the subordinate to respond to positive proactive and retroactive initiatives of the leader for ensuring organizational discipline, becomes a reason for the leader to resort to negative discipline.

Harding (1987) suggests that the researchers should also be subjected to critical examination the way it is done in case of overt subject matter. We also tried to observe and document our own assumptions, perspectives and biases while analysing the data. For example, it was our assumption that a servant leader will be more empathetic, caring and compassionate while ensuring maintenance of organizational discipline. This assumption could affect our interpretation so an extra effort was made to constantly observe and guard against any intrusion by these biases and assumptions. Three stages followed in the development and evaluation of codes and themes i.e. (1) by individual researchers (2) by team of researchers and (3) by auditors (teachers trained in qualitative research) is an example of this effort.

RESULTS

Coetzer, Bussin and Geldenhuys (2017), in their exhaustive systematic review listed various characteristics of servant leadership such as service, humility, listening, compassion, courage, accountability and integrity. All these characteristics could determine the way a servant leader proceeds to shape and influence the conduct of employees and handle employees' indiscipline/behaviour transgressions.

The following conceptual model (Figure 1) suggests that in their effort to maintain discipline, the servant leaders in our study do not directly resort to punishment. With their inherent pro-follower orientation, their first endeavour is to take certain proactive measures to create an organizational environment where employees do not feel the need to indulge in indiscipline. Some employees may, however, indulge in undisciplined behaviour despite all pro-active initiatives of the leader. In such a situation, servant leaders' orientation for listening still propels them to listen patiently and actively to transgressors' perspective in an effort to: (a) diagnose the situation, (b) provide employees an opportunity to give catharsis to their feelings, and (c) persuade them to adopt disciplined behaviour. After putting initial efforts at understanding and counselling the employees, the servant leaders proceed to monitor employee behaviour to observe the impact of these efforts. Employees failure to respond to all pro-active and retro-active measures may ultimately force the servant leader to take recourse to punishment as a last resort. This, however, is an extremely rare step, as reported by our respondents.

The above model suggests that our respondents use a combination of proactive and retroactive measures to maintain organizational discipline and check behavioural transgressions of employees. A description of various proactive and retroactive measures adopted by the servant leaders can be seen in the following sections:

PRO-ACTIVE MEASURES

By using proactive measures, servant leaders endeavour to mitigate the chances of future behavioural transgressions in the organization. Various proactive measures used by our servant leaders can be listed as follows:

1. Open door policy
2. Leading by example
3. Well laid down and well communicated rules and systems
4. Flexibility in implementation of rules

Figure 1. Servant leadership and organizational discipline

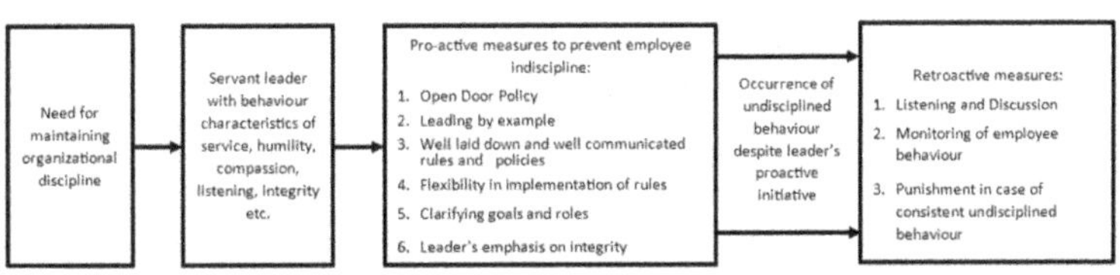

5. Clarifying goals and roles.
6. Leader's emphasis on integrity: Ensuring objectivity, fairness and transparency

Open Door Policy

Open door policy can be used by organizations as a means to improve communication with employees as well as to enhance employee trust and motivation (Aaron Snehar, 1993). Organizational leaders can play a pivotal role in removing the constraints that could discourage the followers from expressing their ideas and concerns (Edmondson, 1999). Servant leaders with their orientation for patient listening, trust and follower wellbeing, create a pervasive social context where followers would feel less risk in voicing their concerns, views and opinions. For example, in the present study respondent 3 felt 'absolutely no need of formal permission' if her subordinates wanted to meet her to discuss personal and professional issues. Respondent 6 also kept his door open to everyone. He stated, "That helps check errors and solve grievances and interpersonal issues in the initial stage itself." Respondent 14 always kept the green light of her room 'on' unless she was in a meeting. She said, "it was much easier to get the work done if the other person was not brow-beaten. A cordial environment encourages discipline and sense of responsibility". Respondent 15 said when "people are frank with you they trust you more and give you better output".

These statements illustrate the importance of open door policy as a pre-emptive measure to understand and manage employees personal and professional concerns as well as to inculcate a sense of discipline and responsibility amongst employees.

Grenny, Switzler and Mc Millan (2002) suggest that instead of always waiting for the employee to come and talk, the leader himself or herself can sometimes initiate one-to-one casual conversations with employees to help them express their views and ask for their feedback. This small step can win the confidence of people and make them more committed.

Some of the respondents in our study manifested a clear orientation for encouraging informal interactions with their subordinates in individual or group settings. This according to them helped them to "understand the needs, desires and problems of their subordinates," (Respondent 6), "to know the strength and weaknesses of the subordinates and manage their aspirations and behaviour better," (Respondent 11), and "to build a rapport which could have a positive impact on organizational climate" (Respondent 13).

Such informal leader-member exchanges play an important role in raising awareness of leader about employees' problems. It could also help to prevent escalation of minor employee concerns into major problems (Grenny et al., 2002).

Leading by Example

According to Social Learning Theory (Bandura, 1977), individuals tend to learn and mimic the values, attitudes and behaviours of credible role models in their environment. Extending this hypothesis to leader-follower relations, Wood and Bandura (1989) suggested that followers might mimic the behaviour of their leaders especially when leaders are perceived to be credible role models. Followers respond strongly to the example set by the leader (Gachter & Renner, 2007; Guth, Levati, Sutter & Van der Heijden, 2007).

The data analysis in the present study indicated that many of our respondents had a strong orientation for leading-by-example i.e., doing what they expect their followers to do. For example, respondent 3, emphasized that the most important determinant for maintaining discipline in the organization is leading by example. She said, "If I myself come late to office and go out in between, or gossip, then my colleagues will not say anything but they will observe, learn and end up becoming undisciplined. It is your integrity which commands respect and nothing else." These views were firmly supported by respondent 11 and respondent 15 as well. Respondent 8 stated, "First and foremost set an example … About 80 per cent people will follow your example." Respondent 14 also said, "People who can't be changed normally, you know, it appeals to them somewhere if you practice what you preach."

These statements suggest that by doing what they say the leaders communicate values that they personally and professionally believe to be important. Through their actions leaders set an unspoken standard about appropriate and inappropriate behaviour.

Well Laid Down and Well Communicated Rules and Policies

Employee behaviour with no ethical basis can generate risks for the organization – both external as well as internal – and requires norms, procedures and controls (Santos, Guevera, Amorim & Neto, 2012). Santos et al. (2012) also suggest that a leader has an important role in influencing individual ethical behaviour in the organization. A leader can focus on forced obedience/compliance to rules or try to raise employee awareness about the need for adherence to code of conduct/ethics. A servant leader with a pro-follower approach and servant orientation may prefer the latter over the former.

Our data analysis revealed adoption of this pro-active approach by many of our respondents. Eight out of fifteen respondents stated that they had well laid out and clear-cut policies, rules and systems for every process and operation. Largely these rules are used as reference points for decision-making. We find confirmation of these assertions in the following statements: "One thing I always believe in is that rules and regulations should be carefully drafted, transparent and widely conveyed" (Respondent 2). "Our rules are fairly well laid down. And employees are made aware of these rules." (Respondent 5). "We have a very simple process wherein all the rules, policies and processes are well defined. Various checks and measures have been placed at different points to prevent escalation of undisciplined behaviour" (Respondent 6). "We have discipline machinery in place, for example, biometric entries for logging in and logging out, giving report of day's work, number of leaves permissible etc. (Respondent 10)"

Four other respondents, all government officials, categorically pointed out that there were well chartered government rules, policies and procedures for keeping a check on employee conduct and being in Government set up normally their decisions were guided by these rules. However, they did not neglect the humane aspect while taking decisions.

Flexibility in Implementation of Rules

In the present study, our respondents indicated an orientation for flexibility in implementing organizational rules especially if employees had a genuine problem. Some of our servant leaders felt that "employees were not robots or cogs in the organizational machinery" (Respondent 3, Respondent 11). Sometimes they could have their genuine problems and compulsions which have to be taken into consideration before implementing the rules. Some other respondents felt that "if rules stifle creativity" (Respondent 2) and "stop employees' progress or harm them directly or indirectly" (Respondent 4), such rules need to be broken or rectified. Respondent 5 emphasised that if a leader wants to help somebody by tweaking the rule a little and thinks that there is genuine cause, he needs to take the step as a leader.

Respondent 10 stated that even though his company ensured quality of deliverables, for other parts, i.e. time and rules employees are given flexibility. Respondent 11 felt that discipline was not the only thing which runs an organization. So, he advocated some flexibility in the implementation of rules. However, Respondent 13 suggested that while giving flexibility leader should also keep on persuading the employees to be more disciplined for their better future.

Leaders can thus play a significant role in humanizing the implementation of code of conduct. They have the scope of formally waiving off or less formally ignoring ethics codes. (Tonge, Greer & Lawton, 2003)

Clarifying Goals and Roles

Tammy Erikson (2012) suggested that without clear role description employee may waste their energy and time negotiating their roles within their teams. In the process, they could lose focus on their productive tasks, and even get involved in unnecessary politics and dysfunctional behaviour.

The servant leaders in the present study are well aware of the value of role and goal clarification. For example, Respondent 1 noted, "In our company we clarify the key responsibility areas (KRAs) and key performance indicators (KPIs). This, takes care of any confusion or resentment as well as enhances efficiency." Clear definition and communication of goals and roles was also considered important for preventing "interpersonal grievances" (Respondent 6) or "confusion and conflict." (Respondent 7), and reducing "chances of wasting time in interpersonal conflicts and organizational politics that may result in unachieved targets."

It is interesting to note here that most of the above respondents are from the corporate. It seems that clarifying roles, goals and career paths is a theme that is most strictly adhered to in the corporate world as it faces more competition and challenge as compared to government sector.

Leaders Emphasis on Integrity: Ensuring Objectivity, Fairness, and Transparency

Integrity emerged as an important characteristic of servant leadership in various studies (Coetzer et al., 2017). In research literature integrity is described as being honest, fair (Russell & Stone, 2002), having strong moral principles (Pekreti & Sendjaya, 2010; Melchar & Bosco, 2010) and behaving ethically (Liden, Wayne, Zhao & Henderson, 2008). Liberman (1981) discussed integrity as an important antecedent to trust.

In the present study leaders' emphasis on integrity emerged as an important proactive approach for prevention of indiscipline. Respondent 2, for example, is very transparent in disseminating information and believes in purity of action. He emphasised that there should be no fear or favour and no discrimination when one takes some decision.

A policy of fairness and objectivity "develops trust and keeps employees satisfied and committed." (Respondent 1) and "helps in establishing credibility and integrity and makes everybody willingly follow the rules." (Respondent 4).

Respondent 6 also believes in total transparency regarding communication of policies and standards of performance evaluation and career progress "to minimize any kind of conflict or grievance." Similarly, Respondent 7 stated that "the clarity of goals, processes and career paths is very important for employees. Else you will have issues and challenges irrespective of who is your leader." This statement found support in respondent 9 who emphasised the need of fairness and clear communication of parameters to avoid any discontentment.

Respondent 12 always made sure that every employee in his organization gets an opportunity for training and work-related seminars and conferences. He stated that in 90-95% cases people will be satisfied with leader's fair approach and will not indulge in any undesirable behaviour. Respondent 13 also tried to be fair and just "as a matter of principle."

Servant leader's orientation for transparency, fairness, justice, honesty, neutrality, open communication, and purity of action etc. is easily discernible in the above statements. These components have the potential to enhance trustworthiness and credibility of the leader.

RETROACTIVE MEASURES

Diagnosis of the Situation: Patient Listening and Discussion

Servant leaders' orientation for ensuring greater good of their followers and their eagerness to help the followers lead themselves, make them more receptive to employees' voice. Instead of forcibly imposing discipline they might first listen to the concerns of the employees.

In the present study, listening and discussion were used as a first step of retroactive approach to maintenance of discipline. Respondent 1 used 'listening' and 'discussion' to understand employees' perspective. while Respondent 2 preferred to "talk it out" whenever he found some employee indulging in indiscipline. He said that almost 90% of the issues got resolved by listening and discussing. Respondent 4 emphasised that listening to the provocative/undisciplined employee demands sincerity and sensitivity. This helped the leader understand the reason for employee's behaviour. Respondent 7 suggested that listening to the provocative employee gives him/her time to cool off. This can be followed by mutual discussion. Respondent 12 also gives cooling off time to the agitated employee and then discusses the issues agitating him/her. Such discussions, according to him, provided a deeper understanding of the issue. He preferred discussing the issue with the erring employee 3-4 times and persuaded them to modify their behaviour before taking any further action. Respondent 10 reported that he has formed some special committees to handle indiscipline issues. "The employees are given ample opportunity to explain their side of things.... Even if they are guilty, most often we politely ask them to modify their behaviour", he stated. Respondent 15 also felt that leader needs to understand why an employee was directly challenging or disobeying him/her. He recommended handling each case differently instead of using same yardstick to everyone.

Patience, Composure, and Temperance

Seven out of fifteen respondents used patience, composure, and self-restraint as a strategy to manage provocative/undisciplined behaviour of their subordinates.

Respondent 3 faced lot of insubordination from one of her immediate subordinates when she joined a new office. But she retained her patience, took decisions by building consensus and used a persuasive approach. "All these things, I think, made him approachable", she said. And at the end when she left that office, they parted as friends. Respondent 4 remarked that even subordinates have certain aspirations and certain problems. "You need immense patience to listen to them". Respondent 7, respondent 8 and respondent 11 also reportedly retained their composure when they faced undisciplined / provocative behaviour and patiently listened to employee's perspective. Respondent 11 remarked that the best way to deal with any situation of indiscipline/ provocation was to act in a sensible and mature manner rather than get provoked and take a hasty step.

Monitoring of Employees' Behaviour

"Active leaders monitor follower behaviour, anticipate problems and take corrective action before behaviour creates serious difficulty (Judge & Piccolo, 2004, p.756)." In our study also, 'monitoring' has been used as a modus operandi to deal with employees' behavioural transgressions. Our respondents, however, used monitoring more as a retroactive measure. When their proactive measures failed to prevent behavioural transgressions, they first tried to patiently and calmly listen, discuss and understand the reason for transgression. They also used counselling to mend employee behaviour and avert recurrence of transgression. After putting all these efforts, monitoring was used to observe employee behaviour over a period of time. According to Respondent 1 the duration of monitoring gives time to the leader to observe whether post-counselling deviance is a consistent phenomenon or an occasional one. Respondent 8 used this strategy for the perennial offenders – a miniscule proportion of the total subordinate staff. He stated, "there will be some two per cent persons who still shirk .... who can also be brought in line by being told that you are being monitored".

Respondent 12 accommodates the erring employee 2-3 times, giving him an opportunity to improve. His behaviour is then monitored for some time before planning any further action. Respondents 14, 15 and 16 resort to surprise checks so that there is constant monitoring of errant officials.

The above statements show that monitoring of employee behaviour is used by servant leaders for various purposes. It sends a message to the erring employee that the leader is observing their behaviour. Monitoring also helps the leader to check whether employees are putting in sincere efforts to change their undesirable behaviour after leader's positive initiatives.

Punishment

Despite various proactive initiatives of the servant leaders some employees may still indulge in improper behaviour or unlawful activities. In such a situation, even servant leaders may administer negative discipline/punishment. Grote (2001) introduced the concept of 'progressive discipline', which, according to him, is employed by many businesses as a sequential process for implementing punishment. Through progressive discipline an employer may try to modify an employee's undesirable behaviour by enforcing a series of disciplinary steps based on the history and nature of employee behaviour (Johnson, 1998).

In the present study, the data analysis indicates that our respondents do sometimes resort to negative discipline whether it happens as a multi-step process (i.e. progressive discipline) or as one- step action.

Respondent 1 in our study aptly puts it, "See, being part of the organization, we can't be soft every time. We have to make some harsh decisions as well...... But such decisions and actions have to go an ethical way". Respondent 2 also resorts to punishment only when situation goes totally out of his hand. In such a case, he politely asks the erring employee to put in his/ her papers. He said, "we make it a point to give him/her a dignified exit saying that he/she has resigned and never say that he/she has been shunted out".

Respondent 6 makes every effort that deviation of behaviour does not escalate. However, if the person fails to mend his/her behaviour despite repeated efforts of the respondent, then the HR may issue a verbal warning followed by written warning. In case the person admits the transgression, and promises not to repeat in future, he is given a chance to again prove himself. Otherwise senior management takes the final decision, depending upon the gravity of the offense. Respondent 9 also has a disciplinary committee in place. In case of any behavioural transgression, the committee investigates the case giving full freedom to the employee to explain his case and mend his ways. Removal of employee happens only if he refuses to respond despite repeated efforts on the part of employer. As respondent 10 puts it, "sometimes we have to punish people, though very rarely, to set an example for others that this type of behaviour will not be tolerated."

Respondents 3, 8, 11, 12 and 14, all government officials reported that there are very clear norms about disciplinary action in government. Some of our respondents who were very high-ranking government officials, reported resorting to strict punishment, but they did it only once or twice in whole of their service tenure. Respondent 11 found himself in a situation where he had to compulsorily retire an employee who had only three months left to his retirement. Dismissal would have meant a negative impact on pensionary benefits and a lot of harassment for the employee. He mitigated his punishment by giving him compulsory retirement. But at the same time, he got all his papers of pensionary benefits prepared in advance. He said, "due to his serious transgression I did not want him to continue in office as he could have created more harm. I could have dismissed him but that does not serve the purpose, you know"

Respondent 14, faced little indiscipline from her employees due to a good rapport she shared with them. But "if some very serious transgression" was there, she "had to follow the disciplinary procedure prescribed by the government", though she will like to "minimise the loss to the employee".

Respondent 15, another senior government official, had to actually get one extremely undisciplined employee fired from service. She said, "this is an extreme case of course. But sometimes you have to take hard decisions."

These statements suggest that even servant leaders, who may like to avoid any strong punishment to their subordinates, may be forced to resort to this mode of corrective measures in the case of unrepentant and incorrigible 'serious' transgressors.

DISCUSSION

Analysis of the statements of servant leaders in this study confirms our assumption that despite their characteristic pro-follower and non-punitive approach, servant leaders do ensure that in their pursuit for employee welfare, the organizational discipline is not compromised beyond a harmless limit. Their approach to achieve this goal is both preventive as well as curative. Proactive measures taken by the servant leaders are aimed at preventing the occurrence of employee indiscipline. Open door policy, adopted by our leaders, is one such proactive measure. Edmondson (1999) reported that organizational leaders can play a pivotal role in removing the constraints that could discourage followers from expressing their ideas and concerns.

An experience of constraints/restrictions could result in bottled up anger, resentment and other negative emotions which could lead to deviant or undisciplined behaviour. Various research studies have reported association of anger with low resistance to frustration (Barrs, Muller, Galhofer & Netter, 2011), reduced emotional self-control capacity (Denson, Pedersen, Friese, Hahm & Roberts, 2011) and destructive behaviour (Noussair, Offerman, Suetens, Van de Ven, Van Leeuwen & Van Veelen, 2014) etc. Our results show that by adopting open door policy servant leaders try to remove constraints on subordinates' free expression of ideas and concerns thus reducing the possibility of any behavioural transgression or indiscipline.

Open door policy, however, implies subordinate 'coming to the leader' for a hearing. Some subordinates may be reluctant to do so. If a leader takes the initiative, reaches out to the subordinates and holds informal conversations with the employees to get their feedback and understand their concerns, he could avert many misunderstanding and behavioural transgressions. Grenny et al. (2002) suggest that informal conversations may prevent escalation of minor issues into big problems.

Our respondents, at times, used informal interactions and conversations with their employees to understand their needs, concerns and problems. Some of our respondents arranged group outings, while others ensured group lunch and tea breaks so that subordinates feel free to discuss and share their problems with the leader.

Another powerful proactive measure adopted by our respondents is 'role modelling' or 'leading-by-example'. Extending Bandura's (1977) social learning theory to leader-follower relationship, Wood and Bandura (1989) reported that followers might mimic the behaviour of their leaders, especially when they are perceived as credible. Giving credence to this suggestion, our respondents also noted that the example set by them was largely followed by their subordinates. Maintaining punctuality, signing in and signing out in biometric machine, avoiding time-out or gossiping during working hours, were some of the examples set by our respondents. Their actions automatically established unspoken standards about appropriate and inappropriate behaviour in the organization.

Contrary to the popular myth that servant leaders are too soft and spineless to formulate and enforce rules, regulations and code of conduct, our respondents did resort to use of policies, rules and code of conduct prevalent in the organization. They normally used these rules and code of conduct as reference point for decision making, for preventing escalation of undisciplined behaviour and for keeping a check on employees conduct. They were, however, very particular about clear communication and transparency as their main objective was not to focus on forced obedience and compliance but to raise employee awareness about the need for adhering to these rules and regulations.

Our respondents, however, manifest a flexible approach while implementing the rules, especially if they felt that employee had a genuine problem or compulsion. They were ready to bend or rectify rules which stifled employees' creativity or compromised the dignity of the employee. The approach of our respondents is in consonance with the approach exhibited by other moral leaders who recognize that growth and welfare of others is more important than rules, and being kind and caring supersedes being right (Pava, 2003). Flexibility of leaders has been reported to be positively related to the effectiveness of leader (Hooijberg, 2005). Being flexible, however, does not entail being oblivious to the deliverables or quality of deliverables. Similarly, even when our respondents showed flexibility, they persuaded their subordinates to inculcate disciplinary habits for their better future.

Many of our respondents emphasised the importance of a leader's integrity and transparency to prevent behavioural transgressions. These characteristics and behaviours of servant leaders fit in well with the concept of leader as a 'moral person' and a 'moral manager' (Trevino, Hartman & Brown, 2000).

In this study, many of the respondents made it a point to clearly set the goals and clarify the key responsibility areas so that there is no scope for confusion and interpersonal conflicts. Clarification of roles and goals is reported to actually facilitate organizational citizenship behaviour (Podsakoff, Nichoff, MacKenzie & Williams, 1993). Podsakoff, MacKenzie and Boomer (1996) also found that a high performance-expectation (that is task and role clarity) and a clear vision of the future were positively linked to OCB.

Along with the above proactive measures used with a view to prevent indiscipline or behavioural transgressions, our respondents reported the use of various retroactive measures for diagnosing and correcting behavioural transgressions. The first step in this direction was diagnosing the whole situation by using listening and discussion as diagnostic tools. Jit, Sharma and Kawatra (2016; 2017) reported that listening and discussion can be used by a servant leader to take perspective of the situation as well as to help the follower give catharsis to their feelings. In this study also, leaders used 'listening' and 'discussion' to understand the perspective of the employee and help the employee give outlet to his/her grievances. During discussion, they also tried to persuade and cajole the employee to modify his/her undesirable behaviour.

Other important characteristics emerging from the statements of our respondents were those of patience, composure and temperance in response to provocative or undisciplined behaviour. Our leaders exercised self- restraint, retained patience and composure to give cooling-off time to the aggressive/ undisciplined employee. This not only helped them hold discussion in a cordial environment but also turn insubordination into a lasting relationship of positivity.

Contrary to the reported proactive use of monitoring by leaders to anticipate problems and take corrective action (Judge & Piccolo, 2004), our respondents used monitoring as a retroactive measure for regulating behavioural transgressions. It helped the leader observe whether undisciplined behaviour persisted despite leader's positive initiatives and then decide whether it needs some further action.

Interestingly, the theme of 'monitoring' emerged in case of four out of five government officials interviewed. Extra need for 'monitoring' in government sector could be attributed to some factors like (1) employee-friendly government policies, sometimes making employees take rules lightly; (2) better employee access to grievance redressal system and (3) more accountability for the action taken by the superior against the subordinates.

Our respondents manifested a stricter side of their leadership when they observed subordinates' unabated and unrepentant indulgence in undisciplined behaviour. The process of progressive discipline, as advocated by Grote (2001), has been followed by some respondents, though not in its exact form. When the erring employee fails to respond to all positive efforts, a milder form of punishment in the form of verbal and written warnings was used as a deterrent to his undesirable behaviour. Removal or retrenchment of the employee happened very rarely and only if the transgressions were recurrent and serious in nature.

But in a servant leader's punishment also there is a trace of care, concern and compassion. As one respondent stated that he did not terminate the incorrigible employee, he just asked him to resign thus giving him a dignified exit. Another respondent who was forced to compulsorily retire one employee due to his serious transgression, not only mitigated his punishment but also got all his papers of pensionary benefits prepared in advance. This compassionate behaviour of our respondents fits in the statement of Kouzes and Posner (2003) who remarked, "credible leaders are compassionate and prefer supporting and sustaining their team members rather than simply punishing them" (2003:56).

Our results also find support in the statement of Peck who says, "although love may disapprove of another's actions and hold others accountable for those actions, the intention of love is always to do that which ultimately achieves the best interests of the other party." (2004:150-158).

LIMITATIONS

As stated by Reissman (1993) 'stories are inherently multi-layered and ambiguous, so narrative analysis is bounded by the constructed nature of truth' and hence present subjective and individual perspectives. While this is a weakness, if we are able to decipher the commonalities in the aspects of perspective it helps at generalization. Identification of themes from the narratives is actually a process of deciphering these commonalities.

Some empirical scientists may argue that for our study we should have used a standardized tool for identifying servant leaders based on self-report. We followed an equally rigorous method of selecting our sample. As mentioned in the section on 'method' the servant leaders in the present study were selected based on the feedback by eight to ten present and past subordinates and colleagues of each leader. This feedback was derived from in-depth interviews of these people. Selecting servant leaders on the basis of feedback has its own strength. When people working with the leader testify the existence of given behavioural dimensions in the leader, it could be a more valid measure of a leader's behaviour.

The interview questions used for getting the feedback were based on Executive Servant Leadership Scale developed by Reed, Cohen and Colwell. (2011). This questionnaire has been specifically developed to measure key dimensions of top executive behaviour – a sample used in this study.

There is no doubt that in a qualitative study there is increased danger of researchers' bias at various stages of data analysis. A conscious effort was thus made to minimize the potential bias by taking various measures to ensure maximum objectivity while identifying themes and categories.

Lack of a comparison group may be another limitation of the study. It may constrain our conclusion that the emergent orientation for predominantly non-punitive handling of employees' undisciplined behaviour is associated with servant leadership only. Here we will like to emphasize that the present research study was designed to elucidate servant leaders' approach to maintenance of discipline, not to compare and contrast it with the approach of other leadership styles. Our results can become a basis for future qualitative research wherein, a comparison can be made between discipline-maintaining strategies of servant leaders and other value-laden leaders.

FUTURE RESEARCH DIRECTIONS

The researchers will like to point out two main gap areas in the present study which could be explored by the future researchers.

1. Exploring attitude of employees towards measures and strategies adopted by servant leaders in maintenance of discipline. This will help evaluate the success of the strategies and measures adopted for the above said purpose.
2. Using a comparison group of any other value-laden leadership style such as ethical leadership, transformational leadership, benevolent leadership, compassionate leadership and spiritual leadership to find out similarities/ differences, if any, in the modus operandi of two leadership styles in ensuring organizational discipline.

CONCLUSION

Discipline is often misinterpreted as menacing and negative. This misconception has resulted in a myth that servant leaders cannot enforce discipline as the very concept of servant leadership entails serving the followers, not punishing them. Our results have burst this myth. It is true that with their pro-follower approach, servant leaders always strive to ensure growth, development and wellbeing of their subordinates, but it is not at the cost of ethical conduct and discipline required for organizational effectiveness. A servant leader's love for followers entails a 360-degree version of love. It incorporates grace as well as accountability, autonomy as well as clear boundaries and forgiveness as well as discipline. Their approach to enforce discipline is, however, different. They believe more in prevention than in cure. With their proactive approach of accessibility, informality, objectivity, transparency, credibility, integrity, fairness and flexibility etc. they try to create a mutual sense of trust with their subordinates. Most of the above said proactive behaviours adopted by our respondents seem to emerge from their care, concern and love for the followers. Such a culture of care and concern may lead to substantial reduction in employees' undisciplined behaviour and behavioural transgressions. The rare cases of such transgressions are managed by our servant leaders retroactively wherein punishment is used only if it is unavoidable. However, this punishment also carries an intention of care and concern for the subordinates.

As suggested by Peck (2004), it is leader's characteristic of love that teaches self-discipline, self-respect and self-control to the followers. It creates a culture where people naturally do the right thing when faced with dilemmas (Back, 2006). With his/her love, care and concern the servant leader may inculcate in his/her followers a conviction for ethical conduct and commitment and loyalty towards work. As suggested by Covey (2004); Cameron, Dutton and Quinn, (2003); and Pfeffer (1998), when leaders consistently exhibit forgiveness, love and trust in relationships, their employees may respond with increased loyalty and commitment.

The so called 'Soft' attitude of a servant leader may be more of an asset than a liability for the organization. A culture of forgiveness, love and trust created by a servant leader may help overcome the withheld commitment, unlock and empower the untapped capabilities and dissipate the employee distrust that is prevalent in command-and-control type of leadership models that show lack of genuine interest in the welfare of employees (Benni & Nanus, 2007; Kouzes & Posner, 2003). With committed, empowered and trusting employees, the need for establishing external mechanisms of discipline may be substantially reduced.

REFERENCES

Aaron, S. (1993). Keeping management's door open: How to establish an open-door policy that works? *Leadership and Organization Development Journal*, *14*(2), 8–12. doi:10.1108/01437739310032665

Agboola, A. A., & Salawu, R. O. (2011). Managing deviant behaviour and resistance to change. *Journal of Business and Management*, *6*(1), 235–243.

Andersen, J. (2009). When a servant leader comes knocking.... *Leadership and Organization Development Journal*, *30*(1), 4–15. doi:10.1108/01437730910927070

Anderson, J. (2008). *The writings of Robert, K. Greenleaf: An interpretative analysis and the future of servant leadership.* Paper presented at servant leadership research round table, Virginia Beach, VA.

Back, P. F. (2006). Principles or Rules? *Public Money & Management, 26*(1), 5–9. doi:10.1111/j.1467-9302.2005.00493.x

Bandura, A. (1977). *Social learning theory.* Eaglewood Cliffs, NJ: Prentice Hall.

Barrs, M. Y., Muller, M. J., Galhofer, B., & Netter, P. (2011). Depressive and aggressive responses to frustration: Development of a questionnaire and its validation in a sample of male alcoholics. *Depression Research and Treatment, 2011*, 352048. doi:10.1153/2011/352048 PMID:21461367

Benni, W., & Nanus, B. (2007). *Leaders.* New York: Collins Business Essentials.

Bolin, A., & Heatherly, L. (2001). Predictors of employee deviance: The relationship between bad attitudes and bad behaviour. *Journal of Business and Psychology, 15*(3), 405–418. doi:10.1023/A:1007818616389

Bowie, N. E. (2000). Business ethics, philosophy and the next 25 years. *Business Ethics Quarterly, 10*(1), 7–20. doi:10.2307/3857690

Brumback, G. B. (1999). The power of servant leadership: A critical but appreciative review. *The Leadership Quarterly, 13*(4), 397–419.

Cameron, K. S., Dutton, J. E., & Quinn, R. E. (2003). *Positive organizational scholarship.* San Francisco, CA: Berrett- Koehler Publishers.

Chin, D. T., & Smith, W. A. (2006). *An inductive model of servant leadership: The considered difference to transformational and charismatic leadership.* Melbourne, Australia: Monash University.

Coetzer, M. F., Bussin, M., & Geldenhuys, M. (2017). The functions of a servant leader [review]. *Administrative Sciences, 7*(5), 1–32.

Covey, S. R. (2004). *The 8th Habit: From effectiveness to greatness.* New York: Free Press.

Crippen, C. (2005). Inclusive education: A servant leadership perspective. *Education Canada, 45*(4), 19–22.

D'Cruz, P., & Noronha, P. (2013). Breathers, releases, outlets, and pauses: Employee resistance in the context of depersonalized bullying. *Qualitative Report, 18*(72), 1–24.

Davis, J. H., Schoorman, F. D., & Donaldson, L. (1997). Toward a stewardship theory of management. *Academy of Management Review*, *22*(1), 20–47. doi:10.5465/amr.1997.9707180258

Day, R. C., & Hamblin, R. L. (1964). Some effects of close and punitive styles of supervision. *American Journal of Sociology*, *69*(5), 499–510. doi:10.1086/223653

De Pree, M. (1992). *Leadership Jazz*. New York: Dell Publishing.

Denson, T. F., Pedersen, W. C., Friese, M., Hahm, A., & Roberts, L. (2011). Understanding impulsive aggression: Angry rumination and reduced self-control capacity are mechanisms underlying the provocation – aggression relationship. *Personality and Social Psychology Bulletin*, *37*(6), 850–862. doi:10.1177/0146167211401420 PMID:21421767

Di Stefano, J. J. (1995). Tracing the vision and impact of Robert A. Greenleaf. In L. C. Spears (Ed.), *Reflections on Leadership: How Robert. K Greenleaf's theory of servant leadership influenced today's top management thinkers* (pp. 61–78). New York: John Wiley.

Dunlop, P. D., & Lee, K. (2004). Work place deviance, organization citizenship behaviour and business unit performance: The bad apples do spoil the whole barrel. *Journal of Organizational Behavior*, *25*(1), 67–80. doi:10.1002/job.243

Edmondson, A. (1999). Psychological safety and learning behaviour in work teams. *Administrative Science Quarterly*, *44*(2), 350–383. doi:10.2307/2666999

Eva, N., & Liden, R. C. (2019). Servant leadership: A systematic review and call for future research. *The Leadership Quarterly*, *30*(1), 111–132. doi:10.1016/j.leaqua.2018.07.004

Ferch, S.R. (2003). *Servant leadership, forgiveness and social justice*. Retrieved from Ferch Essay 5/03 guweb 2.gonzaga.edu/orgl/orgl532/essays/ferch.pdf

Gachter, S., & Renner, E. (2007). *The role of leadership and beliefs in the voluntary provision of public goods*. University of Nottingham.

Giampetro-Meyer, A., Timothy-Brown, S., Browne, M.N., & Kubasek, N. (1998). Do we really want more leaders in business? *Journal of Business Ethics*, *17*(15), 1728–1738.

Grenny, J., Switzler, A., & Mc Millan, R. (2002). *Crucial conversations: Tools for talking when stakes are high*. US Mc Graw Hill Publications.

Grojean, M. W., Christian, J. R., Marcus, W. D., & Smith, D. B. (2004). Leaders, values, and organizational climate: Examining leadership strategies for establishing and organizational climate regarding ethics. *Journal of Business Ethics*, *55*(3), 223–241. doi:10.100710551-004-1275-5

Grote, D. (2001). Discipline without punishment. *Across the Board*, *38*(5), 52–58.

Guth, W., Levati, M. V., Sutter, M., & Van der Heijden, E. (2007). Leading by example with and without exclusion power in voluntary contribution experiments. *Journal of Public Economics*, *91*(5), 1023–1042. doi:10.1016/j.jpubeco.2006.10.007

Hamilton, F., & Bean, C. J. (2005). The importance of context, beliefs and values in leadership development. *Business Ethics (Oxford, England)*, *14*(4), 336–347. doi:10.1111/j.1467-8608.2005.00415.x

Harding, S. (1987). *Feminism and methodology*. Bloomington, IN: Indiana University Press.

Hooijberg, R., & Lane, N. (2005). *Leader effectiveness and integrity: Wishful thinking?* Retrieved from https://www.imd.org/research/publications/upload/Hooijberg_lane_wp_2005_1_level_

Jit, R., Sharma, C. S., & Kawatra, M. (2016). Servant Leadership and Conflict Resolution: A qualitative study. *International Journal of Conflict Management*, *27*(4), 591–612. doi:10.1108/IJCMA-12-2015-0086

Jit, R., Sharma, C. S., & Kawatra, M. (2017). Healing a broken spirit: Role of servant leadership. *Vikalpa*, *42*(2), 80–94. doi:10.1177/0256090917703754

Johnson, B. T. (1998). *Implementing progressive discipline policies to minimize liability and improve employee performance*. Retrieved from www.fwlaw.com/news/146-implementing-progressive-discipline-policies-minimize-liability-improve-employee-performance

Johnson, C. E. (2001). *Meeting the ethical challenges of leadership* (3rd ed.). Thousand Oaks, CA: Sage.

Johnson, N. J., & Klee, T. (2007). Passive-aggressive behaviour and leadership styles in organizations. *Journal of Leadership & Organizational Studies*, *14*(2), 130–142. doi:10.1177/1071791907308044

Judge, T. A., & Piccolo, R. F. (2004). Transformational and transactional leadership: A meta analytic test of relative validity. *The Journal of Applied Psychology*, *89*(5), 755–768. doi:10.1037/0021-9010.89.5.755 PMID:15506858

Kelley, T., & Littman, J. (2001). *The art of innovation: Lessons in creativity from IDEO America's leading design firm*. London, UK: Profile Books.

Kouzes, J. M., & Posner, B. Z. (1995). *The leadership challenge: How to keep getting extraordinary things done in the organizations*. San Francisco, CA: Jossey-Bass Publishers.

Kouzes, J. M., & Posner, B. Z. (2003). *Encouraging the heart: A leader's guide to rewarding and recognizing others*. San Francisco, CA: Jossey-Bass Publishers.

Levering, R., & Moskowitz, M. (2000). The 100 best companies to work for in America. *Fortune, 141*(1), 82–110. PMID:10176767

Liberman, J. K. (1981). *The litigious society*. New York: Basic Books.

Liden, R. C., Wayne, S. J., Zhao, H., & Henderson, D. (2008). Servant leadership: Development of a multidimensional measure and multi-level assessment. *The Leadership Quarterly*, *19*(2), 161–177. doi:10.1016/j.leaqua.2008.01.006

Lowe, L., Mills, A., & Mullen, J. (2002). Gendering the silences: Psychoanalysis, gender and organization studies. *Journal of Managerial Psychology*, *17*(5), 422–432. doi:10.1108/02683940210432655

Magoni, F. (2003). Pyramids and paradigms. *Pastoral Forum, 20*(1), 14.

Martin, R. (2009). *The design of business: Why design thinking is the next competitive advantage?* Harvard Business School Press.

Martin, S., Liao, H., & Campbell-Bush, E. (2012). Directive versus empowering leadership: A field experiment comparing the impact on task proficiency and proactivity. *Academy of Management Journal*. doi:10.5465/amj.2011.0113

Mayer, D. M., Bardes, M., & Piccolo, R. F. (2008). Do servant leaders help satisfy follower needs? An organizational justice perspective. *European Journal of Work and Organizational Psychology*, *17*(2), 180–197. doi:10.1080/13594320701743558

McGee-Cooper, A., & Looper, G. (2001). *The essentials of servant leadership: Principles in practice*. Waltham, MA: Pegasus Communications.

Melcher, D. E., & Bosco, S. M. (2010). Achieving high organisation performance through servant leadership. *Journal of Business Enquiry*, *9*(2), 74–88.

Neuschel, R.P., (1992). *The servant leader: Unleashing the power of your people*. East Lansing, MI: Vision Sports Management.

Noussair, C., Offerman, T., Suetens, S., Van de ven, J., Van Leeuwen, B., & Van Veelen, M. (2014). Predictably angry: Facial cues provide a credible signal of destructive behaviour. *IAST Working Papers from Institute of Advanced Study in Toulouse (IAST).*

Nwogu, O. (2004). The role of follower self-esteem, emotional intelligence, and attributions on organisation effectiveness. The 2004 proceedings of the servant leadership round table at Regent University.

Ouellette, P. M., Lazaer, K., & Chamber, K. (1999). Action leadership: The development of an approach to leadership enhancement for grassroots community leaders in children's mental health. *The Journal of Behavioral Health Services & Research*, *26*(2), 171–184. doi:10.1007/BF02287489 PMID:10230145

Patterson, K. A. (2003). *Servant leadership: A theoretical model* (Doctoral Dissertation). Regent University.

Pava, M. L. (2003). *Leading with meaning: Using covenantal leadership to build a better organization.* New York: Palgrave, MacMillan.

Peck, M. S. (2004). *The road less travelled, 25th anniversary edition: A New Psychology of Love, Traditional Values and Spiritual Growth.* New York: Touchstone.

Pekereti, A. A., & Sendjaya, S. (2010). Exploring servant leadership across cultures: Comparative study in Australia and Indonesia. *International Journal of Human Resource Management*, *21*(5), 754–780. doi:10.1080/09585191003658920

Pfeffer, J. (1998). *The human equation: Building power by putting people first.* Cambridge, MA: Harvard Business School Press.

Plowman, D. A., Baker, L. T., Beck, T. E., Kulkarni, M., Solansky, S. T., & Travis, D. V. (2007). Radical change accidentally: The emergence and amplification of small change. *Academy of Management Journal*, *50*(3), 515–543. doi:10.5465/amj.2007.25525647

Podsakoff, P., Mac Kenzie, S., & Boomer, W. (1996). Transformational leader behaviours substitute for leadership as determinants of employee satisfaction, commitment, trust and organizational citizenship behaviour. *Journal of Management*, *22*(2), 259–298.

Podsakoff, P., Niehoff, B., Mac Kenzie, S., & Williams, M. (1993). Do substitutes for leadership really substitute for leadership? An empirical examination of Kerr and Jermier's situational leadership model. *Organizational Behavior and Human Decision Processes*, *54*(1), 1–44. doi:10.1006/obhd.1993.1001

Reed, L. L., Cohen, D. V., & Colwell, S. R. (2011). A new scale to measure executive servant leadership: Development, analysis, and implications for research. *Journal of Business Ethics, 101*(3), 415–434. doi:10.100710551-010-0729-1

Reinke, S. J. (2003). Does the form really matter? Leadership, trust and acceptance of the performance appraisal process. *Review of Public Personnel Administration, 23*(1), 23–37. doi:10.1177/0734371X02250109

Reissman, C. K. (1993). *Narrative Analysis*. London: Sage Publications.

Ruschman, N. L. (2002). *Servant leadership and the best companies to work for in America. In Focus on Leadership: Servant leadership for the 21st century* (pp. 123–140). New York: Wiley.

Russell, R. F., & Stone, A. G. (2002). A review of servant leadership attributes. Developing a practical model. *Leadership and Organization Development, 23*(3), 145–157. doi:10.1108/01437730210424

Santos, R. A., & Guevera, A. J. (2012). Compliance and leadership: The susceptibility of leader to the risk of corruption in the organizations. *Einstein (Sao Paulo, Brazil), 10*(1), 1–10. doi:10.1590/S1679-45082012000100003 PMID:23045818

Schein, E. H. (2010). *Organizational culture and leadership* (Vol. 2). New York, NY: Wiley.

Sendjaya, S., & Pekerti, A. (2010). Servant leadership as antecedent of trust in organizations. *Leadership and Organization Development Journal, 31*(7), 643–663. doi:10.1108/01437731011079673

Steinbauer, R., Renn, R. W., Taylor, R. R., & Njoroge, P. K. (2014). Ethical leadership and followers' moral judgement: The role of followers perceived accountability and self-leadership. *Journal of Business Ethics, 120*(3), 381–392. doi:10.100710551-013-1662-x

Tammy, E. (2012). The biggest mistake you (probably) make with teams. *Harvard Business Review*, (April): 5.

Tonge, A., Greer, L., & Lawton, A. (2003). The Enron story: You can fool some of the people some of the time.... *Business Ethics (Oxford, England), 12*(1), 4–22. doi:10.1111/1467-8608.00301

Trevino, L. K., Hartman, L. P., & Brown, M. (2000). Moral person and moral manager: How executives develop a reputation for ethical leadership. *California Management Review, 42*(2), 128–142. doi:10.2307/41166057

Van Dierendonck, D. (2011). Servant leadership: A review and synthesis. *Journal of Management, 37*(4), 1228–1261. doi:10.1177/0149206310380462

Van Manen, M. (1998). *Researching lived experience: Human science for an action sensitive pedagogy*. Albany, NY: SUNY Press.

Whetstone, J. T. (2002). Personalism and moral leadership: The servant leader with a transforming vision. *Business Ethics (Oxford, England), 11*(4), 385–392. doi:10.1111/1467-8608.00298

Wood, R. E., & Bandura, A. (1989). Social cognitive theory of organizational management. *Academy of Management Review, 14*(3), 361–384. doi:10.5465/amr.1989.4279067

Yukl, G. (2010). *Leadership in organizations* (7th ed.). New York: Prentice Hall.

ADDITIONAL READING

Anderson, M. (2017). Transformational leadership in education: A review of existing literature. *International Social Science Review, 93*(1), 4.

Boddy, C. R. (2011). The corporate psychopath's theory of global financial crisis. *Journal of Business Ethics, 102*(2), 255–259. doi:10.100710551-011-0810-4

Brown, M., & Trevino, L. (2006). Ethical leadership: A review and future direction. *The Leadership Quarterly, 17*(6), 595–616. doi:10.1016/j.leaqua.2006.10.004

Brown, T. (2009), Change by design – how design thinking transforms organizations and inspires innovation, Haper Collins e-books, Haper Collins 2009, UK.

Crawford, J., Dawkins, S., Martin, A., & Lewis, G. (2017). Understanding the organizational climate of unethical leadership in the Australian Football League. *The Journal of Leadership Studies, 11*(2), 52–54. doi:10.1002/jls.21525

Crawford, J. A., Dawkin, S., Martin, A., & Lewis, G. (2019). Putting the leader back into authentic leadership. Reconceptualizing and rethinking leaders. *Australian Journal of Management*; Advanced online publication. doi:10.1177/0312896219836460

Jooseten, A., Van Dijke, M., Van Hiel, A., & De Cramer, D. (2014). Being "in control" may make you lose control: The role of self-regulation I unethical leadership behaviour. *Journal of Business Ethics, 121*(1), 1–14.

Rego, P., Lopas, M. P., & Simpson, A. V. (2017). The Authentic-Machiavellian leadership grid: A Typology of leadership styles. *The Journal of Leadership Studies*, *11*(2), 48–51. doi:10.1002/jls.21524

KEY TERMS AND DEFINITIONS

Behavioural Transgression: Behavioural transgression refers to a behaviour that exceeds a limit or boundary especially of social acceptability. It is a behaviour that is against a rule, command, or law.

Industrial/New Modernist Leadership: It is also referred to as traditional leadership. It relies heavily on a hierarchical structure, designated change of command, centralised decision making, and dominance and control to regulate employee conduct and increase organizational effectiveness.

Organizational Discipline: Organizational discipline is the practice of self-restraint and learning to follow the best course of action which may not be according to one's desire. Discipline is important as it binds the employees and also motivates them to respect their organization.

Pro-Active and Retro-Active Disciplinary Measures: Pro-active measures are taken by the leader to promote good discipline or to prevent occurrence or escalation of indiscipline. Retro-active disciplinary measures are taken after a person has already indulged in indiscipline or transgression of rules.

Pro-Follower Approach: A leader-follower relationship wherein leaders are attentive to the concerns of the followers and put the needs of the followers first.

Progressive Discipline: This discipline system involves a series of penalties with increasing severity every time an employee breaks the rules. Progressive discipline actions may include verbal counselling, verbal warning, written warning, monetary deductions, suspension and then termination as a last option.

Punitive Leadership and Non-Punitive Leadership: A punitive leader resorts to various negative control methods like belittling, bullying, threatening, humiliating, levying fines and punishing to ensure compliance of rules. A non-punitive leader manifest attributes such as empathic understanding, kindness, active listening, patience, humility, and persuasion to deal with indiscipline issues.

Qualitative Research: It is a scientific method used to gather non-numerical data. Qualitative research tries to answer questions such as why and how a certain phenomenon may occur rather than when and how often it may occur. Qualitative researchers prefer inductive and hypothesis-generating research processes and procedures and use in-depth observation towards investigation of meaning.

Servant Leadership: Servant leadership is a leadership philosophy in which the main goal of the leader is to serve the followers. The need to serve is more important to a servant leader than the need to power. A servant leader feels a moral responsibility to bring success to the subordinates, the organization, the customers and other stake holders.

Value-Laden Leadership: In value-laden leadership style, one finds leaders with strong values and principles about managing their businesses/organizations. Such leaders display attributes such as fairness, openness, honesty, integrity and moral principles.

Chapter 7
Developing Strategic Leadership and Innovation Capability for Manufacturing SMEs Transitioning to Digital Manufacturing Technology

Roderick Walden
University of Technology Sydney, Australia

Stefan Lie
https://orcid.org/0000-0002-3512-2185
University of Technology Sydney, Australia

Berto Pandolfo
https://orcid.org/0000-0003-0179-8862
University of Technology Sydney, Australia

Anton Nemme
https://orcid.org/0000-0002-7846-9971
University of Technology Sydney, Australia

ABSTRACT

Strategic leadership and the development of innovation capability is of critical importance for product design and manufacturing as emerging digital technologies increasingly challenge conventional practice. This chapter presents research intended to provide manufacturing SMEs and product designers with an understanding of the complexities associated with the introduction of advanced manufacturing technologies as part of the transition to Industry 4.0. An included case study describes

DOI: 10.4018/978-1-7998-1108-4.ch007

a university-industry collaboration design project articulating product designs role in developing strategic leadership and innovation capability for SMEs in an era of digitalisation. The chapter provides tangible ways forward for manufacturing SMEs transitioning to additive manufacturing (AM) technology, operationalising a combination of theoretical positions proposed in academic discourse. These include the need to adopt empowering leadership styles, the value of collective entrepreneurship, and the nature of creative process engagement for the generation of ideas and alternatives for AM transition.

INTRODUCTION

The research presented in this chapter investigates the relationship between strategic leadership, innovation and design practices as manufacturing small-medium enterprises (SMEs) begin to transition to advanced manufacturing technology. Additive Manufacturing (AM), sometimes referred to as 3D printing, is an emerging advanced technology that today enables autonomous production of both large and small format parts, directly from computer models (Ghaffar et al. 2018) created using computer aided design (CAD) software. Transitional Integration of AM is by no means straightforward. AM challenges some of the fundamental operating principles of product design practice and manufacturing management. SMEs are typically cautious in the technological upgrading of their organisation as their capacity to absorb errors in decision making is often very limited. Various Government funding programs have encouraged SMEs and external research units, most predominantly, university research units to conduct collaborative (often) practice-led research to investigate the various challenges associated with the implementation of advanced manufacturing technologies including AM (Walden et al. 2018a).

The research describes a university-industry collaboration project between an Australian based SME and a university research unit specialising in additive manufacturing and product design research. The SME is a company with over 30-years of experience in the design and manufacture of urban street furniture for local government both nationally and internationally. The SME is interested in integrating AM technology with their current manufacturing system for end-part production. A key reason stated for wanting to make a change to AM is for improved inventory management. The SME may only require small batch production of specialised parts for custom designs, smaller scale installations or repair of previously installed furniture. Typically, these parts are heavy, made of cast metal that requires tooling and consequently production runs far exceeding the actual number of parts required. Alternatively, limited-run custom parts may be created by modifying existing parts using laborious cutting, shaping and welding techniques. AM technology allows for limited production runs, efficient material usage and the freedom to make specialised forms. As such it is deemed to be a way for the SME to overcome some of the issues associated with conventional manufacturing including wasteful practices and improved inventory control. The SME approached the research unit with these concerns and agreed to work with them on research to determine the feasibility of AM technology for production. The university-industry collaboration adopted a practice-led research methodology to investigate the viability and feasibility of introducing AM technology for the manufacture of parts currently made in cast alloys with a particular focus on urban street furniture.

The AM technologies investigated ranged from metals for the smaller items to polymers and polymer composites for the larger structural items. The project involves the design, optimisation, printing and testing of selected parts currently manufactured by the SME. The process was carefully documented and analysed to understand where changes to a typical design practice occur by working with this new technology and how these 'new' practices inform designs leadership role inside the broader innovation strategy of the organisation.

BACKGROUND

Advancing digitization of the manufacturing sector is in large part referred to as a transition to Industry 4.0. Where Industry 3.0 represents the era of automation and computing, Industry 4.0 represents digitization of the wider manufacturing system - a connected network of operations that is designed to be far more flexible, adaptable and responsive to changes across a range of factors that includes, but not limited to, evolving consumer needs and technology advances. The transition to Industry 4.0 is considered important for maintaining competitive advantage in a globalised market. Additive manufacturing (AM), also known as 3D printing is an emerging group of digital technologies that produce physical 3D objects directly from computer models in polymers, ceramics or metals (Ford, 2014). AM has been described as an 'intelligent production system' and part of the move to Industry 4.0 (Dilberoglu et al. 2017). Typically, manufactured products are made from parts moulded or cast in large expensive machines, designed to produce the part repeatedly in the most efficient way. AM does not require this machinery and provides a seamless link between the design as created on the computer and holding the physical item in your hand.

Used originally as a prototyping tool, AM has improved and can now produce cost efficient, low volume customised parts that can contain previously unachievable levels of complexity and functionality (Huang et al. 2015). Improvements in material performance, efficiency and quality means that AM is increasingly being accepted as an end-part production or a "direct production" process (Wang, 2018). And these developments have encouraged small-medium manufacturing enterprises to research the potential of the technology further (Walden et al. 2018a&b). However, it's important to note that AM represents much more than an alternative fabrication method as it has the potential to concurrently affect design processes and supply chain logistics (Wang, 2018). AM technology can restructure and redefine the operational relationship between design research and design for manufacture due to its capacity to respond to consumer needs so quickly (Walden & Lie, 2019). AM has been described as a "hyper-flexible technology that can provide highly customised and personalised products and production" that will challenge conventional business models (Bogers et al. 2016). Therefore, integrating AM is likely to be unique for each company based on, among other factors, technology selection, improving product and service quality; and implementation strategy. Wang (2018) suggests that in practical terms, future manufacturing systems will integrate AM with existing and emerging production methods, rather than replace all machine processes. Technological transitions are dependent on identifying strategic windows of opportunity (Geels, 2005) and manufacturing SMEs may be the optimum enterprise for AM adoption. The technology can bridge the resource, knowledge and absorptive capacity gap that SMEs typically struggle with, resulting in efficiency and waste reduction (Ghobadian et al. 2018).

TRANSITIONING TO AM

The transition to AM technology encompasses many concerns in ensuring quality standards can be achieved. Chua, Wong & Yeong (2017) has developed a Quality Management Framework in Additive Manufacturing that recommends that top management in an AM organisation play a critical role in ensuring that the production systems are safe and that the parts produced meet quality standards. The Framework developed by the authors outline a series of guidelines to support manufacturers transitioning to AM. Key to this research, there are particular recommendations that are important to refer to here. The authors recommend the development of benchmarks that include the measuring of the performance of materials and material testing. They also recommend committing resources to knowledge retention and the development of new design methodologies to optimise products for the AM process that include 'validation experiments' and 'continuous process verification' to ensure part performance.

Transitioning to AM is a shift to a more complex system because of the pace design changes can be made with the technology. The organisation naturally moves from a conventional manufacturer of product 'repeats' to one that can create many variables. Consequently, the company transitions to a product-service integration where the system becomes buyer (or supplier) centric. Research on the management innovation of complex products and systems has found that managers should experiment and innovate ways to develop relationships with their contractors and that joint activities between external and internal 'change agents' may identify new performance improvements (Roehrich et al. 2018). The research advocates the use of cross-functional teams to better integrate specialised knowledge and expertise.

Factors Causing a Change in Leadership Style

One factor that must be managed when companies transition to AM is the speed of new product development. Where in conventional manufacture an SME may take advantage of the time taken for the production of parts because of the need to develop new tooling, etc. an AM organisation will be expected to supply product (often customised) very efficiently. Cheng & Yang (2019) examine new product innovativeness and its relationship to the speed of production through the lens of creative process engagement (CPE) (see Amabile, 1993; Gibson & Shalley, 2004). Cheng & Yang find that leadership encouragement of proactive, creative development of 'ideas and alternatives' positively impact new product development speed and performance, highly contingent on facilitation full communication of innovation viewpoints between departments. Manufacturing SMEs are limited by size and resources so by seeking the support of external research units expand their organisation for a short time and enable idea generation and cross-functional team support. The research by Cheng & Yang indicates that while it may seem counterintuitive to commit time to the development of new ideas and the exploration of new avenues of practice, it can potentially be an important activity for both identifying innovation pathways and stabilising conditions for fast new product development. These are critical goals for improving the competitive advantage of an SME. The development of university-industry partnerships that adopt design-led research to explore alternative and emerging technological opportunities has proven beneficial in the past as in the case of the project titled, 'Digifacture' by the Integrated Product Design Research (IPDR) Unit at the University of Technology Sydney (Walden et al., 2018a). Digifacture worked closely with a local AM Service Provider to test the limits of and explore opportunities for AM integration in a range of industries to provide AM application insights for SMEs. The project produced prototypes and in doing so, was able to provide implementable guidelines for practice that can ultimately reduce the risk and increase the pace of transitioning to AM technology. Developing innovation (process and products) by organisations working collaboratively, or *collective entrepreneurship* has been identified as significant factor in shaping the systems and strategies associated with the future incorporation of AM technology, particularly when engaging universities where academic research and business insights converge (Lavoie & Addis, 2018).

Research analysing entrepreneurial strategies associated with AM adoption identifies that the 'diffusion' of printing technology develops a trend towards shorter, more diverse value chains that essentially encourage localised manufacturing that operates close to the end-user (Laplume et al. 2016 in Lavoie & Addis, 2018). Consequently, the advantages that low cost, low wage countries currently enjoy may lose that advantage in the future if AM is broadly integrated (Lavoie & Addis, 2018). The adoption of AM by SMEs can therefore strengthen their production agency and for SMEs that have a diverse and localised customer base, the adoption of AM ought to be seriously considered.

A study of business models for AM finds that adoption of the technology promotes the transition from manufacturer-centric to consumer-centric value logic and a shift toward decentralised supply chains (Bogers et al. 2016). Further, the study indicates that a hybrid approach that organises centralised AM production for specilaised, low volume production or special additions; combined with localised (decentralised) AM supply chain can optimise a consumer-centric approach and the reduction of inventory levels. However, 'platforms' need to be created to ensure part reliability and address any IP issues. Any move toward decentralised, consumer-centric models of AM production strongly indicates that organisations must continuously research and review technology parameters, part design and customisation boundaries. Adoption of manufacturing decentralisation and transition toward more customer-centric approaches raises interesting perspectives on leadership. Agentic leadership (visionary) leadership orients the specific future direction of the company and exerts control. And communal (empowering) leadership grants autonomy and control from leaders to followers by imparting trust. Research has indicated that these two styles complement each other to improve the performance of the organisation and 'goal clarity'. The research acknowledges that in the context of advancing digitalisation and its effect on the future work, the synergy between visionary and empowering leadership is something to be further researched (Kearney et al. 2019).

Knowledge Management of Product Design for AM

In connection to leadership, knowledge management is also an important function particularly for companies interested in adopting new technologies that require new practices. The modern form of the product design discipline was born out of the industrial revolution (from about 1760 – 1840) and the industrialisation of consumer products. The profession developed significantly after WW2 particularly in Europe and the United States with designers such as Henry Dreyfuss and, later, Dieter Rams who developed methods and approaches that both strengthened the bond between 'good' design and 'good' business; as well as demonstratively expanded the strategic influence of product design in an organisational context, that has been recognised globally. Essentially, the product design process is concerned with developing new products for manufacturing, though the product designer must take into consideration a wide set of concerns throughout the process, simultaneously. To represent the dynamically interrelated issues that must be addressed in designing a new product, W. H. Mayall (1979) developed 'the principle of totality' that has 'performance', 'serviceability', 'manufacturability', 'size', 'styling', 'comfort', 'ease of control', and 'safety'; all connected to each other. This mode of product design is highly process driven so that products can be developed 'on time' and 'on budget'; and meet the marketing goals of the company or client, the needs of the consumer and the requirements set by the manufacturing technologies available. The knowledge generated through this highly prescriptive practice often only emerges in the form of prototypes – essentially becoming the 'carriers of knowledge' – and then the knowledge is generally implicit or hidden. Such is the nature of product design operating in the service of manufacturing. However, as globalisation increases and as manufacturing becomes decentralised, so does the complexity of the problems that product designers need to address through their work. The problem of how to transition to AM technology must be addressed through research enquiry, because each company is different and the range of different types of AM technology is vast and always developing. It may be classified as a modern design problem in that it is 'open' (without boundaries), 'complex' (having many interrelated elements), 'dynamic' (changing over time) and 'networked' (across organisations and globally) (Dorst, 2015).

A conventional design for manufacture approach is too limited. A design-led research approach is needed because 'organisational learning' is needed to provide the right means of integrating the technology. Prototyping with various versions of the technology, through an external partner, without committing to the purchase of expensive machines would form part of a design-led research approach for this knowledge-directed enquiry. Knowledge management research has indicated that the complexities associated with globalisation may be addressed by better accommodating human-values and a more holistic approach into knowledge management, inviting creative cooperation and the adoption of responsibility for interdependence (Ženko et al. 2017). These findings serve to reinforce the value of empowering leadership styles and the granting of a degree of autonomy to teams for the conduct of research. Additionally, research conducted by Turulja & Bajgorić (2018) finds that knowledge management processes can be improved by promoting "innovation and experimentation" assisted by the development of ideas with external sources. This in-turn develops the firms learning culture. The research further indicates that organisational knowledge management can be positively impacted when the employees "consider themselves as partners in mapping the direction of the organisation", as it serves to promote organisational learning in connection with knowledge management capability.

Increased digitisation and the changing nature of work encompasses the practices and processes of AM and Industry 4.0. As discussed earlier a transition to AM by a manufacturing organisation will require the development of new business models that account for the decentralisation of manufacture and higher-level consumer-centric value logic. The transition must naturally place greater demands on the design products and platforms for product codesign, but if negotiated correctly has the potential to improve the production agency and competitive advantage of the organisation. Design-led research cooperatively pursued through industry partnerships with universities has been shown to be effective in supporting creative process engagement through collective entrepreneurship. Case analysis of these factors combined may provide practical insights for leadership and innovation for SMEs transitioning to AM.

PRODUCT DESIGN AND AM

A manufacturer transitioning to AM are most likely interested in redesigning their original products into new forms compatible with AM printing technology. Oh, Zhou & Behdad (2018) identify a number of approaches to this including material optimisation (*topology optimisation* - discussed later in this section), *part consolidation* where an assembly of components are merged into a single part and, the opposite of this, *part decomposition* which is partition of an original part into several smaller component parts so that it can be printed. The research notes the importance of *part decomposition* for AM adoption given the limited build envelope of an AM machine and the need to maximise printability and productivity while also acknowledging other advantages including the opportunity to increase product flexibility (designing in interchangeable parts for product variability).

The advance of 3D printing as a legitimate technology for end-part production (AM) has progressed alongside of, and in association with, computer aided design software tools. AM production allows for the production of complex forms in durable materials free from the geometric constraints conventional manufacture (e.g. injection moulding, die-casting, etc.) imposes. This allows for the creation of highly optimised parts that perform structurally with the smallest amount of material volume. The most efficient combination of shape, structure and material required is determined by computer software compatible with (sometimes built into) computer aided design (CAD) software used by product designers to ultimately generate production data for the making of physical parts and assemblies. The process is known as *topology optimisation* which is achieved by a *generative design* process that generates optimised variations of a part based on design inputs such as boundary volumes and load-case data. Generative Design may be described as a process whereby the designer creates a set of algorithms (non-ambiguous instructions) that enable concept forms to self-generate. Generative Design has been described as a "generic" term for using computation in design or as computation-driven design that uses 'the cloud', GPUs and artificial intelligence to autonomously generate part variances based on the set-up of an algorithmic framework and instructions from the designer (Altair, 2019, Desktop Metal, 2019). Generative systems can be useful in "sparking new design ideas ... solving difficult tasks ...(which) extend designers' current capabilities" (Shea, Aish and Gourtovaia 2005) and can allow "greater exploration of possibilities" and create "exuberant forms" (Schumacher and Krish 2010). McCormack et al. (2004) provide a good description of the designer's role in Generative Design. They state, rather than a direct relationship between the designer and the artefact as with conventional design practice, Generative Design creates the artefact autonomously based on creating and varying instructions noting that the role of the human designer remains central to the design process but transitioning toward *meta-design* (their term) where it's important to master relationships between process specification, environment and generated artefact. Developing a meta-design approach to responsibly incorporate these new digital technologies and, as McCormack et al. (2004) described it in their research, 'master relationships' between production, environment factors and part features suggests that these relationships should be stable enough to reliably and rapidly generate multiple variants, that are compatible with the organisations systems and vision. Adoption of Generative Design into product design for consumer products has found that it can be useful for both ideating part forms and for technical refinement of the selected form, however, the 'raw' shapes typically require stylised surface treatment worked through by the designer using more conventional practices (Nemme & Walden, 2019).

The processes discussed above account for only some of the ways product design and manufacturing leadership will be affected by AM. Management of and for product design for an SME transitioning to AM requires adoption of new tools (computer software) and new methods that operationalise connections between the practical work of designing products for manufacture and the strategic goals of the (developing) AM production organisation. The company itself needs to conduct this research themselves, through practice as the way one company may integrate

AM technology is likely to be very different from another. Integration of AM is to establish ways for the technology to complement traditional manufacturing (see Oh et al. 2018) and design for manufacturing process. Recent research proposes a framework that places the role of (product) design in an AM organisation on a spectrum of practice ranging from design as research (DAR) through to design for AM (DFAM) with design for manufacture (DFM) in the centre (Walden & Lie, 2019). Design as Research is an enquiry-based concept design that examines future opportunities in part or process design through the application of practical design methods. The outcomes are created to generate knowledge about technology application, cross-functional teams and future directions rather than arrive at commercialisable product solutions. The framework considers that the role of product design expands beyond Design for Manufacture (a conventional yet still relevant role) into design-led research enquiry to proactively define how the organisation defines (for itself) design for AM. Performing all of these roles cohesively is difficult. A divide between line-up design (Design for Manufacture) and next-gen design (more research focused) is evident in large Korean companies. Acknowledging this; separate research by the authors argues that for SMEs there are ways of bridging that divide by focusing on research prototyping (Wensveen & Matthews, 2015) and engaging university-industry collaboration (UIC) projects (Walden et al. 2018b).

RESEARCH QUESTION

What is the role of product design in strategic leadership and the development of innovation capability when transitioning to digital manufacturing technology, and the integration of AM, particularly for manufacturing SMEs?

HYPOTHESIS

That for manufacturing SMEs to transition from conventional manufacturing technology to digital manufacturing technology the incorporation of design-led research is required to continuously innovate and investigate part design variability and performance as technology evolves. Additionally, that to incorporate design-led research practice, a combination of visionary (agentic) and empowering (communal) leadership styles, guided by a knowledge management process that promotes "innovation and experimentation", is necessary.

RESEARCH METHOD

The chapter presents a case study of a prototyping method. Case studies are a form of qualitative research that provides a description and analysis of (in this case) a process and concept (Bloomberg & Volpe, 2016). The case study analysis focuses on aspects of the process that are related to the research question and hypothesis presented above. Prototyping as a part of design research can be used for a number of different forms of enquiry. Referring to research by Wensveen and Matthews on the role of prototypes in design research (2015), they have identified four roles. 1) As an 'experimental component', where the prototype form is created to test the aesthetic and functional performance of the intended product; 2) as a 'means of inquiry', where the prototype may (strategically) not take on the form of the intended product as they are 'sent out' as open-ended explorations for design discourse about use or users; 3) as 'research archetypes' where the prototype essentially functions as the physical embodiment of a research perspective; and 4) prototyping as a 'vehicle for inquiry', where the construction of the prototype is equally (and often primarily) concerned with the making of the prototype and where the process is documented, analysed and critiqued. In this project, the prototyping research is somewhat of a hybrid between type 1 and 4, based on Wensveen and Matthews description. There is the construction of parts that resemble production versions, for the purpose of testing the performance of those parts under various conditions (type 1). And there is also the documentation and analysis of the process of developing those parts that challenge standard design and production practices. It is via this combined approach that we are able to investigate how the product designer's role might change with regard to strategic leadership and the development of innovation capability with the integration of AM.

The Integrated Product Design Research (IPDR) Unit at the University of Technology Sydney was approached by a local public furniture manufacturer (SME) to investigate the potential of additive manufacturing (AM) as a fabrication method for some of their parts. The manufacturer supplies benches, seats, tables and other types of public furniture to councils around Australia and in other parts of the world. The company values design, always upgrading and advancing their products to ensure their reputation as a leader in the field, capable of delivering high performing and stylish furniture for public spaces. The company has been in operation for 30-years and has succeeded in the design, manufacture and supply of public furniture, by focusing heavily on ensuring that their products are strong and reliable for decades after installation. They primarily use press formed steel and cast aluminum for the structural elements that are concreted into the ground. The cast aluminium parts require tooling and minimum production volumes of between 7-12,000 units as well as post machining and polishing operations per part. Years after installation, a council may request an upgrade or refurbishment of an existing product, or for more of the same product to be installed in the area. These orders generally mean either committing to a production quantity far beyond what is actually needed or, custom-making new parts - both options are not optimal. This situation aside, the minimum production volumes for many new product types often tend to exceed the order quantities anyway given the rate of change the market today experiences. Consequently, AM presents an attractive alternative if the parts can be printed at a reasonable cost and also be strong and reliable.

Significantly, the company had appointed an external design expert to work with company management on a vision for the organisation's innovation agenda, before the start of this research project. The design expert appointed has extensive industry experience in the product design field having founded and directed a very successful design consultancy in Australia as well as holding an eminent professorial position at a leading Australian University. The IPDR Unit was appointed to the project as a result of the recommendations of the design expert combined with collaboration with company management to formulate a project objective and methodology. The action by the company to form partnerships with external expertise and external academic research units is an important step for SMEs to adopt in order to inspire innovation change in their organisation.

An objective of the project was to investigate the potential of AM as an end-part production technology for the fabrication of structural parts typically made from cast aluminium alloy. An additional objective was to identify how a transition to advanced (digital) manufacturing technology may impact the role of product design inside the organisation, particularly regarding strategic leadership and the development of innovation capability. Therefore, processes were documented carefully. Design-led research is set up to objectively investigate the overarching consequences of integrating this emerging technology, quite differently from the research that the company designers may conduct to shape design features for a product for manufacture with a predetermined function, production method, budget and release date. The IPDR research team operates in a more speculative realm where research prototyping, concepting and scientific testing converge to provide knowledge insights rather than commercial solutions. However, the significance of conducting design-led research is that it is also an effective way of determining how to *leverage* embedded product design techniques mastered by the company's design team, toward knowledge-directed methods of enquiry.

CASE ANALYSIS

The IPDR Unit collaborated with the industry partner early in the project to determine the best approach for investigating ways of adopting AM technology into their production system. The industry partner is a manufacturer of a diverse range of parts, many that need to be customised for various clients. The component parts and the assembled products they ultimately form are outdoor public structures, subject to weather as well as the stresses and strains of public use (and occasional misuse). As public structures they must perform safely and reliably for years after installation to ensure that the client (the council responsible for their purchase) can be confident that their citizens are not harmed. The industry partner for this project is an excellent company with a very good reputation for high quality design and engineering. They take great care in the development of new products and it was from this position (this vision) that the project team formed an approach forward.

A decision was made to concentrate the research around three types of parts. Firstly, a joining part that has a critical structural role in dispersing load through a frame structure for outdoor benches that must take human weight and withstand other types of impacts from public use. Additionally, the original part is an aluminium alloy casting often left unfinished and exposed to the weather. The selection of this part was to examine redesign and material optimisation processes for achieving suitable strength and durability performance using AM technology. The second type of part is an armrest originally made in aluminium alloy but adopting a more 'fluid' complex form. Being a less critical part structurally, it enables the examination of more 'aggressive' topology optimisation and generative structures using larger format AM printing. The third part type was a smaller badge-cap component that is open for high levels of customisation as AM would definitely allow the client can have their council's logo embossed on the cap. The investigation of part type three was related to part consolidation / decomposition for customisation, build envelope analysis to determine production efficiency and profitability. As a smaller part, it was the type of part most likely to adapt to AM in the short to medium term. This Chapter will concentrate on part type 1 and 2 since these provide the richest insight into changing leadership and strategy for new product development using AM technology.

Design Strategy: Internal Material Optimisation

The post and beam junction is a joining part subject to significant loads and impact. It distributes the load through the structure and into the ground to make the bench stable. The original part is made in diecast aluminium alloy - a process and material composite combination that has been confirmed as being able to reliably produce strong, durable parts through hundreds of years of research and development by companies all over the world, tested via the millions of different forms its adopted and conditions its been applied to over very long periods of time. Over that time these tried and tested process and material combinations adopt signature features that themselves are replicated time and time again by companies learning from each other. The product designer and engineer working on the design of such a part draw on embedded knowledge to confidently give shape to the part with emphasis on the exterior surface form as the metallurgical structures of the solid interior no longer require in-depth scientific analysis to confirm the parts durability. The industrialisation of the process and of the formulation of the material are so compatible that design of parts can happen quickly and confidently because part performance can be assured by these factors alone.

This is not the case with AM. Exterior form, internal structure, material type, material density and process dynamically interrelate to produce an enormous number of potential variables, only a small number of which are likely to perform nearly as well as the original metal part. To ‘stabilise’ the research experiment, the team decided to keep the exterior form exactly the same as the original part. The experiment would examine internal (lattice) structure, material type, material density and process for a select range of AM machines for printing polymers and polymer composites. Below (Figure 1) are two samples of over sixteen test variations produced and tested (see Table 1).

Metal printing was excluded for this test because as of the writing of this chapter metal printing except for very small and critical parts is a process generally too expensive for consumer products. The IPDR unit arranged for the parts to be tested with a material scientist under controlled conditions to determine if a printed polymer version of the part may suitably replace the metal version. The testing procedure for each printed part was set up in the same way each time. The testing comprised of bend load and compression load testing (Figure 2) under controlled laboratory conditions on a Shimadzu Autograph AGS-X testing machine. Testing revealed that some versions of the part performed very well in printed polymers in both tests (bend and compression) printed on accessible and affordable AM technology. Analysis of the process identifies opportunities for integrating AM and demonstrates practical ways for further testing to occur. As can be seen from this example the product design and production process for AM requires a much more in-depth and scientific analysis of part design. The case recommends that the AM organisation should adopt R&D strategies that enable product design teams to work with other specialists inside (and potentially outside) the organisation to conduct experimental design-led research that develops internal knowledge about the integration of AM technology and its relationship to strategic and innovation goals. Design teams (represented in this case by the IPDR unit) must be able to work with degrees of autonomy so that creative

Figure 1. Redesigned (optimised) and printed versions of the original Post & Beam Junction. The images show the part printed in a nylon with carbon fibre inlays that has been printed on a Markforged printer (left) and the same part printed in 'Alumide' - an aluminium filled polyamide - that has been printed on an EOS printer (right). A range of samples were printed on different printers and with different lattice densities for testing.

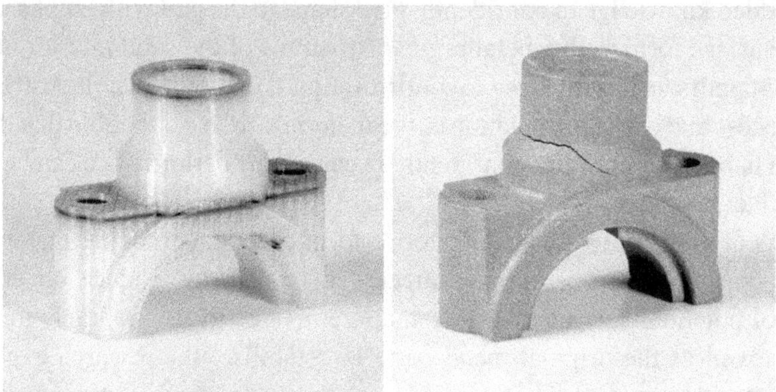

exploration can take place. In conventional settings a product design department in a manufacturing organisation (particularly an SME) would typically receive briefs outlining clear design objectives and project scope for the creation of a product (or prototype) for manufacture and commercialisation. Whereas, design for AM requires the adoption of 'research prototyping' as a part of knowledge-directed, design-led research; to establish design processes and guidelines.

Design Strategy: Generative Design and Part Decomposition

The next series of tests examined the potential for AM to produce a large format part by generating topologically optimised variations and investigating the use of part decomposition. The armrest investigated is originally cast in aluminium alloy and then polished (with some post-machining). For larger components such as this, it is a very common method of manufacture that has been in place for over 150-years. The development of AM for the production of large parts is an emerging and developing technology. Large format printing is less accessible because most commercially available AM machines have limited construction platform dimensions though research into multi-axis robot assisted AM is a promising way forward (Urhal et al. 2019). A part such as an armrest is considered a large part for AM and can only be made on some printers though there are options available for SMEs. Investigation of how a larger component may be produced using AM by the industry partner involved

Table 1. Machine, process and material variations for redesigned (optimised), printed and tested Post & Beam Junction variations. Processes include SLS (Selective Laser Sintering), Multi Jet Fusion, FFF (Fused Filament Fabrication) and FDM (Fused Deposition Modeling).

Part Description	Supplier / Machine	Process	Material	Test Performed
Post and Beam Junction	i.Materialise EOS	SLS	Alumide: Nylon + Aluminium Particles (hollow / 3mm wall)	- Bend - Compression
	UTS Protospace HP Jet Fusion 3D 4210	Multi Jet Fusion	Nylon (Polyamide12) Powder (solid)	- Bend - Compression
	UTS Protospace HP Jet Fusion 3D 4210	Multi Jet Fusion	Nylon (Polyamide12) Powder (hollow / 3mm wall)	- Bend - Compression
	UTS Protospace Markforged X7	FFF	Onyx (Black) Nylon + Carbon Powder (infill)	- Bend - Compression
	UTS Protospace Markforged X7	FFF	Nylon (translucent) + Carbon Fibre Inlay (Horizontal) (infill)	- Bend - Compression
	UTS Protospace Markforged X7	FFF	Nylon (translucent) + Carbon Fibre Inlay (Vertical) (infill)	- Bend - Compression
	UTS Product Design StrataSys UPrint	FDM	ABS (infill)	- Bend - Compression
	UTS Product Design StrataSys UPrint	FDM	ABS (solid)	- Bend - Compression

the testing of topology optimisation software - an emerging suite of technologies that augment the 3D CAD software tools that product designers use to design and generate production data (see Figure 3). A large part printed on an expensive machine produces expensive parts though the cost can be reduced by optimising the part, i.e. reducing the amount of material used while maintaining its structure. Calculating the performance potential of an AM part that has been redesigned for material optimisation is the responsibility of the designer and new software tools that use algorithmic modelling to generate part variants based on information provided by the designer (including volume boundary, load case and lattice density) were tested. The research produced a series of prototypes of the armrest in a range of materials (an example is shown in Figure 4). Part decomposition was revealed as a promising outcome. The original part was modified using the software to essentially become

Figure 2. Testing procedure of the Post & Beam Junction in various printed polymers and polymer composites. Parts were tested in a controlled laboratory setting by an engineering specialist using a Shimadzu Autograph AGS-X testing machine. Test data was produced revealed the best combination of AM material type, internal structure and printer.

a polymer (nylon) shell augmented with a simple plate metal spine to provide structure. The result was achieved through design-led practice-based research and experimentation and can potentially set design guidelines for the development of large parts using AM for this industry partner. The investigation of software provided insights for the adaption of workflow in the design process for AM.

The project led to the development of thirty-two prototypes (Figure 5) and a comprehensive report that outlined process and procedures conducted as well as a detailed overview of the current state of the technology with a list of machines and printing capabilities that can be easily referred to for future research work. The project has helped the industry partner to understand the practical measures involved in the development of strategic leadership and innovation capability for its organisation as it transitions to AM. The need for knowledge driven enquiry conducted through practice by the designersinside the organisation cannot be understated. The strategic adoption of AM technology must allow for experimentation, idea generation and creative engagement as demonstrated in the case presented. There is no more reliable way of determining what type of AM technology should be purchased and how it will be incorporated into the organisation because the technology (the hardware and

Figure 3. A close-up of the 'dual edge delaunay' pattern algorithmically generated using topology optimisation software, Autodesk Meshmixer

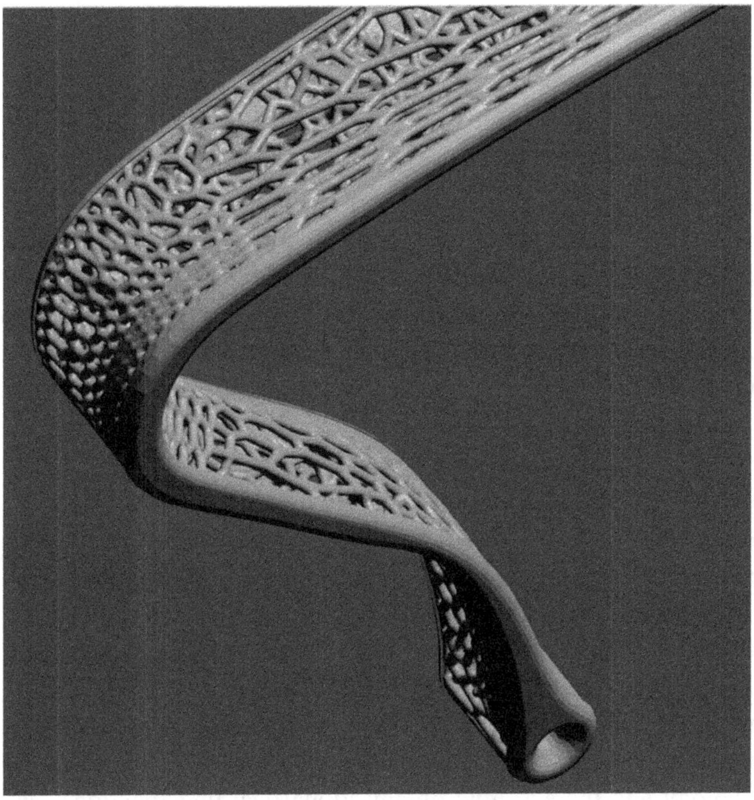

software) present too many avenues of development. The stable and reliable nature

Figure 4. Armrest printed in a nylon on the HP Jet Fusion printer

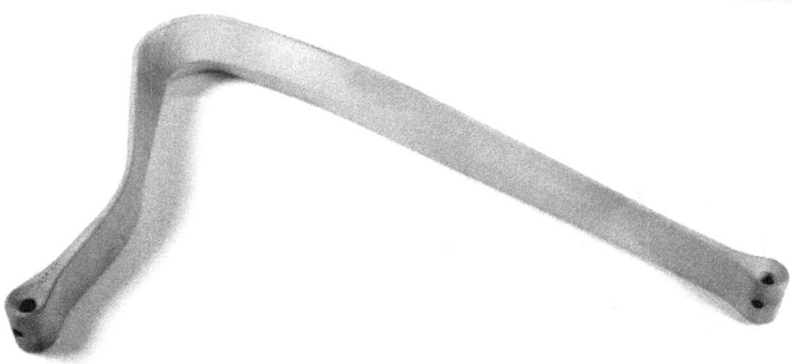

of traditional manufacturing technologies does not apply for AM and therein lies both tremendous challenges but also tremendous opportunity.

FUTURE RESEARCH DIRECTIONS

As discussed earlier in this chapter under the heading titled *Design for AM and Leadership* any organisation intending to move toward decentralised, consumer-centric models of AM production must continuously research and review technology parameters, part design and customisation boundaries with clear positioning in relation to their consumers. To assist organisations in this area, a more in-depth investigation of how a holistic and human-values approach to knowledge management may impact knowledge-directed enquiry in product development research, may provide valuable insights.

University-industry projects with a product design focus depend on knowledge management, creative co-operation, the integration of all participants and the relationships between them. These relationships are complex, and the knowledge generated through them has potential value for future projects. It would therefore be beneficial to find ways of documenting the relationship between knowledge management, creative co-operation and participant integration, which could be done as a form of action research.

Future research should also be conducted into the role of research prototyping that considers the specific complexities associated with the integration of emerging AM technology and the systemic organisational learning functions that may be impacted. This research could also address questions related to developing innovation capability in technology dependent organisations that must consistently update processes and practices.

Figure 5. Final set of parts delivered to the client

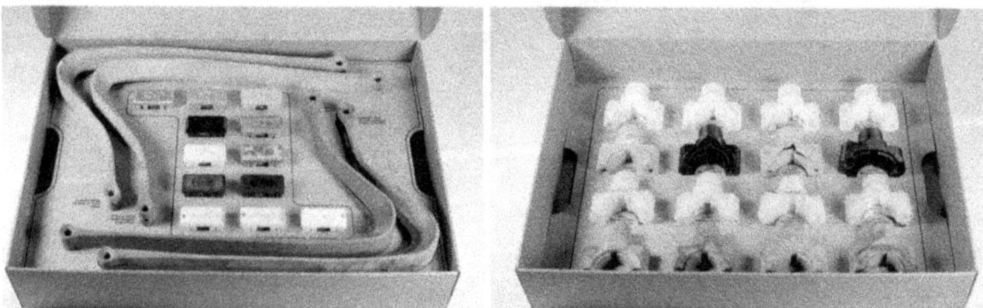

CONCLUSION

The research led to the identification of issues concerning the wider manufacturing process including the need to upgrade to new CAD software for optimising parts to be 3D printed, testing requirements for printed parts and; the opportunities and limitations associated with design for customisation. Additionally, the research informs knowledge management and development strategies that the SME should consider given the significant way in which the technology impacts conventional practices. The adoption of technology such as AM is to acknowledge that it operates as part of a network of operations digitally connected in a dynamically interrelated way, widening innovation capability but impacting strategic leadership. We find that product design must assume a greater level of responsibility in the conduct of research, knowledge development and strategic innovation planning. Digitisation of the manufacturing organisation means that the product design team becomes a critical node along lines of communication between key sections of the organisation in more ways than with conventional manufacture. As mentioned earlier, the modification of an existing part to create a custom version of that design, might have been managed primarily by the manufacturing department as there are only a very limited number of ways that those modifications could be made. AM parts cannot (typically) be modified and in any case, the technology makes it possible to manufacture a unique custom part in very limited quantity. However, this requires design and engineering teams to be centrally involved with not only the design, but the manufacture and requisite testing that must be done. We determine that design-led strategies for parts customisation need to be carefully developed at all levels of the organisation to ensure that the manufacturing system remains stable.

From a management perspective, research indicates that the transition to AM is a transition to decentralised, consumer-centric and localised production systems. This is good news for SMEs because it can return production agency and associated R&D work necessary to develop unique and innovative knowledge that connects the organisations new product development goals and the manufacturing system. However, the challenge for SMEs is to develop new models of leadership that combine visionary (agentic) and empowering (communal) leadership styles, that are guided by a knowledge management process that promotes “innovation and experimentation”. Currently we don’t know what they may look like for many manufacturing SME AM organisations though we may be able to predict that due to the very nature of AM technology conditions, that each SME must address this paradox of styles in their own unique way. The case study presented offers some practical insights as to how a degree of autonomy granted to the product development team can allow for examinations of practice and the testing of how the AM technology might in both the medium and longer term improve the company’s performance. Notably, the set up for the project presented in the case study involved firstly, the appointment of an external design expert to lead the development of an innovation vision for the advancement of the company technologically, and then the appointment of an external research unit (IPDR) operating with degrees of autonomy to develop practical in-roads toward that vision. Another project that developed concepts to examine the potential future applicability of AM titled Digifacture (referred to in Walden et al. 2018a) identifies the importance of university-industry collaboration for research and experimentation work that generates knowledge and returns it back to inform practice. At a technical level the case presented in this chapter demonstrates the value of (product) design-led research in cross-functional teams that mediate between knowledge generation and creative process engagement; and supports the development of new ideas and alternatives for positive and proactive AM integration.

ACKNOWLEDGMENT

This research was supported by the NSW Government TechVoucher Scheme and the Industry Partner, Street Furniture Australia (SFA). The authors would also like to acknowledge the contributions of key people including Dr Barry Liu, Technical Officer in the Faculty of Science at the University of Technology Sydney; June Boxsell, Marketing Manager at SFA and Prof. Mark Armstrong, Practice Prof. in Design at Monash University and design advisor at SFA.

REFERENCES

Altair, S. R. (n.d.). *Generative Design and Topology Optimisation by Altair*. Retrieved from https://cdn2.hubspot.net/hubfs/47251/Altair_Generative_Design_Report.pdf

Amabile, T. M. (1983). The social psychology of creativity: A componential conceptualization. *Journal of Personality and Social Psychology*, *45*(2), 357–376. doi:10.1037/0022-3514.45.2.357

Archer, B. (1995). The nature of research. *Co-Design Journal*, *2*(11), 6–13.

Bloomberg, L. D., & Volpe, M. (2016). *Completing your qualitative dissertation: A road map from beginning to end*. Sage Publications.

Bogers, M., Hadar, R., & Bilberg, A. (2016). Additive manufacturing for consumer-centric business models: Implications for supply chains in consumer goods manufacturing. *Technological Forecasting and Social Change*, *102*, 225–239. doi:10.1016/j.techfore.2015.07.024

Cheng, C., & Yang, M. (2019). Creative process engagement and new product performance: The role of new product development speed and leadership encouragement of creativity. *Journal of Business Research*, *99*, 215–225. doi:10.1016/j.jbusres.2019.02.067

Chua, C. K., Wong, C. H., & Yeong, W. Y. (2017). *Chapter Nine-Quality Management Framework in Additive Manufacturing*. Academic Press.

Dilberoglu, U. M., Gharehpapagh, B., Yaman, U., & Dolen, M. (2017). The Role of Additive Manufacturing in the Era of Industry 4.0. *Procedia Manufacturing*, *11*, 545–554. doi:10.1016/j.promfg.2017.07.148

Dorst, K. (2015). *Frame innovation: Create new thinking by design*. MIT Press. doi:10.7551/mitpress/10096.001.0001

Ford, S. L. (2014). Additive manufacturing technology: potential implications for US manufacturing competitiveness. *Journal of International Commerce and Economics*. Retrieved from http://www.usitc.gov/journals/Vol_VI_Article4_Additive_Manufacturing_Technology.pdf

Geels, F. W. (2005). *Technological Transitions and System Innovations: a Co-evolutionary and Socio-technical Analysis*. Edward Elgar Publishing. doi:10.4337/9781845424596

Ghaffar, S. H., Corker, J., & Fan, M. (2018). Additive manufacturing technology and its implementation in construction as an eco-innovative solution. *Automation in Construction*, *93*, 1–11. doi:10.1016/j.autcon.2018.05.005

Ghobadian, A., Talavera, I., Bhattacharya, A., Kumar, V., Garza-Reyes, J. A., & O'Regan, N. (2018). Examining legitimatisation of additive manufacturing in the interplay between innovation, lean manufacturing and sustainability. *International Journal of Production Economics*. doi:10.1016/j.ijpe.2018.06.001

Gilson, L. L., & Shalley, C. E. (2004). A little creativity goes a long way: An examination of teams' engagement in creative processes. *Journal of Management*, *30*(4), 453–470. doi:10.1016/j.jm.2003.07.001

Huang, Y., Leu, M. C., Mazumder, J., & Donmez, A. (2015). Additive manufacturing: Current state, future potential, gaps and needs, and recommendations. *Journal of Manufacturing Science and Engineering*, *137*(1), 014001. doi:10.1115/1.4028725

Kearney, E., Shemla, M., van Knippenberg, D., & Scholz, F. A. (2019). A paradox perspective on the interactive effects of visionary and empowering leadership. *Organizational Behavior and Human Decision Processes*. doi:10.1016/j.obhdp.2019.01.001

Laplume, A. O., Petersen, B., & Pearce, J. M. (2016). Global value chains from a 3D printing perspective. *Journal of International Business Studies*, *47*(5), 595–609. doi:10.1057/jibs.2015.47

Lavoie, M., & Addis, J. L. (2018). Harnessing the potential of additive manufacturing technologies: Challenges and opportunities for entrepreneurial strategies. *International Journal of Innovation Studies*, *2*(4), 123–136. doi:10.1016/j.ijis.2018.11.001

Mayall, W. H. (1979). *Principles in design*. Design Council.

McCormack, J., Dorin, A., & Innocent, T. (2004). *Generative design: a paradigm for design research. Proceedings of Futureground*. Melbourne: Design Research Society.

Nemme, A., & Walden, R. (in press). *Integrating generative design and topology optimisation with product design values*. Paper presented at the Technology Education Conference (TENZ), Auckland, New Zealand.

Oh, Y., Zhou, C., & Behdad, S. (2018). Part decomposition and assembly-based (Re) design for additive manufacturing: A review. *Additive Manufacturing*, *22*, 230–242. doi:10.1016/j.addma.2018.04.018

Roehrich, J. K., Davies, A., Frederiksen, L., & Sergeeeva, N. (2018). Management innovation in complex products and systems: The case of integrated project teams. *Industrial Marketing Management*.

Schumacher, P., & Krish, S. (2010). *Teaching Generative Design Strategies for Industrial Design*. Academic Press.

Shea, K., Aish, R., & Gourtovaia, M. (2005). Towards integrated performance-driven generative design tools. *Automation in Construction*, *14*(2), 253–264. doi:10.1016/j.autcon.2004.07.002

Turulja, L., & Bajgorić, N. (2018). Knowing Means Existing: Organizational Learning Dimensions and Knowledge Management Capability. *Business Systems Research Journal*, *9*(1), 1–18. doi:10.2478/bsrj-2018-0001

Urhal, P., Weightman, A., Diver, C., & Bartolo, P. (2019). Robot assisted additive manufacturing: A review. *Robotics and Computer-integrated Manufacturing*, *59*, 335–345. doi:10.1016/j.rcim.2019.05.005

Walden, R., & Lie, S. (2019). *Design Knowledge Development and Additive Manufacturing Systems: How Does Design Knowledge Change With Design for AM?* Academic Press. doi:10.4018/978-1-5225-7152-0.ch011

Walden, R., & Lie, S. (2019). Design Knowledge Development and Additive Manufacturing Systems: How Does Design Knowledge Change With Design for AM? In *Standards, quality control, and measurement sciences in 3D printing and additive manufacturing*. Academic Press. doi:10.4018/978-1-5225-7152-0.ch011

Walden, R., Lie, S., Pandolfo, B., Lee, T., & Lockhart, C. (2018a). Design Research Units and Small to Medium Enterprises (SMEs): An Approach for Advancing Technology and Competitive Strength in Australia. *The Design Journal*, *21*(2), 247–265. doi:10.1080/14606925.2018.1427913

Walden, R., Pandolfo, B., Lie, S., & Nemme, A. (2018b). *Adaptable practices for next-generation design in manufacturing SMEs informed by university-industry collaboration*. Paper presented at the Design Thinking Research Symposium DTRS, UNIST.

Wang, B. (2018). The Future of Manufacturing: A New Perspective. *Engineering*, *4*(5), 722–728. doi:10.1016/j.eng.2018.07.020

Wensveen, S., & Matthews, B. (2015). *Prototypes and prototyping in design research. In The Routledge Companion to Design Research*. Taylor & Francis.

Ženko, Z., Mulej, M., & Potočan, V. (2017). Knowledge-cum-values Management belongs to the Way out from Global Crisis. *Business Systems Research Journal*, *8*(1), 113–123. doi:10.1515/bsrj-2017-0009

ADDITIONAL READING

Chu, C., Graf, G., & Rosen, D. W. (2008). Design for additive manufacturing of cellular structures. *Computer-Aided Design and Applications*, *5*(5), 686–696. doi:10.3722/cadaps.2008.686-696

Diegel, O., Nordin, A., & Motte, D. (2019). *A Practical Guide to Design for Additive Manufacturing*. Springer. doi:10.1007/978-981-13-8281-9

Gibson, I., Rosen, D. W., & Stucker, B. (2010). Design for additive manufacturing. In *Additive manufacturing technologies* (pp. 299–332). Boston, MA: Springer. doi:10.1007/978-1-4419-1120-9_11

Gibson, I., Rosen, D. W., & Stucker, B. (2014). *Additive manufacturing technologies* (Vol. 17). New York: Springer.

Loy, J., Canning, S., & Haskell, N. (2016). *'3D Printing Sociocultural Sustainability', Handbook of Sustainability in Additive Manufacturing* (pp. 51–72). Springer. doi:10.1007/978-981-10-0549-7_4

Rael, R., & San Fratello, V. (2018). *Printing Architecture: Innovative Recipes for 3D Printing*. Chronicle Books.

Walden, R., & Lie, S. (2019). Design Knowledge Development and Additive Manufacturing Systems: How Does Design Knowledge Change With Design for AM? In *Handbook of Research on Industrial Advancement in Scientific Knowledge* (pp. 205–222). IGI Global. doi:10.4018/978-1-5225-7152-0.ch011

Wohlers, T., Caffrey, T., Campbell, R. I., Diegel, O., & Kowen, J. (2018). [Printing and Additive Manufacturing State of the Industry; Annual Worldwide Progress Report. Wohlers Associates.]. *Wohlers Report, 2018*, 3D.

KEY TERMS AND DEFINITIONS

3D Printing: Three-dimensional printing.

CAD: Computer aided design. Software applications used to create accurate, virtual three-dimensional representations of products and parts thereof. In product design and manufacturing, CAD software can be used to generate production data so that the parts can be fabricated directly using computer-controlled machines, as is the case with additive manufacturing technology.

Design for Manufacture: The practice of designing and optimising products and parts thereof in such a way that they are cost effective to be manufactured in high volumes.

Design-Led Research: A knowledge-directed research methodology that integrates design practices, processes and tools to investigate what can be learned through practitioner action (direct intervention often by the researcher themselves). As an example, the construction of prototypes, in connection with other methods, as a means of knowledge-directed enquiry is a common form of design-led research evident in the design research literature.

End-Use Product: A product that is to be used as a real product for its intended use.

End-User: The group of people for whom the product is intended to be used by.

Generative Design: A CAD process that through software applications generates optimised variations of a part based on design inputs such as boundary volumes and load-case data.

IPDR: Integrated Product Design Research Unit at the University of Technology Sydney.

Lattice Structure: A three-dimensional space-filling unit cell that can be tessellated along any axis with no gaps between cells, similar to biological open-cell structures such as bones. Additive manufacturing is the only manufacturing method able to produce cost effective lattice structures by infilling a part rather than printing it as a solid.

Low Volume Production: This refers to number of units produced during a production run where low volume production can be between 1 and 1000 units produced.

Polymer: A substance which has a molecular structure built up chiefly or completely from a large number of similar units bonded together, e.g. many synthetic organic materials used as plastics and resins.

Prototypes: The construction of parts that resemble production versions, for the purpose of testing the performance of those parts under various conditions.

R&D: Research and development.

SME: Small to medium enterprise. The Australian Bureau of Statistics defines a small to medium sized business as employing between 5 and 199 employees that has annual revenue turnover of less than $10 million Australian dollars.

Topology Optimisation: A computer-aided design process that uses mathematical methods to optimise material layout within a given design space, for a given set of loads, boundary conditions and constraints with the goal of maximizing the performance of the system or parts thereof.

Chapter 8
Interconnected Areas of Research:
Collaborations in Social Innovation

Derya Fındık
https://orcid.org/0000-0002-3002-4391
Ankara Yıldırım Beyazıt Üniversitesi, Turkey

Erdal Akdeve
Ankara Sosyal Bilimler Üniversitesi, Turkey

Gülsen Kaya Osmanbaşoğlu
Ankara Sosyal Bilimler Üniversitesi, Turkey

ABSTRACT

The authors aim at revealing themes in this research field throughout the years between 1970 and 2016 by using the terms of social innovation and collaboration together. They apply co-citation analysis to find out the theoretical foundations of this recently emerged field. Thus, they obtain six clusters with different attributes, such as cross-sector partnerships in social innovation, the definition of social innovation, transition studies, social entrepreneurship, innovation studies, and inter-organizational relations.

INTRODUCTION

The number of studies on the concept of social innovation has increased overwhelmingly in recent years. This concept, unlike the traditional definition of innovation emphasizing value creation activities such as idea generation, prototyping,

DOI: 10.4018/978-1-7998-1108-4.ch008

and commercialization, deals with the social benefits of the created value and aims at sustaining these benefits. While the profit-oriented innovation paradigm does not place much concern for the environmental impacts of production, social innovation activities focus more on the production of environmentally friendly technologies, installation of energy-saving facilities, and the roles of local actors in finding solutions to problems emerging at local levels.

Achieving predetermined societal goals requires the inclusion of actors from different sectors such as firms, non-profits, local actors, and society at large. These parties form various types of collaborations such as cross-sector partnerships, social alliances, and social partnerships. Cross-sector partnerships, being used interchangeably with the concept of social alliances (Le Ber and Branzei, 2010), emerge between firms and non-profit organizations in order to achieve a common societal goal, while both parties sustain their efforts over maximizing their own benefits. In social alliances, unlike strategic alliances, non-profit organizations or, broadly speaking, the whole of society is one of the parties. When commercial firms become party to social innovations, they tend to see themselves as 'good corporate citizens' (Berger et al, 2004). This type of partnership enables each party to focus on social problems and find appropriate solutions with the help of non-profit organizations. In other words, all parties including both non-profits and firms benefit from these partnerships. Another type of partnership is called social partnerships (Seitanidi and Crane, 2009). Similar to social alliances, all partners, being composed of public and private sector institutions, share their resources in order to accomplish a predetermined goal. Complementarity among partners' resources is desired to achieve success. Since each party has different motivations, a detailed examination is required to obtain successful outcomes. Such partnerships sometimes arise from necessity. For instance, firms may not have access to local markets due to their distinctive features and need to get in touch with local communities to clearly identify their needs and expectations. Moreover, developing the skills required to solve problems is costly both in terms of time and money. Thus, firms outsource some assets such as knowledge and skills due to production costs. These partnerships, on the other hand, commonly end by mutual miscommunication in a short time. Each party may have a fear of losing control and power, or mistrust may develop among parties. Considering all these examples, we aim at understanding the structure of collaborations in social innovation activities focusing on the studies that terms of collaboration and social innovation intersect. There has not been any study dealing with these practical issues. This study contributes to the current literature in at least three ways. Firstly, to our knowledge, collaboration in social innovation literature is only now elaborated on in this present study. Secondly, core themes and potential diversification among research streams in the field are determined by using co-citation analysis. Thirdly, we propose a set of implications based on the current structure of the network. The following section elaborates on data and methodology. The third section focuses on results. We discuss conclusions and implications for further research in the last section.

DATA AND METHODOLOGY

Sample Selection

Bibliometric data is collected from Thomson Reuters Web of Science database by using the topic search option. We use the phrase "social innovation" in the first stage and constrain our sample with the keywords "collaboration," "network," or "partnership" in the second stage in order to find papers falling into the areas where social innovation and collaboration intersect. We also limit our sample to journals in selected Web of Science categories[1]. Additionally, we select document type article, and the time span of the study covers the years between 1970 and 2016. The first stage of our topic search options results in 335 articles. The sample reduces to 97 articles in the second stage. We eliminate 16 irrelevant articles through carefully reading the abstracts. The final sample size is composed of 81 articles with 3865 papers in the references. During the data cleaning procedure, we eliminate 114 duplicate reports in the reference list (for example, mistyped author names) and we transform the data into a standard format. Looking at the distribution of the main articles in the collaboration-social innovation field, except for some years such as 2003, 2009, 2013, we observe an increasing trend in the number of articles during the period investigated. As shown by Figure 1, the first papers in this field first appeared in 2003.

Table 1 demonstrates the occurrence and co-occurrence values of the author keywords in publications. As expected, the phrase social innovation appears most frequently in the list and collaboration comes next. Further, as observed in the examples of innovation, sustainability, and urban generation, variation in the keywords on the list attracts attention.

As a generic field of study, social innovation has been studied in journals mainly related to business and society, organization management, and business ethics. Among these, *Business & Society,* with an extensive outlet within a range of societal impacts and intersections with business, receives the highest citation record to a single article (see Table 2). The journal started publication activities in 1960. Since then, mainly society-oriented business articles have been published in the journal. In recent years, corporate social responsibility is the focus[3]. Another field journal, *Group & Organization Management,* has the second highest number of citations to a single article. Other journals belonging to other scientific fields such as sociology

Figure 1. The distribution of web of science publications on social innovation and collaboration (2003-2016)

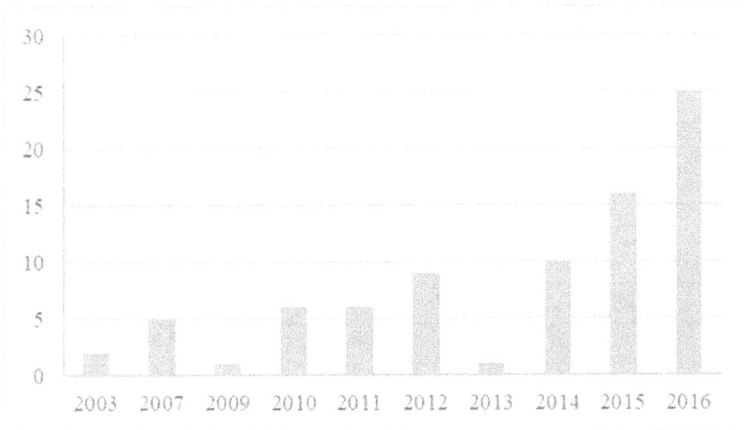

or urban studies also attract attention. The School of Urban studies is represented by two well-known journals: *European Urban and Regional Studies* and *Urban Affairs Review*.

Methodology

There are three commonly used citation analysis approaches, namely direct citation, bibliographic coupling, and co-citation analysis (Boyack and Klavans, 2010). The first one demonstrates the direct link between two papers, while bibliographic coupling and co-citation go further. Having at least one common reference is an indication of a link between two articles for bibliographic coupling (Kessler, 1963). Co-citation occurs between two articles when both are cited by another article (Crane 1972; Small, 1973). Each technique could be used depending on the objective of the study conducted. Bibliographic coupling is an appropriate method when the objective is to determine recent trends in the field. Co-citation, on the other hand, demonstrates the theoretical foundations of the field (Kovacs et al, 2015; Loi et al, 2016; van der Have and Rubalcaba, 2016). The more often two articles are cited together, the more strengthened the link becomes between these articles as an indication of belonging to the same scientific discipline.

Co-Citation Analysis

Social innovation emerged as an interdisciplinary field of research that attracts the attention of many scholars from various fields. However, there are only a few

Table 1. Co-occurrence of the author keywords[2]

Label	Occurrences	Co-occurrences
social innovation	36	22
Collaboration	9	6
co-creation	4	3
İnnovation	4	3
social entrepreneurship	4	3
social networks	4	3
Innovations	4	3
corporate social responsibility	3	3
Learning	3	2
actor network theory	2	2
collaborative consumption	2	2
cross-sector partnerships	2	2
grassroots innovation	2	2
grassroots innovations	2	1
human services	2	2
open innovation	2	1
public-private partnerships	2	2
regional development	2	1
renewable energy	2	0
Resilience	2	2
social innovation	2	0
socio-technical systems	2	0
Sustainability	2	2
Transformation	2	2
urban regeneration	2	0
Vulnerability	2	2

studies dealing with the knowledge base of this area. Cluster analysis of the links between the main articles fromVan der Have and Rubalcaba (2016) reveal four academic communities: community psychology, creativity research, social and societal challenges, and local developments. These clusters show the main themes observed in the area of social innovation. In this study, we focus on papers studying collaborations in social innovation research. In order to reveal the foundations of this field, we conduct co-citation analysis (see Figures 2 and 3) of which nodes

Table 2. Citation statistics of sources[45]

Label	Documents	Citations
Business & Society	1	7
Group & Organization Management	1	3
Journal of Business Ethics	4	3
Academy of Management Perspectives	1	2
Human Service Organizations Management Leadership & Governance	3	2
Journal of Business Research	4	2
Sociologia Ruralis	2	2
Energy Policy	1	1
European Urban and Regional Studies	1	1
Global Environmental Change-Human and Policy Dimensions	1	1
Health Promotion and Chronic Disease Prevention in Canada-Research	1	1
Journal of Publlic Policy & Marketing	1	1
Journal of Social Policy	1	1
Organization Science	1	1
Paths of Convergence for Agriculture, Health, and Wealth	1	1
Urban Affairs Review	1	1

represent papers in the reference list of the main articles. It refers to the frequency with which two documents are cited together.

We use a technique called Vos algorithm to visualize the similarities and determine clusters in the co-citation network that are anchored in minimizing the weighted sum of squared distances between all pairs of items (van Eck et al, 2010). Vos places items in space with multidimensional scaling, so that the distance between any two items reflects the degree of similarity or association between these items as precisely as possible. The similarities are calculated by the association strength shown in the following equation where i and j represent two papers in our case.

$$as_{ij} = \frac{c_{ij}}{c_i c_j} \tag{1}$$

The association strength (as_{ij}) is thus calculated by the ratio of the i and j papers to the co-occurrence numbers (c_{ij}) of the observed co-occurrence numbers ($c_i c_j$). As emphasized by van Eck et al. (2010), both scaling and VOS algorithm give similar results, but scaling has shortcomings, such as locating the most important items in the center and less important items in the periphery. However, Vosviewer is not influenced by these weaknesses.

RESULTS

We download 97 cited references as full-text and extract title, keywords, and abstracts, carefully read all the papers in each cluster, and eliminate 16 papers that do not fall into our area of interest. We only include references that have been cited a minimum of two times (Kovacs et al., 2015). We determine 6 clusters based on cited references we derive from 81 papers. The maps of the clusters are shown in Figures 2 and 3. Each cluster is represented by a different color. For each cluster, papers having high values of the link, total link strength, and more than 3 citations are used in order to determine the theme of each cluster.

Theme 1: Cross Sector Partnerships in Social Innovation

Figures 2 and 3 demonstrate the clusters assigning different colors. Cluster 1 is shown in red and includes the most influential papers in terms of their citations, links, and total link strengths in the network. It is composed of 24 articles and is named *Cross-sector partnerships in the context of social innovation*. In this cluster, there are two methodological studies, Yin (2003) and Glaser and Strauss (1967), that are based on case study analysis and grounded theory approaches. These two methodologies are heavily used in social innovation literature due to the fact that revealing the dynamics of collaborations for social innovation activities requires detailed analysis based on in-depth interviews or other methodologies of case analysis. Through close reading of the remaining papers in this cluster, we realize that this field targets the essential factors of cross-sector partnerships (Le Ber and Branzei, 2010; Austin et al, 2006) and formulation of effective corporate social and environmental responsibility policies (Seitanidi and Crane, 2009; Berger et al., 2004; Rondinelli and London, 2003; King, 2007) with the help of collaborations.

The term social innovation is defined as finding a new solution to a social problem that is more efficient, effective, and sustainable than the existing one (Le Ber and Branzei, 2010). In order to achieve this goal, partners from different sectors come together, which is known as cross-sector partnerships. Elsewhere, Berger et al. (2004) argue that cross-sector partnerships are the engine of collective action. Accordingly, firms recently became aware of the fact that their conventional vision giving emphasis to profits is no longer valid, so they started to position themselves as good corporate citizens. This recent movement is called social alliances, which can be important value generators for companies, non-profits, and society at large.

Contrary to strategic alliances, which necessitate resource sharing among similar business parties, social alliances, oriented towards non-economic objectives, are composed of at least one non-profit partner. Another attempt to define essential factors for cross-sector collaborations, Bryson et al (2006) treat inter-sectoral collaborations, referring to partnerships including government, business world, and non-profit organizations, as a structure that enables each partner to respond to complex social problems. They mainly focus on the factors that may have an impact on collaboration formation, constraints, process, and governance of all actors being involved in the collaboration. One of the crucial factors is the presence of a turbulent environment that forces cross-sector collaboration. In other words, collaboration stems from a necessity, due to a single party's failure at solving the problem.

This cluster also has papers on collaborating for the achievement of corporate social responsibility (CSR) activities. Seitanidi and Crane (2009) emphasize the importance of CSR processes emerging through non-profit organizations and business called *social partnerships*, 'social problem-solving mechanisms among organizations' (Waddock, 1989). The definition refers to the coordination of organizational resources to solve societal problems, which in turn provides mutual benefits both for private sector partners and for society. Austin (2006) treats social partnerships as a formal agreement and adopts an approach that combines social and business entrepreneurship using an analytical model, aiming to explore the similarities and differences between these two types of entrepreneurship and to develop a more systematic approach to social innovation, both for policymakers and for researchers. In a similar vein, Rondinelli and London (2003) focus on how

Figure 2. Co-citation network of papers in collaborations in social innovation (1970-2016)

Figure 3. Co-citation network of papers in collaborations in social innovation (1970-2016) using density visualization

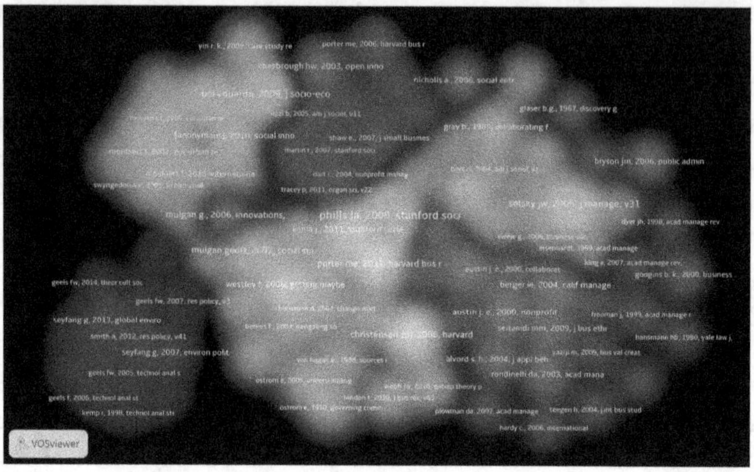

companies and nonprofit organizations effectively share knowledge and expertise on environmental issues through strategic partnerships. They focus on the strategic criteria that help executives evaluate the feasibility of cross-sector collaborations. Alliances are the preferred type of collaboration for firms due to the fact that internal development of skills required for environmentally friendly solutions is too costly. For managers, a desired cross-sector collaboration plays a dual role in the company's internal operations, exploiting internal skills on the one hand, and gaining competence in new skills on the other. Focusing on collaboration for environmental protection, Rondinelli and London (2003) observe three types of collaboration between companies and non-profit organizations. These are *arm's length, interactive collaboration,* and *intensive alliances.* The first one encourages workers to participate in non-profit organization activities. Interactive collaborations include initiating a project such as a certificate programme in collaboration with non-profit organizations. Both parties play an active role in managing the project. The last one, intensive alliances, differs from the other two collaboration types with respect to its character. Thus, companies may need to set up formal alliances with non-profit organizations depending on the extent of the problem, or non-profit organizations may need the help of companies to prevent a societal problem; for instance, environmental pollution before it generates dramatic outputs for society.

For all types of partnerships mentioned above, the collaboration process does not work smoothly from the very beginning to the end. Collaboration between private and non-profit sectors may fail due to differences in their motivations. While business partners must focus on maximizing profits, social benefit is the focal point of their non-profit counterparts (Le Ber and Branzei, 2010). Additionally, as observed by Berger et al (2004), the collaboration may fail due to various reasons, such as talking in different languages, being inexperienced in working with the non-profit sector or a company, mistrust among partners, misallocation of costs and benefits, and problems arising from power and time issues. In addition to these factors, Rondinelli and London (2003) argue that both parties may have concerns such as fear of losing control and reputation that generate undesired results for collaboration. Each partner, therefore, needs to understand the nature of the relationship and to clarify the conditions of achieving mutual benefits before forming a partnership (Googins and Rochin, 2000). Besides, each partner should consider these issues during various stages of the partnership, such as partner selection, risk assessment techniques, partnership design, and institutionalization. How well they deal with these issues determines the degree of success.

Theme 2: Defining Social Innovation

This cluster is shown in green and is composed of 17 studies that focus on *the concept of social innovation and its content.* Pol and Ville (2009) define social innovation and distinguish social innovation from business innovation by focusing on its effects on both quality and quantity of life. The concept of social innovation, unlike business innovation, refers to the generation of new ideas without profit maximization motives. They propose a new term to understand what social innovation is: *a desired social innovation* targeted at improving living conditions both quantitatively and qualitatively. Likewise, Phills et al (2008), in their paper "Rediscovering social innovation," argue that social innovations differ from technological innovations in terms of distribution of benefits throughout society, but describe social innovation by using a similar terminology. Accordingly, social innovation should be novel, as is expected for technological innovations. However, social innovations put society at the center and are concerned with social problems that couldn't be solved with a profit-oriented approach. Similarly, Howaldt and Schwarz (2010) define the concept of social innovation by focusing on the differences between social and technological innovation. Thus, the nature of the outcome is the core difference. For social innovations, the outcome is to meet the social needs that existing commercial and public sectors fail to satisfy. However, they highlight that it is difficult to distinguish social innovation from technological innovation by using this argument because technological innovation may also aim at meeting social needs. Also, there are papers dealing with the definition of the field by applying a similar approach observed in technological innovations. To illustrate, Murray et al (2010) focus on processes of social innovation including stages as prompts, idea generation, prototyping, sustaining, scaling, diffusion, and systemic change, respectively. In the first stage, initial conditions, which could be either a catastrophic event or an inspiration leading to changes in future stages, are established. Following the idea generation stage, the outcome (e.g., a novel project idea that targets improving the health conditions of refugees) is tested during the prototyping stage. The remaining phases are implemented respectively. Brown and Wyatt (2010), using a similar approach, apply the three-stage model in conducting design-led social innovation, including inspiration, ideation, and implementation.

Mulgan et al (2007) deal with the definition of innovation, activities, actors, and stages of social innovation and build up a framework called 'connected difference theory'. Accordingly, social innovation activities could only be achieved under certain conditions that trigger communication among different actors. Moulaert and Nussbaumer (2005), in attempting to define the field of social innovation, discuss the main limitations of the territorial innovation model and its failure to meet regional needs, and propose an alternative model called 'community development based on social innovation'. Moulaert et al (2007), in a similar vein, examine the institutional change patterns in the context of urban regimes and find that institutional change is path dependent. From a social innovation point of view, in the generation of alternative actions, local communities work more effectively than their counterparts at the national or global level due to two main reasons. Firstly, local communities are more knowledgeable about the needs of the local society so they can govern this process more efficiently. Secondly, at the local level, values (identity and vision) are regenerated through social interaction. Swyngedouw (2005), introducing a new perspective in social innovation, examines the relationship between government and society in the context of governance. Beyond the traditional roles of government and citizens, known as governing body and citizenship, new rules of the game depend on designing institutional arrangements that give a greater role to business stakeholders and civil society in policymaking. Governance mechanism in his paper is defined as socially innovative institutional arrangements organized as a network of relationships including government and civil society organizations.

Theme 3: Defining Social Entrepreneurship

This cluster is shown in light blue on the map and includes 14 papers. It is labelled *the definition of social entrepreneurship*. The most influential papers in this cluster are Mair and Marti (2006), Mulgan (2006), Porter and Kramer (2006), and Shaw and Carter (2007). The scope of social entrepreneurship is discussed by Mair and Marti (2006) with special emphasis on its potential to create social value. Shaw and Carter (2007) discuss the similarities and differences between social entrepreneurs and business entrepreneurs. They both carry similar characteristics with respect to individual features such as leadership, charisma, and inspiring others; but social entrepreneurs behave differently from their counterparts in the business world in that they focus more on ethical values. Additionally, their mission is different. For instance, while profit maximization is the primary objective for business entrepreneurs, meeting social objectives is the primary motive for social entrepreneurs. Another difference between social and business entrepreneurs is related to innovativeness. While it is possible to become an entrepreneur without any innovation effort, for social entrepreneurs, innovation is essential. In this cluster, we also observe that Mulgan (2006)[6], placing emphasis on the *processes of social innovation*, argues that primary social innovation activities are performed by the great efforts of social entrepreneurs. Porter and Kramer (2006), unlike these two articles, deal with social value creation ideas from the corporate social responsibility perspective and focus on the successful results of activities driven by society's demands as observed in the Toyota case. The company's early response to customer complaints about car emissions led to the production of the hybrid-engine. In a similar vein, Chesbrough (2003) discusses closed and open innovation in the case of Xerox PARC. While the vertical integration model based on closed innovation works for a large, established company, PARC necessitated a new business model that required collaborating with other parties, which is called open innovation.

Among these papers, we also mention Granovetter (1973), namely *Strength of Weak Ties*, one of the papers cited as a methodological paper in this cluster. He argues that weak ties developed through the relationships among members of different groups led to the emergence of new and different ideas, which in turn generates greater value for all members. Social entrepreneurs play a critical role in establishing relationships among different groups such as business, non-profits, and society.

Theme 4: Transition Studies

There are 15 papers in this cluster shown in blue, which belong to *transition studies* emphasizing the important role of niche activities in satisfying the needs of society. Seyfang and Smith (2007) use the term 'grassroots innovation' to describe networks of activists and organizations serving this purpose. They describe the role of innovation and community action in sustainable development and explore the link between technological innovation, which is based on a market-oriented approach placing emphasis on profits for rental seekers and community action. Technological innovation is classified as product, process, marketing, and service innovations, the supply side of which is composed of firms and the demand side of consumers. The main purpose is to sustain market-based exchanges, without considering societal needs and ideologies. Community action, on the other hand, reflects the other side of the coin, putting society at the center. *Grassroots* efforts provide an alternative way of encouraging innovative behavior in areas where societal needs are unmet. For instance, Smith and Raven (2012) highlight the three important areas of shielding, nurturing, and empowerment that need further consideration because they lack a systematic approach to mobilize resources in favor of niche improvements targeting these areas. They also deal with the main constraints of a linear approach to mobilize resources and propose a dynamic approach attempting to reveal complex relations among variation, selection environments, and existing technological regimes[7].

Considering existing applications in niche development fields, the learning process is not well developed and there is no clear link between innovative niche activities and the existing regime (Seyfang and Longhurst, 2013). In a similar vein, Geels and Schot (2007) discuss transition pathways as transformation, reconfiguration, technological substitution, de-alignment, and re-alignment. They introduce a multilevel perspective claiming that transition paths are not deterministic and relations among pathways are not automatically established, but rather follow their own paths due to their dynamic nature.

Theme 5: Innovation Studies

This cluster is shown in yellow and consists of 15 papers on *innovation studies.* The main papers with respect to their links, total links, and citations, are Christensen et al (2006), Kania and Kramer (2011), Porter and Kramer (2011), and Westley et al (2006). Christensen et al (2006) use the term catalytic innovation, which is introduced as a new way of meeting needs of underrepresented societal groups. Organizations, in general, tend to have control over their existing capabilities and resources and maintain the status quo. However, there are some other ways of providing services to a large population as observed in education, health, and other related sectors. To illustrate, community colleges are educational programmes providing lower cost options to four-year degree programmes. Moreover, there are some alternative credit-lending institutions willing to provide credit to people who do not have the assets required to launch a new business. All these products and services are included in the definition of catalytic innovation. In a similar vein, Porter and Kramer (2011) discuss capitalism and its value creation rationale with emphasis on its constraints. Accordingly, value creation activities are narrowly defined, focusing on short-term financial performance that does not provide sustainable gains when societal needs are not met.

The current situation of value creation activities needs further consideration to fulfil societal needs and cope with challenges. It is different from social responsibility or philanthropy, but rather a new way of achieving economic success by putting externalities at the center. There are various issues that the firm needs to consider throughout its production processes, such as environmental impact, energy use, water use, employee health, worker safety, employee skills, supplier access, and viability.

Further, Kania and Kramer (2011) focus on the issue of creating collective impact in collaboration with different actors to meet societal needs. The success factors of the system are based on sharing a common agenda, using the same measurement system, following common standards to collect data at the community level, conducting complementary activities, continuous communication among partners, and presence of supporting organizations.

Theme 6: Formations of Inter-Organizational Collaborations

This cluster is shown in light blue and consists of 13 papers. The main papers having the highest citations, links, and total link strength in this cluster include Austin and Seitanidi (2012a), Gray (1989), Austin and Seitanidi (2012b), Ring and van de Ven (1994), Selsky and Parker (2005), and Waddock and Post (1991). These papers focus on different formations of inter-organizational relations. Selsky and Parker (2005) and Austin and Seitanidi (2012a) discuss different types of inter-organizational relations such as strategic alliances, partnerships, joint ventures, franchises, research consortia, and various forms of network organizations with special emphasis on the development process of collaboration. They analyze this process through dynamic relations among negotiations, commitments, assessments, and executions. All the trust and risk issues are discussed in the negotiation process and the output of this stage affects commitments to future plans and the role of interaction at the execution stage. All these processes are assessed in terms of efficiency and equity conditions.

Selsky and Parker (2005) discuss project-based cross-sector partnerships applying two platforms, such as resource dependency and social issues. As for the first one, organizations need to collaborate with partners having complementary resources and skills. In a similar vein, Austin and Seitanidi (2012a;b) use a collaborative value creation framework and evaluate the success of value creation activities at different levels including type of activity, actors, and time period. These papers use the same framework as observed in collaboration on technological innovation activities for choosing the right partners. This process follows steps including the definition of the (social) problem, identification of required resources to achieve a common goal, and determination of partners' contributions to the final outcome. Additionally, identification of partners' motivations and their relations in the past helps actors build trust through collaboration. Likewise, Waddock and Post (1991) touch upon a new form of collaboration called social partnership as a symbol of collective action, and introduce a sound typology to reveal its dynamic nature and uniqueness. Unlike previous typologies, they apply an integrated framework including mechanisms of collaboration and exchange through which social partnerships reduce the costs of the transaction and strategic dependencies.

Burt (2004) and Granovetter (1985) are two methodological papers cited in this cluster. Burt (2004), with his well-known paper, "Structural Holes and Good Ideas," mentions the institution of a brokerage connecting different actors, and brokers' strategic roles in the creation of social capital. Elsewhere, Granovetter (1985) discusses the problem of embeddedness in relation to economic action and social structure, and argues that economic behavior is embedded in social relations. In this cluster, papers on different forms of inter-organizational collaborations heavily use the assumptions of social network theory with special emphasis on structural holes as connectors and behavioral factors embedded in the society.

CONCLUSION AND DISCUSSION

In this study, six types of clusters are identified based on co-citation analysis. The main themes for each cluster are *cross-sector partnerships, innovation studies, the definition of social innovation, transition studies, social entrepreneurship,* and *types of inter-organizational relations,* respectively. According to the results, these six clusters are connected to each other except for the *transition studies* cluster including papers on grassroots, which is located differently on the map. In other words, articles studying social innovation in the context of grassroots are less connected to articles of other clusters. As far as the remaining clusters are concerned, the first cluster, red, is connected to the light blue cluster. Selsky and Parker (2005) on project-based partnerships as a connector is linked to almost all papers in the first cluster. In other words, these two clusters are intertwined.

The reason why these two clusters are so close is that both clusters focus on collaboration types. While cross-sector partnerships are the focus for the first one (Austin and Seitanidi, 2012a), Selsky and Parker (2005) deal with project-based partnerships. Further, Cluster 2 on the definition of social innovation and Cluster 3 on social entrepreneurship are also connected to each other. The closeness of these clusters stems from similarities in topics. This result is in line with the findings of Tracey and Stott (2017), which use the term social entrepreneurship to explain social innovation typology. Accordingly, the term social entrepreneur refers to a person whose main goal is to resolve social problems regardless of their identity. In other words, regardless of whether they are profit seekers or actors in the non-profit community, social entrepreneurs are included in social innovation typology.

The green cluster including studies on the definition of social innovation and the blue cluster on innovation studies are closely linked to each other. As mentioned by the two papers by Christensen et al (2006) and Porter and Kramer (2011), the outdated definition of innovation with a special emphasis on short-term value creation takes on a new meaning, which touches upon the environmental effects of production, energy and water use, and health conditions of workers.

The main implication of this study is that literature on collaboration in social innovation is not fragmented, contrary to the studies treating it that way (Cajaiba-Santana, 2014; Dawson and Daniel, 2010; Pol and Ville, 2009). Based on the results of this study, it is rather integrated. However, the literature on transition studies has the potential to become a distinctive research area. As for the remaining clusters, themes of social innovation, social entrepreneurship, cross-sector partnerships, and inter-organizational relations are intertwined (e.g., key papers playing brokerage roles are Phills et al (2008) in the green cluster; Mulgan (2006) in the purple cluster; Selsky (2005) in the light blue cluster; and Austin (2006) in the red cluster).

The second implication is that there has not been a systematic framework that enables researchers to measure social innovation. Further, existing literature highlights that social innovations differ from technological innovations in terms of their motives. Accordingly, social innovations are carried out to satisfy societal needs, while the other is profit-oriented (Howaldt, 2010; Le Ber and Branzei, 2010, Phills et al., 2008, Pol and Ville, 2009). However, these studies remain insufficient to reveal other critical differences between social innovation and technological innovations. In these articles, a similar framework including resource complementarity and organizational fit as observed in technological innovation is applied to explain the dynamics and nature of social innovation activities. The measurement problem in the context social innovation studies, therefore, could be eliminated by developing comprehensive methods including both quantitative and qualitative techniques. Although studies in social innovation predominantly apply case study analysis, network approach, and grounded theory, collecting survey data at the local, regional, and national level will enable social innovation researchers to conduct quantitative analysis and reveal cross-country variations. As the number of studies in collaborations in social innovation increases, changes in research streams could be observed throughout the years.

Thirdly, the definition of the field is still in progress. Although one cluster is assigned to the definition of social innovation, we observe that there are papers dealing with the definition of social innovation (e.g., Mulgan, 2006; Mulgan et al, 2007) in other clusters. Further, social entrepreneurship and social innovation are twin fields because almost all clusters are connected to each other except for one.

Through examining collaborations in the field of social innovation, this study provides a comprehensive perspective using co-citation analysis. However, this study has some limitations that require further consideration in future studies. The number of articles dealing with collaborations in social innovation is limited due to the newness of the field. Thus, as the number of papers increases over time, collaborations in social innovation will elaborate in a broader sense.

REFERENCES

Austin, J., Stevenson, H., & Wei-Skillern, J. (2006). Social and commercial entrepreneurship: Same, different, or both? *Entrepreneurship Theory and Practice*, *30*(1), 1–22. doi:10.1111/j.1540-6520.2006.00107.x

Austin, J. E., & Seitanidi, M. M. (2012a). Collaborative value creation: A review of partnering between nonprofits and businesses: Part I. Value creation spectrum and collaboration stages. *Nonprofit and Voluntary Sector Quarterly*, *41*(5), 726–758. doi:10.1177/0899764012450777

Austin, J. E., & Seitanidi, M. M. (2012b). Collaborative value creation: A review of partnering between nonprofits and businesses. Part 2: Partnership processes and outcomes. *Nonprofit and Voluntary Sector Quarterly*, *41*(6), 929–968. doi:10.1177/0899764012454685

Berger, I. E., Cunningham, P. H., & Drumwright, M. E. (2004). Social alliances: Company/nonprofit collaboration. *California Management Review*, *47*(1), 58–90. doi:10.2307/41166287

Boyack, K. W., & Klavans, R. (2010). Co-citation analysis, bibliographic coupling, and direct citation: Which citation approach represents the research front most accurately? *Journal of the Association for Information Science and Technology*, *61*(12), 2389–2404.

Brown, T., & Wyatt, J. (2010). Design thinking for social innovation. *Development Outreach*, *12*(1), 29–43. doi:10.1596/1020-797X_12_1_29

Bryson, J. M., Crosby, B. C., & Stone, M. M. (2006). The design and implementation of Cross-Sector collaborations: Propositions from the literature. *Public Administration Review*, *66*(s1), 44–55. doi:10.1111/j.1540-6210.2006.00665.x

Burt, R. S. (2004). Structural holes and good ideas. *American Journal of Sociology*, *110*(2), 349–399. doi:10.1086/421787

Cajaiba-Santana, G. (2014). Social innovation: Moving the field forward. A conceptual framework. *Technological Forecasting and Social Change*, *82*, 42–51. doi:10.1016/j.techfore.2013.05.008

Christensen, C. M., Marx, M., & Stevenson, H. H. (2006). The tools of cooperation and change. *Harvard Business Review*, *84*(10), 72–80. PMID:17040041

Crane, D. (1972). *Invisible Colleges: Diffusion of Knowledge in Scientific Communities*. Chicago, IL: University of Chicago Press.

Dawson, P., & Daniel, L. (2010). Understanding social innovation: A provisional framework. *International Journal of Technology Management*, *51*(1), 9–21. doi:10.1504/IJTM.2010.033125

Geels, F. W., & Schot, J. (2007). Typology of sociotechnical transition pathways. *Research Policy*, *36*(3), 399–417. doi:10.1016/j.respol.2007.01.003

Glaser, B., & Strauss, A. (1967). *The discovery of grounded theory.* London: Weidenfield & Nicolson.

Googins, B. K., & Rochlin, S. A. (2000). Creating the partnership society: Understanding the rhetoric and reality of cross-sectoral partnerships. *Business and Society Review*, *105*(1), 127–144. doi:10.1111/0045-3609.00068

Granovetter, M. (1985). Economic action and social structure: The problem of embeddedness. *American Journal of Sociology*, *91*(3), 481–510. doi:10.1086/228311

Granovetter, M. S. (1973). The Strength of Weak Ties. *American Journal of Sociology*, *78*(6), 1360–1380. doi:10.1086/225469

Gray, B. (1989). *Collaborating: Finding common ground for multiparty problems.* Academic Press.

Hargreaves, T., Hielscher, S., Seyfang, G., & Smith, A. (2013). Grassroots innovations in community energy: The role of intermediaries in niche development. *Global Environmental Change*, *23*(5), 868–880. doi:10.1016/j.gloenvcha.2013.02.008

Howaldt, J., & Schwarz, M. (2010). *Social Innovation: Concepts, research fields, and international trends.* Sozialforschungsstelle Dortmund.

Kania, J., & Kramer, M. (2011). *Collective impact.* Academic Press.

Kessler, M. M. (1963). Bibliographic coupling between scientific papers. *Journal of the Association for Information Science and Technology*, *14*(1), 10–25.

King, A. (2007). Cooperation between corporations and environmental groups: A transaction cost perspective. *Academy of Management Review*, *32*(3), 889–900. doi:10.5465/amr.2007.25275680

Kovacs, A., Van Looy, B., & Cassiman, B. (2015). Exploring the scope of open innovation: A bibliometric review of a decade of research. *Scientometrics*, *104*(3), 951–983. doi:10.100711192-015-1628-0

Le Ber, M. J., & Branzei, O. (2010). (Re) forming strategic cross-sector partnerships relational processes of social innovation. *Business & Society*, *49*(1), 140–172. doi:10.1177/0007650309345457

Loi, M., Castriotta, M., & Di Guardo, M. C. (2016). The theoretical foundations of entrepreneurship education: How co-citations are shaping the field. *International Small Business Journal*, *34*(7), 948–971. doi:10.1177/0266242615602322

Moulaert, F., Martinelli, F., González, S., & Swyngedouw, E. (2007). *Introduction: social innovation and governance in European cities: urban development between path dependency and radical innovation*. Academic Press.

Moulaert, F., & Nussbaumer, J. (2005). The social region: Beyond the territorial dynamics of the learning economy. *European Urban and Regional Studies*, *12*(1), 45–64. doi:10.1177/0969776405048500

Mulgan, G. (2006). The process of social innovation. *Innovations: Technology, Governance, Globalization*, *1*(2), 145–162. doi:10.1162/itgg.2006.1.2.145

Mulgan, G., Tucker, S., Ali, R., & Sanders, B. (2007). *Social innovation: what it is, why it matters and how it can be accelerated*. Academic Press.

Murray, R., Caulier-Grice, J., & Mulgan, G. (2010). The open book of social innovation. London: National Endowment for Science, Technology and the Art.

Phills, J. A., Deiglmeier, K., & Miller, D. T. (2008). Rediscovering social innovation. *Stanford Social Innovation Review*, *6*(4), 34–43.

Pol, E., & Ville, S. (2009). Social innovation: Buzz word or enduring term? *Journal of Socio-Economics*, *38*(6), 878–885. doi:10.1016/j.socec.2009.02.011

Porter, M. E., & Kramer, M. R. (2006). Strategy and society: The link between corporate social responsibility and competitive advantage. *Harvard Business Review*, *84*(12), 78–92. PMID:17183795

Porter, M. E., & Kramer, M. R. (2011). The Big Idea: Creating Shared Value. How to reinvent capitalism—and unleash a wave of innovation and growth. *Harvard Business Review*, *89*(1-2).

Ring, P. S., & Van de Ven, A. H. (1994). Developmental processes of cooperative inter-organizational relationships. *Academy of Management Review*, *19*(1), 90–118. doi:10.5465/amr.1994.9410122009

Rondinelli, D. A., & London, T. (2003). How corporations and environmental groups cooperate: Assessing cross-sector alliances and collaborations. *The Academy of Management Executive*, *17*(1), 61–76. doi:10.5465/ame.2003.9474812

Seitanidi, M. M., & Crane, A. (2009). Implementing CSR through partnerships: Understanding the selection, design, and institutionalization of nonprofit-business partnerships. *Journal of Business Ethics*, *85*(2), 413–429. doi:10.100710551-008-9743-y

Selsky, J. W., & Parker, B. (2005). Cross-sector partnerships to address social issues: Challenges to theory and practice. *Journal of Management*, *31*(6), 849–873. doi:10.1177/0149206305279601

Seyfang, G., & Longhurst, N. (2013). Desperately seeking niches: Grassroots innovations and niche development in the community currency field. *Global Environmental Change*, *23*(5), 881–891. doi:10.1016/j.gloenvcha.2013.02.007

Seyfang, G., & Smith, A. (2007). Grassroots innovations for sustainable development: Towards a new research and policy agenda. *Environmental Politics*, *16*(4), 584–603. doi:10.1080/09644010701419121

Shaw, E., & Carter, S. (2007). Social entrepreneurship: Theoretical antecedents and empirical analysis of entrepreneurial processes and outcomes. *Journal of Small Business and Enterprise Development*, *14*(3), 418–434. doi:10.1108/14626000710773529

Small, H. (1973). Co-citation in the scientific literature: A new measure of the relationship between two documents. *Journal of the Association for Information Science and Technology*, *24*(4), 265–269.

Smith, A., & Raven, R. (2012). What is protective space? Reconsidering niches in transitions to sustainability. *Research Policy*, *41*(6), 1025–1036. doi:10.1016/j.respol.2011.12.012

Swyngedouw, E. (2005). Governance innovation and the citizen: The Janus face of governance-beyond-the-state. *Urban Studies (Edinburgh, Scotland)*, *42*(11), 1991–2006. doi:10.1080/00420980500279869

Tracey, P., & Stott, N. (2017). Social innovation: A window on alternative ways of organizing and innovating. *Innovation*, *19*(1), 51–60. doi:10.1080/14479338.2016.1268924

van der Have, R. P., & Rubalcaba, L. (2016). Social innovation research: An emerging area of innovation studies? *Research Policy*, *45*(9), 1923–1935. doi:10.1016/j.respol.2016.06.010

van Eck, N. J., Waltman, L., Dekker, R., & van den Berg, J. (2010). A comparison of two techniques for bibliometric mapping: Multidimensional scaling and VOS. *Journal of the American Society for Information Science and Technology*, *61*(12), 2405–2416. doi:10.1002/asi.21421

Waddock, S. A. (1989). Understanding social partnerships: An evolutionary model of partnership organizations. *Administration & Society*, *21*(1), 78–100. doi:10.1177/009539978902100105

Waddock, S. A., & Post, J. E. (1991). Social entrepreneurs and catalytic change. *Public Administration Review*, *51*(5), 393–401. doi:10.2307/976408

Westley, F., Zimmerman, B., & Patton, M. Q. (2006). *Getting to maybe: How the world has changed*. Toronto, Canada: Random House Canada.

Yin, R. K. (2003). *Applications of case study research*. Applied social research methods series Thousand Oaks, CA: Sage Publications.

Yu, K. H. (2013). Institutionalization in the context of institutional pluralism: Politics as a generative process. *Organization Studies*, *34*(1), 105–131.

ENDNOTES

1 Management or psychology applied or business or multidisciplinary sciences or education special or social sciences interdisciplinary or economics or industrial relations labor or environmental studies or history of social sciences or planning development or health care sciences services or education educational research or behavioral sciences or architecture or sociology or hospitality leisure sport tourism or anthropology or health policy services or social work or energy fuels or nutrition dietetics or urban studies or psychology multidisciplinary or medical ethics or public administration or area studies or geography or social sciences biomedical or environmental sciences or literary reviews or social issues or linguistics or public environmental occupational health or international relations or law or education scientific disciplines or art or communication or family studies or geriatrics gerontology or agricultural economics policy or forestry or green sustainable science technology or women s studies or folklore or business finance or water resources or film radio television or political science or transportation or ethnic studies or psychology social or psychology educational or philosophy or humanities multidisciplinary or history philosophy of science or archaeology or gerontology or history or agriculture multidisciplinary or ethics or genetics heredity or ecology.

2 Occurrences show the number of occurrences of the keyword. Co-occurrences show the occurrence of the keyword with other keywords.

3 In 2017, this journal published four special issues on corporate social responsibility. The theme of the first issue is 'SMEs and CSR in Developing Countries'. A related concept, Corporate Sustainability, is the theme of the second special issue with a specific emphasis on new perspectives, research methods, and theories in CSR. Another issue is composed of papers dealing with CSR applications in China. Additionally, *Business & Society* included measurement of CSR applications as a themed issue in July 2017.

4 Documents represent the number of papers published in the journal on the list. Citations show the number of citations documents received.

5 Sources in table belong to main papers that we obtained based on our WoS search.

6 As seen in Figures 2 and 3, Mulgan (2006) is connected to Mulgan et al (2007). Both papers. in an attempt to build up a framework for social innovation, focus on the definition of social innovation and the role of connectors on successful social innovation activities.

7 Hargreaves et al (2013), similarly, find that existing niche-building theories do not explain much about the diversity in societal systems and do not meet the needs of society due to its dynamic structure. With specific emphasis on sociotechnical niches in protected spaces, they concluded that most niche-related activities are implemented at the national level and in the diffusion of new projects in niche areas; national networks play a crucial role.

Chapter 9
Building Theories-in-Practice on Social Innovation in Disability Nonprofit Organizations

Rachel Taylor
Federation University, Australia

Nuttaneeya (Ann) Torugsa
Mahidol University, Thailand & University of Tasmania, Australia

ABSTRACT

This chapter discusses the key theoretical and empirical steps undertaken throughout the authors' previous-but-related mixed methods studies on social innovation in nonprofit organizations (NPOs) in the Australian disability sector with the aim of using the key findings of these studies to develop 'theories-in-practice' in disability NPOs. In this chapter, the authors summarize the associated theory-building processes deployed to explain how disability NPOs develop and implement social innovations and the societal 'system-level' impacts of such innovations. These theory-building processes involve two broad phases, and the culmination of these phases (grounded in the abductive logics of inquiry, complexity theorizing, and set-theoretic methods) leads to the development of several 'theories-to-practice' that not only convey the interactivity of contextual causal mechanisms leading to social innovation by NPOs, but also outline change-oriented solutions for managers who are working to address complex social challenges.

DOI: 10.4018/978-1-7998-1108-4.ch009

RESEARCH BACKGROUND AND DESIGN

The authors' research studies (i.e., Taylor, Torugsa, & Arundel, 2018a,b; Taylor et al., 2019a,b) explore the emerging phenomenon of social innovation within the context of nonprofit organizations (NPOs), with the focus on how NPOs develop and implement social innovations. By developing a systemic understanding of such a perplexing phenomenon through a rigorous-and-innovative research design, the current research chapter revitalizes theories to better capture the contingent and asymmetrical trajectories of NPO-based social innovation.

Social innovation – which has been recognized as a practice-led domain – can be broadly defined as a service and/or process that contributes to generating a novel solution to a social problem that is more effective, efficient, sustainable or just than existing solutions, and for which the value created primarily benefits society as a whole (Phills, Deiglmeier, & Miller, 2008). Despite gaining immense popularity and being viewed as a potential mechanism for addressing complex or wicked problems, academic research on social innovation has failed to establish a coherent conceptual foothold or strong empirical grounding (Ayob, Teasdale, & Fagan, 2016; Phillips, Lee, Ghobadian, O'Regan, & James, 2014). This is evidenced by a dearth of systematic analysis of the organizational capabilities, outcomes and theories associated with this phenomenon (Howaldt & Schwarz, 2017). Within the nonprofit sector, social innovation is considered a nascent area of research that has yet to mark out its own theoretical territory (Taylor et al., 2018a).

To overcome the hurdles presented by this pre-theoretical domain, the authors undertake a context-driven inquiry that explores interlinked practice-theory questions, and employs a multi-staged research design to ensure that experiential knowledge and scholarly understandings are advanced hand-in-hand. The distinctive features of this research design include: (1) employing 'abduction' (Peirce, 1955) as the primary method of inquiry for discovering new knowledge of social innovation within organizational contexts; (2) complexity theorizing (Byrne, 2005; Taylor et al., 2018b) for interpreting the complex, multiple realities of social innovation; and (3) using set-theoretic methods grounded in complexity theory to determine social innovation's patterns of emergence at a broader systems (i.e. sectoral) level.

The critical role of context permeates and informs each of these methodological aspects of the authors' research. This is deemed vital to ensure that the research can capture the complexity of social change processes manifesting in social innovation's contextualized expressions (Westley, Antadze, Riddell, Robinson, & Geobey, 2014). Thus, this research uses the Australian disability sector as an illustrative example and embed this context in the research design, where the aim is to explore organizational capabilities for social innovation and the outcomes of social innovation in nonprofit organizations (NPOs) as part of developing a theory of social innovation in NPOs.

The Australian disability sector is specifically chosen as the context for the study due to the recent introduction of the National Disability Insurance Scheme (NDIS). Considered the most significant social reform in Australia for over 40 years (Connellan, 2014; Goggin & Wadiwel, 2014), the arrival of the NDIS is expected to increase the need and potential for socially innovative responses by NPOs in the sector (Green & Mears, 2014).

To explore social innovation in the context of disability NPOs, the authors' research design incorporates two broad phases. Phase 1 embeds abductive logic in methodological framing. This phase results in: (i) the development of the framework of *Nonprofit Social Innovation (NSI)*, i.e. a social innovation enacted specifically in the disability NPO context, and (ii) the generation of the theoretical propositions, i.e. abductively-derived working hypotheses. The theoretical propositions that are put forward at the end of Phase 1 are then subsequently tested in Phase 2. Phase 2 uses a whole-of-sector lens and Qualitative Comparative Analysis (QCA) to test the NSI framework and verify the working hypotheses.

A complete diagrammatic overview of the steps in this multi-staged research project is depicted in Figure 1 below. Sequential practice-theory building processes can be seen to thread through each stage of this research journey.

In the following sections, the discussion of the key research findings is framed in relation to Phases 1 and 2. Through the culmination of these sequential phases of data collection and practice-theory bridging, several theories-in-practice are developed.

PHASE 1: ABDUCTIVE CASE STUDIES

In tackling an ill-defined problem domain, the authors' research is driven by abductive logics and methods of inquiry. The use of abduction as an operation or process is argued by Peirce (1955) to be the only way to bring about new ideas or scientific theories. Abductive reasoning is used as a promising starting point for considering previously unexplained facts and anomalies (through practical and theoretical observations), creatively generating new plausible hypotheses, comparing competing explanations, and ultimately aims at developing better theories (Patokorpi & Ahvenainen, 2009; Svennevig, 2001). In addition to using abduction as a logic of discovery, the authors adopt complexity theorizing (which views the knowledge and practices shaping social innovation as contextually sensitive rather than universal, and occurring through patterns of emergence that are self-organizing, nonlinear, and sensitive to initial conditions) to better understand the complex and contingent reality under study (Byrne, 2005). Importantly, the use of abductive reasoning is put forward as a methodological stepping-stone between the study's complexity theorizing approach and the practice-based world of NPOs, as it is concerned with the

Figure 1. Summary of research steps undertaken in this mixed methods research project

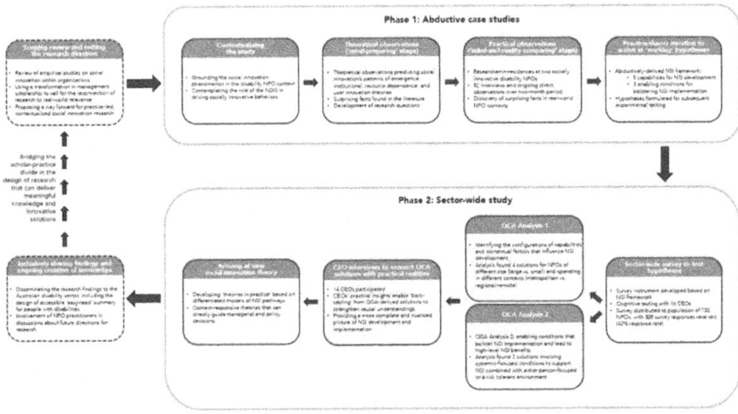

particularities of specific situations and emphasizes the need to make contextualized judgements via a logical process of sense-making.

The specific methodological components of the study's four-step abduction process are detailed in the authors' previous paper (see Taylor et al., 2018a). For the purpose of this chapter aimed at building theories-in-practice, the key steps undertaken are described briefly below, with a focus on the implications of each step as it pertains to the processes of theory development.

Step 1: Grounding the Social Innovation Phenomenon in Context

Due to its highly variable applications across distinct contextual fields, grounding the social innovation phenomenon in a particular context forms the first step in this abductive journey. The initial contextualization stage involves contemplating the interplay of micro/meso/macro elements within the practice domain of interest, i.e. specifically pertaining to NPOs operating within the Australian disability sector. A predominant feature of this context is the newly implemented National Disability Insurance Scheme (NDIS), which is widely regarded as destabilizing for disability NPOs in terms of their immediate and intense focus on organizational adaption and improvement processes (Connellan, 2014; Green & Mears, 2014). Hence, the NDIS policy is predicted in this contextualization phase to be a critical factor influencing NPOs' socially innovative activities.

Step 2: Theoretical (Non-Field) Observations

As well as grounding the social innovation phenomenon in context, the abduction process in the authors' study also involves contemplating the phenomenon in relation to existing theoretical knowledge (Van de Ven, 2015). This ensures that new discoveries logically arise out of an appreciation of the existing body of knowledge, avoids gaps in knowledge about emergent constructs and their implications, and aims to generate sufficient diagnoses of the phenomenon and its context. Making theoretical (non-field) observations thus forms the 'mind-preparing' stage of this abductive study, and this is tightly interwoven with (i.e. emerges out of) the initial contextual grounding process (Step 1 above).

The theories used for comprehending social innovation in disability NPOs are: institutional theory to explain the role of regulatory forces (DiMaggio & Powell, 1983); resource dependence theory to decipher the effects of resource-scarcity, competition, and collaborative opportunities in the sector (Pfeffer & Salancik, 2003); and, user innovation theory to denote the impetus for NPOs to engage in consumer-led design processes (von Hippel, 1986; 2005). The 'surprising facts' encountered in this mind-preparing stage result in the formulation of research questions to be explored in the subsequent direct field stage.

Step 3: Practical (Direct Field) Observations

Two researcher-in-residency case studies were carried out in late 2015 and early 2016. The authors' previous paper – Taylor et al. (2019b) – provides a full account of these abductive case studies. The researcher spent two months full-time in the field, first at a disability NPO situated in the regional city of Hobart, followed by a second disability NPO in the metropolitan center of Melbourne. During this time, 52 semi-structured interviews were conducted with a vast array of organizational actors, including the CEO of each organization, senior managers, service coordinators, clients, board members, and external collaborators. In addition, extensive observational data were recorded. Forming the 'mind-and-reality comparing' stage of the abduction research process, these direct field experiences enabled the researcher to attend to the richness and particularities of practical everyday contexts (Taylor et al., 2018). The case study data were analyzed through thematic coding using NVivo.

Step 4: Practice–Theory Iteration to Arrive at A 'Working' Hypothesis

A deliberate teasing out of practice–theory implications occurs in the fourth step of the abduction research process (see Taylor et al., 2018a). Various 'surprising facts' uncovered in the direct field stage point to inconsistencies between the initial theoretical predictions and the lived organizational processes and complex role relations discovered within the two NPO settings.

The practical evidence gained through the research-in-residencies indicates that the public policy reform of the NDIS is not the leading influence on socially innovative behavior in disability NPOs. While the NDIS is ushering in many changes that these organizations are adapting to, the key drivers of their social innovations derive from a range of factors hinging on the social justice objectives of these disability NPO contexts. It was observed that these disability NPOs develop social innovations to increase opportunities for people with disabilities to make a direct and positive impact in their communities, and to enhance the inclusiveness of spaces and structures within society to lead to the full rights and participation of people with disabilities. This reduces the degree of focus on adaptive capacity at an organizational level as a direct response to NDIS-fueled changes.

Through the iterations between practice and (pre-identified) theories in Step 4, the authors observed the incompatibility of *institutional theory* for understanding social innovation in the context of NPOs. The case study findings show that the institutional pressures in the operating environment (i.e. legislative mandates) do not play a central role in how socially innovative NPOs develop and implement social innovations. Specifically, rather than conforming to the prevailing rules within society's institutions, disability NPOs develop new services and processes that attempt to reorient institutional policies and challenge structural inequalities. In so doing, these NPOs focus their work at local scales to bring about positive life changes, and deal with the on-the-ground 'felt' realities of inclusion which, as stated by (Stainton, 2017, p. 1), "cannot be legislated or mandated…[but] speaks to the way of being and belonging in community; our connections to others, mutual respect, and a deep sense of equality".

With regard to *resource dependence theory*, which was also initially identified as a useful lens to explain how increased competitiveness in resource-constrained environments may intensify the need for innovative response by disability NPOs (Pfeffer & Salancik, 2003), the case study findings, however, indicate that such increased competitiveness does not determine new ways to access the external resources needed for survival. For example, during one of the researcher-in-residences it was discovered that when external (i.e. financial) resources could not be attained to pilot a socially innovative program, the NPO decided to implement the program without funding. In this instance, the deficiency of resources from their external environment was mitigated by mobilizing resources from within the organization: specifically, the relevant manager was deeply committed to the project and undertook the work to implement the program on top of her usual workload. In addition, while active engagement with external stakeholders was found to be a critical factor in the development of social innovations in NPOs, the disability NPOs in the authors' study sample did not purely gain resources through these partnerships. Rather, they made use of their partners' investment in the form of time, expertise, and their willingness to openly share ideas and jointly devise initiatives (through interorganizational learning). This suggests that external engagement in the context of disability NPOs was typically oriented towards the processes of open innovation (Chesbrough, 2006), rather than resource acquisition for organizational survival.

The last theory that was initially expected to hold relevance to NPOs' social innovation development was *user innovation theory* (von Hippel, 1986). During the researcher-in residences, the importance of this theory – particularly the role of direct client input in shaping socially innovative services or processes – was clearly observed in formalized constructs (e.g. being stipulated in an organization's mission statement), and was perceived to take on significant personalized meaning for people through the articulated beliefs of all organizational members. Based on thes findings, user innovation theory was reinstated within the emerging theory of social innovation.

THEORY-BUILDING PROCESSES AT THE END POINT OF THE ABDUCTION PHASE (PHASE 1)

In this section, the authors detail the processes of theory-building that occurred near the conclusion of the abduction phase (in Step 4 of Phase 1), which led to the formulation of new, testable hypotheses based on this study's NSI framework. This framework is grounded in a complexity theory lens and involves a combination of pre-identified theory (i.e. user innovation theory) and post-identified theories i.e. self-determination theory (SDT), inclusive citizenship theory, open innovation, absorptive capacity, organizational ambidexterity, servant organization, and complexity leadership theory. The mutual practice–theory implications behind each of the components within the NSI framework are also discussed in this section.

Conceptualizing Nonprofit Social Innovation (NSI)

Through the abductive process that closely entwines real-world and theory-based observations, this study arrives at a contextually-derived concept of *Nonprofit Social Innovation (NSI)*, defined as 'a new service or process enacted specifically by a disability nonprofit which promotes the broader community's inclusion of people with disabilities, and thereby has the potential to achieve systems-level (not just organization-level) impacts'. Underlying NSI is a palpable social justice objective stemming from the social mission orientations of disability NPOs as they seek to increase the inclusion of people with disabilities in community life and/or society more broadly.

Significantly, the NSI conceptualization can be seen to bridge the 'utilitarian' (Ayob et al., 2016) social innovation conceptualizations of the management discipline, such as the definition put forward by Phills et al. (2008), with a more systematic, empowerment-focused approach. The latter is sensitive to historical and socio-cultural contexts, and contends with the multi-layered constraints imposed by institutional structures that can result in social inequalities (Moulaert, 2009). The NSI construct thus integrates concepts from the management and organizational science literatures, for example emphasizing organizational learning, strategy, and leadership-based creativity (e.g. Chesbrough, 2006; Cohen & Levinthal, 1990), while also synthesizing the complex dynamics occurring across nested scales of organizational agency on the path to social change (Westley & Antadze, 2010). In other words, the theoretical construct of NSI explains how disability NPOs mobilize resources and pursue strategies in novel ways as they address the problem of systemic social exclusion experienced by people with disabilities.

For the detailed explanation of how the authors conceptualized the NSI framework, please refer to the authors' previous paper (see Taylor et al., 2019b).

Capabilities for NSI Development

The culmination of insights from the 52 interviews and direct observations conducted during the research-in-residences evidence a myriad of complex foci in terms of NSI development by disability NPOs. Through complexity theorizing, this study unpacks these interwoven facets within the world of practice by incorporating multiple levels of reality and by using different yet complementary theoretical constructs in careful alignment (Burnes, 2005; Byrne, 2005). There are five 'pivotal capabilities': transformational empathy, place-based relationing, diversity learning, paradoxical change-making, and complexity leadership. The wide range of actors and processes involved means that these pivotal capabilities for social innovation development are multifaceted, based on a combination of internally-oriented capabilities (i.e. encompassing facets of organizational culture and intrinsic modes of relating), together with externally-oriented capabilities (i.e. responding to and influencing the broader environment and co-creating social action). So too, the variety of theoretical concepts underpinning each pivotal capability underscores the complexity of the overall NSI construct.

Transformational Empathy (Pivotal Capability A)

This concept emerged out of observations of NPO staff and managers who were intrinsically motivated to enact consultative practices. Within the socially innovative NPOs in this study, such practices could be seen in the active encouragement of reflexivity and criticism as a means for driving social innovation, and the inclusion of clients not just in planning or design discussions but in the day to day culture-building discussions occurring within the organization. The notion of transformational empathy that originated out of the two NPO case studies encapsulates these participatory and inclusive behaviors. The processes of co-creation with clients and incorporating their ideas through blended back-and-forth relationships is theorized with reference to user innovation theory (von Hippel, 1986), while the importance of self-determination displayed by participants is tied to self-determination theory (Ryan & Deci, 2000).

In related literatures, the role of empathy has previously been linked to socially innovative social work practices (Berzin & Pitt-Catsouphes, 2014), social entrepreneurship theory and practice (Schwartz, 2012), and is integrated into models of empathic design and human-centered design (Brown, 2009). However, this study's transformational empathy construct recognizes the marginalized vantage point of certain individuals and groups involved in user-led processes within disability NPOs and the (dis)empowerment dynamics that shape social innovation agency. Accordingly, the coupling of user innovation and self-determination theories conveys the significance of human relatedness as demonstrated by a shared sense of purpose between those engaging in co-design processes, and implies the importance of each person having a sense of autonomy which motivates them to actively commit to the organization's development of social innovations (Ryan & Deci, 2000).

Place-Based Relationing (Pivotal Capability B)

Many of the approaches adopted by the socially innovative NPOs in this study were driven by an understanding that systems-level change requires the transformation of long-serving structures and ways of being that are entrenched within spatial and relational contexts. As such, exploiting the specificities and potentialities of 'place' was a common feature of the place-based work of these NPOs. One practical example is a program that catalyzes new friendships between members of a community, which enables a person with disability to attend live music events together with a non-disabled person who enjoys the same music. In this instance, the NPO opted not to set up a fee-for-service solution based on the employment of a support worker to accompany the individual; rather, this service is driven by mutual friendship between people within a local community context, making the program freely available, self-sustaining, and transformative for those involved.

Throughout the researcher-in-residences, a wide variety of initiatives were found to harness context-dependent approaches in which the notion of 'inclusion' is understood as lived (daily) experience/s; meaning it cannot be divorced from its context, including its spatial context. In the social innovation literature, a territorial perspective (Neumeier, 2012) emphasizes the role of socio-spatial structures as both a catalyst and constraint. This study's concept of place-based relationing links these structural perspectives with a conceptualization of 'inclusive citizenship' (Kabeer, 2005). This theoretical lens focuses on the situatedness of constraints within social structures and emphasizes the mechanisms used by excluded groups and individuals to (re)define themselves, and in so doing to (re)story contextual realities. This capability requires that disability NPOs work closely with excluded groups who have been disenfranchised by the prevailing structures, to develop context-dependent activities promoting new relationships between citizens (based on a horizontal view of citizenship). Over the long-term, the aim of these activities is to establish or restore democratized (vertical) relationships between citizens and social structures (Kabeer, 2005).

Diversity Learning (Pivotal Capability C)

The core premise of the diversity learning capability is that NSI involves multiple perspectives and voices from both outside and inside the organization. The disability NPOs in this study interacted with diverse networks of stakeholders to learn about the many facets of social exclusion and to collectively develop novel solutions. Such cross-fertilization of knowledge via multi-partied exchanges features prominently in the social innovation literature (e.g. Caulier-Grice et al., 2012; Nicholls & Murdock, 2012; Sanzo, Álvarez, Rey, & García, 2015). This study's pivotal capability of diversity learning draws on the open innovation paradigm (Chesbrough, 2006) to emphasize the importance of these external links and knowledge sources.

Yet this outward-looking orientation is balanced with the need for internally-derived knowledge. During the researcher-in-residences, the researcher found that a valuable source of diverse knowledge for social innovation development was often tacit in nature and was prominently (though not exclusively) derived from managers and staff members with lived experience of disability or mental health issues. As tacit knowledge is closely tied with social context, this form of knowledge integration may be critical to tackling the less visible aspects of social exclusion that have contextualized meanings, defy easy explanation, and may only be understood experientially (Gertler, 2003). Important connection points are made within this pivotal capability by incorporating the concept of absorptive capacity (Cohen & Levinthal, 1990), which denotes an NPO's capacity to take in new forms of information while also utilizing prior related knowledge. As far as can be determined, the role of absorptive capacity as an aspect of social innovation has only been explored within a study by Chalmers and Balan-Vnuk (2012); however, rather than viewing absorptive capacity as an isolated capability for NPOs to exploit, the authors embed it within a multi-theoretical concept (i.e. 'diversity learning') which explains how diverse knowledge sources are captured and used by NPOs to better deal with the multidimensionality and complexity of social change processes.

Paradoxical Change-Making (Pivotal Capability D)

This pivotal capability includes a number of seemingly disparate operational approaches that combine to create socially innovative agency that is both impact-driven and purpose-driven. For example, during the researcher-in-residences the researcher perceived the disability NPOs' tendency to harness a 'productive tension' with other organizations, including government agencies. This involved openly pushing outside agencies to do things differently, and it meant that the NPO was not afraid to ask what were at times confronting (and often seen as radical) questions as a way to disrupt the status quo. Simultaneously, these NPOs were seen to adopt a 'servant culture', in which individual managers and staff were driven by a desire to serve a greater cause in modest or humble ways, rather than in ways that necessarily increased the organization's profile. This outspoken/humble dichotomy was a surprising paradox found in the practical domain that required careful practice–theory linking to explain it. A complexity-oriented view illustrates why occasionally taking on riskier (and more unpredictable) social innovation experiments is necessary, as opposed to always adhering to smoother (easier to predict) social innovation pathways (Goldstein, Hazy, & Silberstang, 2010). The theory of servant organizations is tied to the latter incremental pathway of change (Sousa & van Dierendonck, 2017) to explain the organizational actor behaviors characterized by the virtue of humility.

The disability NPOs in the current study also espoused simultaneously long-term/ short-term outlooks. Being comfortable extending out project timeframes well beyond their initial deadlines, while at the same time constantly shifting in a manner that is responsive and opportunistic, are examples of the exploration/exploitation facet of NPO-based socially innovative behavior (Gupta, Smith, & Shalley, 2006). In this vein, 'slack' is given to projects which apply novel, forward-looking models of exploration and variability, yet exploitation-related projects are more tightly coordinated as they aim to seize on immediate opportunities. While this view of organizational ambidexterity has been applied to social innovation processes within public-private partnerships (Battisti, 2014), its explicit connections to NPO-based social innovation have not previously been outlined in the extant social innovation literature.

Complexity Leadership (Pivotal Capability E)

In this study, the pivotal capability of Complexity Leadership Theory (CLT; Uhl-Bien, Marion, & McKelvey, 2007) is used to explain emergent forms of leadership in response to adaptive challenges, i.e. for disability NPOs, tackling the issue of social exclusion requires exploration, new discoveries, and adjustments on the path to social change. Shifting the emphasis away from the actions of an individual (i.e. a 'leader') and instead focusing on patterns of relational organizing, CLT involves three intertwined forms of leadership: adaptive, administrative, and enabling leadership (Uhl-Bien et al., 2007).

Diverse and interdependent arrangements of agents were found to engage in the social innovation-focused practices within disability NPOs. The role of complex adaptive system (CAS) dynamics (Stacey, 2003; Tapsell & Woods, 2010; Uhl-Bien et al., 2007) was evident as these groups of agents each behaved according to their own principles and aims, yet also adjusted their behaviors in response to conditions and events as they materialized. For example, new synergistic alliances and unplanned projects emerged nonlinearly out of processes of co-influencing among agents, each holding conflicting needs and preferences. According to CLT theory, arriving at these adaptive outcomes within the social system is linked to the concept of adaptive leadership. Yet planned and structured activities were also embedded in the disability NPO context in the form of administrative leadership. Enacted by people in formalized management roles within the organization, administrative coordination was important for overseeing the delivery of disability support services in a predictably safe manner and in accordance with organizational goals. Critically, the role of the third form of leadership within the CLT theory, enabling leadership, was seen to overlap both the adaptive and administrative functions within the disability NPOs in the case studies. Enabling leadership involves catalyzing CAS dynamics such that novel ideas can originate through the horizontal, interactive dynamics of adaptive leadership and be formalized as new order within the hierarchical administrative system (Arena & Uhl-Bien, 2016).

CLT has been linked to social entrepreneurship in the area of human rights (Rhodes & Donnely-Cox, 2008), socially entrepreneurial indigenous cultures (Tapsell & Woods, 2010), and has been used as the basis of a diagnostic tool for tackling 'wicked' social problems by local governments (Zivkovic, 2015). This study's use of CLT to predict the leadership mechanisms of disability NPOs as an aspect of social innovation development extends the relevance of this relatively new leadership theory which is still in its early stages of being operationalized (Hazy & Uhl-Bien, 2015).

Enabling Conditions for Maximizing NSI Impact

During the researcher-in-residences, three enabling conditions within NPO contexts were discovered that promote an organization's successful implementation of social innovations with high levels of societal benefits. The enabling conditions are *'person-focused'*, *'systemic-focused'*, and a *'risk tolerant environment'*. While an NPO may be able to develop NSIs through deployment of its capabilities, the three enabling conditions were found to play a role in bolstering the implementation of NSI strategies leading to the generation of social change within systems (i.e. at community and/or societal scales). Rather than specific arrangements of resource endowments in the form of organizational capabilities, these enabling conditions can imbue (sometimes subtly, at other times conspicuously) most or all facets of the organizational milieu, broadly shaping the underlying ethos and operational mode/s of the NPO.

Person-Focused Enabling Conditions

These conditions may enable impactful NSI implementation and lead to social change outcomes through an intensively person- or client-centered focus, and an overall commitment to valuing the personal experience of individuals. At all levels of organizational activity, it requires the central involvement of people with disabilities as autonomous decision-makers (Stainton, 2000). This condition was observed in both disability NPO case studies, with a person-centered approach permeating their organizational missions, behaviors, structure, and processes. As a practical example, the NPO clients were empowered to lead their own initiatives in line with their personalized goals and ambitions.

Systemic-Focused Enabling Conditions

Systemic-focused enabling conditions rest on the premise that creating systemic change requires the NPO to take a broad-level systems view within almost every action or plan undertaken. It is enacted through collective cross-sectoral approaches in which multiple actors work according to an overarching objective of 'coherent action' (Davies, Mulgan, Norman, Pulford, Patrick, & Simon, 2012), rather than necessarily formalized collaborative arrangements. The abductive case studies revealed how NPOs inhabit an important intermediary institutional space between government, private businesses, and the general community, from which they can coordinate and engage in networks that may lead to important processes of social change. For example, the NPOs in this study demonstrated their full commitment to, and investment in, cross-sectoral partnerships as an important condition for the fostering of systemic solutions.

Risk Tolerant Environment

This final enabling condition involves an NPO maintaining an open posture towards risk. This can be seen in the types of innovative projects that are instigated, with an inclination to engage in ambitious or radical projects signaling a greater risk tolerance. In order to disrupt prevailing institutions and mitigate systemic forms of exclusion in effectual ways (Howaldt & Schwarz, 2017), both disability NPO case studies viewed their willingness to take risks as a necessary ingredient to successful innovative strategy implementation. An acceptance of failure was also important for fueling the implementation of high-risk NSI strategies.

Formulation of 'Working' Hypotheses for Further Empirical Testing

This study's abduction-based NSI model offers an explanation of how particular capabilities are used for the development of social innovations by disability NPOs, and the impacts attributed to these innovative strategies. Stemming from this NSI model, the authors generate four main 'working' hypotheses as follows (also see Taylor et al. 2019a,b):

- **Hypothesis 1**: A complex interplay of pivotal capabilities could lead to NSI development.
- **Hypothesis 2**: Not all the five pivotal capabilities may be required for NSI, and the choice of which pivotal capabilities to deploy is likely to depend on the specificity of organizational and environmental contexts (e.g., size and location).
- **Hypothesis 3**: Successful NSI development is anticipated to be a mechanism by which organizations may alter broader community and societal structures and thus contribute to social change.
- **Hypothesis 4**: Certain enabling conditions (i.e., person-focused, systemic-focused, and risk-tolerant environment) may strengthen the potential for an NSI to generate high levels of beneficial outcomes in terms of promoting social change at community and/or societal scales.

Hypotheses 1 and 2 pertain to the first part of this study's research question (i.e. within the disability NPO context, what organizational capabilities are required for the development of social innovations?), and Hypotheses 3 and 4 are relevant to the second part of the research question focusing on impacts (i.e. what impacts can be attributed to the successful implementation of social innovations by NPOs?).

As the 'end point' of the abduction process, these tentative working hypotheses represent plausible, but not yet logically necessary, conclusions (Kovács & Spens,

2005). In order to verify these hypotheses, a follow-on phase of quantitative and/or qualitative testing must be carried out. This study's research design is structured such that the working hypotheses arising out of the abductive case study phase are directly tested in a subsequent sector-wide study. This next (Phase 2) component of the study, including the results of the hypothesis testing and its fulfilment of practice–theory building, will now be discussed.

PHASE 2: SECTOR-WIDE STUDY

Phase 2 tests the working hypotheses generated from the abductive case studies in Phase 1 (refer to Taylor et al., 2019a). It involved a mixed methods investigation into NPO-based social innovation across the nationwide disability sector. The survey instrument was designed to directly test the use of pivotal capabilities for NSI development and the role of enabling conditions for maximizing NSI impact. Prior to dissemination, the survey instrument was 'cognitively tested' through face-to-face interviews with 16 CEOs from the sector. Following the testing phase, the research team sent a hard copy survey to the CEO or senior manager of all NPOs operating in Australia that were registered with the NDIS (n=735). A total of 308 surveys (representing 42% of the total population) were completed, with 301 NPOs reporting the development of an organization-level social innovation during the previous two years. The survey was followed by the collection of qualitative data through 14 semi-structured interviews with CEOs who had participated in the survey.

Phase 2 of this research project bridges the complexity theorizing and practice-based findings of the abduction process with configurational methods, i.e. fuzzy-set Qualitative Comparative Analysis (QCA), in order to capture the nonlinear, systemic dynamics of the social innovation phenomenon at a whole-of-sector level. QCA is a set-theoretic method that empirically tests for complex causal structures and 'equifinality' (Katz & Kahn, 1978) by examining relationships between a given outcome and all possible combinations of binary states (i.e. presence or absence) of its predictors (Ragin, 2000). Based on the notion of configurations (Fiss, 2007), each separate case can be viewed as characteristic configurations (combinations of capabilities or enabling conditions), thereby making it possible to view the interdependencies of the different parts which form a coherent whole within a given context (Ragin, 1987).

Two fuzzy-set QCA analyses were carried out in the Phase 2 component of this study (the results were reported in Taylor et al., 2019a). The first QCA analysis (QCA Analysis 1) identifies the configurations of the five pivotal capabilities to ascertain the conditions for high-level NSI development. The contextual factors of organizational size (large or small) and geographical location (metropolitan or regional/remote locality) are incorporated into QCA Analysis 1 in order to interpret certain context-dependent elements of social innovation in NPOs (Moulaert, Martinelli, Swyngedouw, & Gonzalez, 2005). The second QCA analysis (QCA Analysis 2) tests for the enabling conditions used to orchestrate NSI implementation in order to generate high levels of societal benefits. In order to implement NSIs successfully and attain benefits at community and/or societal scales, this second QCA analysis focuses on the interactivity of high levels of NSI development together with personalized, systemic, and risk-focused enabling conditions.

Subsequent to the survey phase, 14 semi-structured interviews with CEOs who had participated in the survey were conducted. Roughly equal numbers of CEOs from small regional/remote NPOs, small metropolitan NPOs, and large metropolitan NPOs participated in this aspect of the study. Through the 45-60 minute interviews, the researcher gained rich qualitative insights into the socially innovative practices occurring in particular NPO contexts. The data were thematically coded and integrated with the QCA findings to provide a more in-depth understanding of the contextualized processes linked to NSI.

Key Findings From the QCA Analyses

QCA Analysis 1

The first QCA analysis uncovers four solutions, or *'recipes'*, for capability deployment by socially innovative NPOs, as depicted in Figure 2 below.

- Solution 1 (labeled the **'Regional Responsiveness'** recipe) specifically pertains to small regional/remote NPOs. This solution indicates that the pivotal capabilities of place-based relationing, paradoxical changemaking, and complexity leadership lead to high levels of NSI development by these organizations. However, due to the presence of particular environmental factors, it was found that the capability of diversity learning is not harnessed.
- Solution 2 (labeled the **'High Input – High Impact'** recipe) highlights that small NPOs, irrespective of their location, are able to develop social innovations by deploying all five capabilities.
- Solution 3 (labeled the **'Big Innovators'** recipe) focuses specifically on large metropolitan NPOs, and outlines the importance of using four capabilities in combination (transformational empathy, place-based relationing, paradoxical change-making, and complexity leadership).
- Solution 4 (labeled **'Place-Based Diversity'** recipe) reveals that the two capabilities of place-based relationing and diversity learning are critical for social innovation development by small metropolitan-based NPOs. This

targeted capability arrangement renders the other three capabilities as not required by these organizations.

The results from QCA Analysis 1 highlight a number of important findings that have the potential to enhance an understanding of organization-based social innovation. First, the findings lay emphasis on the organizational capability of place-based relationing. This capability appears in all four solutions, rendering it a 'necessary' condition for social innovation adoption by disability NPOs. The prominence of place-based activities suggests that a territorial or 'place' dimension may be essential for theoretical understandings of social innovation (Butkevičienė, 2009; Neumeier, 2012). This contributes a salient perspective on NPO-based social innovation. Yet, it must be noted that this dimension is not sufficient on its own to drive high levels of NSI development by these organizations; all four solutions in QCA Analysis 1 include additional capabilities in conjunction with place-based relationing. The ramification of this finding for social innovation theory points to a pluralistic explanation; what this means is the influence of localized responses may be undeniably important for social innovation, however it requires the integration of additional organizational capabilities beyond solely community-focused approaches.

Another key take-away from the QCA Analysis 1 findings is that there is no single pathway or one 'best' approach to social innovation adoption by NPOs that can suffice for all contexts and contingencies. Thus, the importance of *'place'* is evidenced not just in terms of the capability of place-based work as described above, but also through

Figure 2. Configurations of organizational capabilities for Nonprofit Social Innovation (NSI) development by NPOs

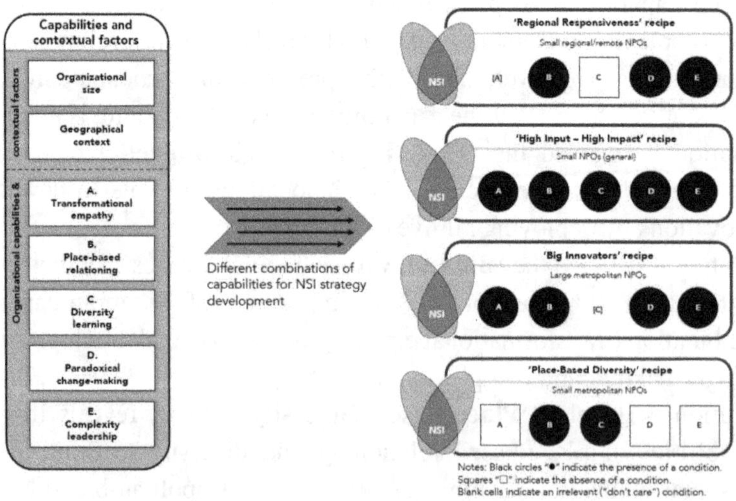

the alternative ways NPOs combine and harness capabilities as they innovate within different contexts. These QCA findings address some pressing questions that have remained largely unanswered in disability and social innovation research: namely, *"what works... for whom, and under what conditions?"* For instance, for regional and/or remote NPOs, the particular combination of capabilities used can be seen to form a vital response to the challenges this context poses for socially innovative service delivery (NRHA, 2013). Prior research into social programs within these localities evidences the effects of geographical-related factors such as distance, isolation, and disadvantage (Matheson, Dew, & Cumming, 2009). An interesting element to the findings, however, is that these NPOs specifically did not harness the capability of diversity learning. Although extant studies emphasize co-evolutionary learning processes (Neumeier, 2012) and knowledge synergies (Butkevičienė, 2009) for fostering social innovations as part of rural development in European contexts, the findings of the current study may reflect the greater levels of remoteness within the Australian context resulting in significant barriers to participation in external networks and limited pools of diverse staff and volunteers for these NPOs to draw upon (NRHA, 2013). This shines a light on the importance of social innovation's contextualized expressions (Howaldt & Schwarz, 2017).

Just as geographical contexts may shape social innovation adoption, different organizational traits are found to give rise to different arrangements of capabilities. For example, according to the QCA Analysis 1 results, large NPOs operating in metropolitan areas demonstrate the use of four capabilities, whereas small NPOs also operating in metropolitan areas are found to develop social innovations through the adoption of just two capabilities (with the only commonality being the place-based relationing capability). Within the extant literature, organizational size has been found to alter the innovative capacity of NPOs (Green & Mears, 2014), and so too the configurational analysis findings allude to such distinctiveness of practices within disparate organizational settings.

Significantly, the QCA Analysis 1 findings provide a holistic, complexity-based explanation of NSI development by uncovering the interrelatedness of causal processes as they manifest in response to localized conditions (Matheson et al., 2009). For instance, in returning to comparison of large- and small-sized NPOs, comprehending the underlying system trajectories is possible based on the QCA analysis. In relation to large NPOs, direct client participation, context responsiveness, ambidextrous modes of operating, and a complexity leadership model must all be explicitly harnessed, i.e. a large variety of actions and systems are instrumental for social innovation adoption by these NPOs. This broad set of capabilities may be important for overcoming the often-inbuilt inflexibility of bureaucratic systems within these particular organizations and a greater distance between key organizational decision-makers and service users (Hammack, 1995). On the other hand, smaller NPOs in metropolitan contexts evidently focus on developing tailored service offerings in partnership with other community actors and networks, which is perhaps due to their community embeddedness (Toepler, 2003). By uncovering these configurational patterns, the QCA findings can ascertain the role of initial conditions and complex causal relationships within different cases. They also offer a view of system behavior which showcases how the different parts (i.e. capabilities) may interact to form a coherent whole within a given context.

In summary, the results from QCA Analysis 1 suggest that high levels of NSI development stems from a complex interplay of pivotal capabilities, thus supporting Hypothesis 1. The context-related factors of geographical location and organizational size influence whether all of the five pivotal capabilities need to be present and deployed in an integrated way for high-level NSI development, which supports Hypothesis 2.

QCA Analysis 2

The second QCA analysis pertains to the enabling conditions that promote an NPO's implementation of an NSI with high-level benefits. The two solutions generated in QCA Analysis 2 are depicted in Figure 3 below.

Both solutions indicate that a high level of NSI development is *necessary but not sufficient* for attaining significant community-level or societal benefits. These QCA results, along with the CEO interview data, imply that high levels of societal benefits are not easily achieved by socially innovative NPOs. To some extent, this supports the existing concept of 'systemic social innovation' (Davies et al., 2012),

Figure 3. Configurations of enabling conditions for high levels of Nonprofit Social Innovation (NSI) benefits

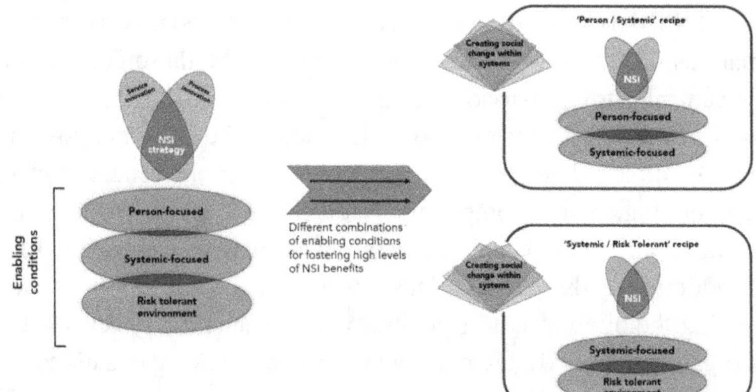

which posits that generating broad social change is beyond the scope of a single organization (Westley, Antadze, Riddell, Robinson, & Geobey, 2014). Yet while an NSI on its own may not lead to these kinds of benefits, QCA Analysis 2 findings extend knowledge of the potential for organization-based social innovation to alter systemic structures in a number of key ways.

Systemic-focused conditions are found to play a crucial role, which means that NPOs must cooperate with a broad range of organizations (including organizations from different sectors), and at the same time draw heavily from those sources (Laursen & Salter, 2006). Yet focusing squarely on a systemic orientation is unlikely to be sufficient for attaining significant community-level or societal benefits. One way to enhance collaboration-based social innovations is by incorporating an intensively personal element (as per solution 1 in QCA Analysis 2, labeled the **'Person/Systemic'** recipe). This suggests that attempting to experiment exclusively at the level of the system (Davies et al., 2012) but not attending to the deep-rooted personalized aspects and the related 'micro moments' of change, may ignore or devalue the substrata of socially transformative agency that may be crucial for NSI (Hallahan, 2013). This finding points towards the critical importance of a complexity orientation within social innovation theory, whereby interconnected 'parts' at different social levels (Anderson, 1999), including the everyday interactions, relationships, and a wide range of 'lower-level' forms of self-organizing, are seen as having potential to lead to emergent 'wholes' within the system (Matheson, 2016).

In the second solution within QCA Analysis 2 (labeled the **'Systemic / Risk Tolerant'** recipe), systemic-focused enabling conditions are combined with a risk tolerant environment to generate social change within systems. While the nonprofit sector is generally recognized for its propensity for risk averse behaviors (Chalmers & Balan-Vnuk, 2012; Chetkovich & Frumkin, 2003), social innovations that can mitigate systemic forms of social exclusion may entail a high degree of risk tolerance (Howaldt & Schwarz, 2017). By combining this with systemic-focused conditions, systemic impact is achieved through collaboration-based projects involving high levels of uncertainty, such as cross-sectoral initiatives that look for emerging strategies with no specific final solution or 'end point' (Davies et al., 2012).

To summarize the results of QCA Analysis 2, high-level NSI development is a necessary but not sufficient condition for high societal impacts to be achieved. NSI needs to be developed from the systemic-focused perspective and implemented in tandem with either person-focused or high-risk conditions in order to attain high levels of societal benefits from NSI. Based on these QCA results, Hypothesis 3 is partially supported and Hypothesis 4 is supported.

The next section of this chapter will discuss the implications of the study's Phases 1 and 2 findings for theory of social innovation. This will be followed by a discussion of the implications for managerial practice, research design, and policy. The study's limitations and directions for future research are provided at the end of this chapter.

BUIDING THEORIES-IN-PRACTICE ON SOCIAL INNOVATION IN DISABILITY NONPROFIT ORGANIZATIONS

Implications for Theory of Social Innovation

In its entirety, this study entails a number of logical steps through abductive (Phase 1) and deductive/inductive (Phase 2) stages of inquiry that are guided by contextually-driven questions and explorations, and that utilize methods capable of measuring complex realities (Woodside, 2014). It is through the culmination of these phases that this study is able to make a number of interesting and important theoretical contributions to the emerging field of social innovation scholarship. These key contributions, which lead to the development of theories-in-practice, are displayed in Figure 4 below.

The Phase 1 'abduction' process, involving research-in-residences at two disability NPOs, laid the groundwork for the generation of new plausible hypotheses for explaining social innovation in NPOs. These hypotheses, which are encapsulated in the Nonprofit Social Innovation (NSI) conceptual framework, incorporate a diversity of constructs and mechanisms linked to social innovation in the particular contextual domain under study, i.e. within disability NPOs in the Australian disability services sector. While this inbuilt context-specificity may limit the NSI model's direct applicability to other sectors or organizational types, it is in fact an important feature

Figure 4. Theories-in-practice on social innovation in disability nonprofit organizations

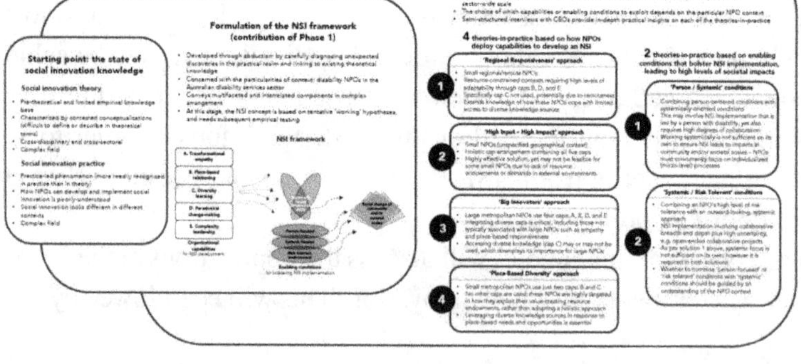

of its theoretical contribution. This is due to the abductivist logic behind the model's formulation – based on a careful decrypting of real-world context/s, the synthesizing the concrete realities (perception) with higher-order reasoning (cognition), and linking these together to form plausible theoretical premises – all of which is vital for properly 'establishing' the social innovation phenomenon. Thus, the abductive foundation of this study's NSI framework, which positions the particularities of the organizational context at its center (George, 2014), avoids grounding new social innovation theory in myths and superstitions (Van de Ven, 2015); or in other words, ensures sufficient diagnosis of the phenomenon.

Based on the extensive observational and interview data gained through the research-in-residences, it is apparent that a plurality of foci and a diverse range of participants act as key drivers of social innovation in disability NPO contexts. The authors draw connections between these practical realities through complexity theorizing, adopting a blended approach to theory construction that mindfully responds to the surprising paradoxes and unique forms of emergence that characterize NSI. Such anomalous empirical patterns are not adequately explained by current social innovation theories (Howaldt & Schwarz, 2017; Phillips et al., 2014; Westley & Antadze, 2010). The study's use of a multi-theory lens to predict the key factors behind the development and implementation of social innovations by NPOs deviates from the typically mono-theoretical and/or a-theoretical approaches found in the literature (e.g. Barraket & Furneaux, 2012; Chalmers, 2012; Weerawardena & Mort, 2012). Complexity theorizing allows for a holistic understanding of multiple levels of analysis (Byrne, 2005; Taylor et al., 2018b), and avoids delineating what are interlinked, emergent forces playing out across complex social systems.

Accordingly, the abductively-derived NSI framework provides a conceptual blueprint for social innovation in NPOs by conveying its multifaceted and interlinked components. While it is difficult to predict the ways in which a multiplicity of actions and conditions may ultimately lead to social change across complex systems (Howaldt & Schwarz, 2017), the NSI model takes a small yet critical step on the road to social innovation theory development by explaining: 1) the diversity of components behind this complex phenomenon – through theoretical pluralism it brings together different inter-disciplinary theories into careful arrangement; 2) how certain patterns emerge – this is outlined through the breadth of behaviors and routines which coalesce to form the capabilities in the NSI framework; 3) the interconnections between different components – the many internal/external sub-capabilities align in particular formations to create the complex 'pivotal' capabilities in the framework; and, 4) the possible influence on long-range outcomes – the NSI framework offers a rudimentary theory of 'initial plausibility' (Haig, 2005) for explaining how organizations may contribute to social change processes through NSI.

Yet, abduction on its own does not yield highly developed or well-validated theories, but instead serves as a starting point for the development of new theories (Kovács & Spens, 2005). Thus, the Phase 2 component of this study builds directly upon the tentative theoretical understandings that emerged out of Phase 1 by bridging complexity theory with configurational thinking and QCA analysis methods (Fiss, 2007; Ragin, 2000). Critically, the second phase repositions the study's lens from the localized scale of the two researcher-in-residence investigations to adopt a sector-wide view. In so doing, it examines the interface between localized NSI trajectories and broader patterns of emergence across the system domain. Based on the testing of the NSI framework and the resultant QCA findings in Phase 2, a number of complexity-based rules of interaction (Burnes, 2005) can be ascertained. These consist of asymmetrical combinations of organizational capabilities and enabling conditions as used by the individual entities (disability NPOs) making up the system or population. From these rules, it is possible to decipher the diverse forms of self-organizing, as well as the nonlinearity and 'equifinality' of NSI, by identifying how different capabilities and enabling conditions are harnessed within different organizational contexts, and to what effect.

Through Phase 2, this study is able to formulate contextualized 'theories-in-practice' (Argyris & Schön, 1974) to understand social innovation in NPOs. This interrelated set of middle-range theories encapsulates the dynamic properties of NSI's patterns of emergence. Theories-in-practice are instrumental in connecting localized and systemic realities. First, at a local level these theories contribute useful knowledge into NSI's complex causal structures and its varied contextualized manifestations. For example, such theories recognize that different geographical and organizational contexts are likely to have different ways of responding, as "no two communities will be identical and small differences matter" (Eppel, Matheson, & Walton, 2011, p. 53). Second, these theories can be understood systemically, according to a whole-of-population perspective consisting of multiple trajectories within broader systems of social transformation.

Thus, in terms of this study's sequential steps of theory development, the theoretical contribution of the Phase 2 investigation is the transformation of the NSI model from an undifferentiated or globalized (Fox, 2003) theoretical construct explaining NPO-based social innovation (i.e. as per the untested, 'tentative' NSI model as it was understood at the end of the abduction phase) into several contextually responsive 'theories-in-practice'. Through both the Phase 1 and Phase 2 theoretical contributions, which are logically and intimately linked, this study is able to contribute holistic new understandings to the young and growing field of social innovation research. In broad terms, the study's theoretical contribution elucidates how NSI may serve as a mechanism by which localized NPO-driven action can be influential in altering the broader temporal-spatial fabric of society.

Implications for Managerial Practice

Through the development of theories-in-practice, this study serves as a change process based on trans-disciplinary problematization, involving researchers and organizational actors collaboratively tackling an issue of shared concern (such as the current lack of knowledge about social innovation), with the aim to jointly contribute perspectives, expectations, and experiences that can increase the broader system's capacity to confront critical challenges. This contribution of change-oriented findings is a hallmark of complexity theory's focus on prescriptive (as in, how to change), rather than purely descriptive forms of knowledge (Morrison, 2005). Based on the complexity theory tenets of self-organization, context-specificity, and equifinality, the social innovation pathways presented in the NSI framework and the formulation of theories-in-practice provide practitioners with knowledge into the specific ways that complex problems can be addressed according to localized conditions and constraints. This is arguably a necessary hallmark of social innovation research that intends to bear fruit for practitioners, as systemic social problems are unlikely to have one 'best' approach to solve them (Eppel et al., 2011).

The study's findings attempt to support managers to become more adept at weaving ambiguity into every action as a way of countering the potential disabling effects that complexity may otherwise have on NPO managers and their teams. This is a requisite mode of thinking and operating for managers seeking to use NSI for dealing with the mess of social change. As such, the authors propose that the strength of the complexity-informed theories-in-practice is their capacity to: 1) avoid being overly prescriptive or uniform, incorporating a multiplicity of perspectives or actions; 2) communicate the subtle patterns and underlying order within the SI phenomenon, but also remaining malleable, fine-tunable, and responsive to different conditions as and when needed; 3) allow for a level of ambiguity via localized and evolutionary pathways; and, 4) encourage continual evaluation of experience as it occurs via actions and reactions through feedback loops.

Implications for Policy

This study is situated in the midst of a turbulent and game-changing period of social reform in Australia (Connellan, 2014; Goggin & Wadiwel, 2014). For decades, the government has relied on NPOs to deliver welfare services and social programs which aimed to benefit people with disabilities (Hallahan, 2013). However, according to the highly influential Productivity Commission (2011) report, 'Inquiry into Disability Care and Support', the disability services sector has been characterized as inequitable, fragmented, and giving people with a disability little choice with regards to service design and delivery (Productivity Commission, 2011). With the NDIS rollout in the

Australian disability sector, a major paradigm shift providing the structural means for socially innovative and rights-based community care services for people with disabilities has now become a real possibility (Green & Mears, 2014).

While the development of a responsive disability services marketplace with effective and efficient service provision is one key aspect which may drive innovation in the sector (Connellan, 2014; Skelton, 2016), at the heart of present-day policy frameworks is an explicitly stated vision for "an inclusive Australian society that enables people with disability to fulfil their potential as equal citizens… [through] innovation in the design of communities and environments to invite participation on a universal and equal footing" (FAHCSIA, 2011, p. 22). This vision suggests an imperative to look beyond quantifiable notions of service delivery, i.e. in terms of observable and 'knowable' processes with measurable outcomes, based on 'how many' or 'how much' (Stainton, 2002), to instead embrace new ways of tackling entrenched forms of social exclusion. Yet an absence of rigorous analysis into how NPOs can play a role in contributing socially innovative solutions may be holding back the social justice elements outlined within current policy agendas.

A lack of knowledge about the practice of social innovation and its sector-wide patterns may make it difficult for governments to shape public policy in ways that can constructively promote socially innovative behavior (Mulgan, 2006). The authors argue that for these new policy directions to be realized, government's recognition that implementation of the NDIS is influenced mainly from 'the bottom up', and not the other way around (Hallahan, 2013), is critical. This means understanding how the social objectives of the NDIS policy may be interpreted, enacted, and ultimately become a reality within local NPO contexts. The study findings reveal that this often occurs in idiosyncratic ways, as NPO practitioners develop contextually-sensitive responses informed by their experiential knowledge of both local conditions and systemic barriers.

Through the study's findings which explore the effects of context, policymakers will be better placed to develop informed policies and interventions that develop, support, and sustain NPO-based social innovation. While current capability building activities in the sector in response to the NDIS rollout typically focus on the ubiquitous needs of all NPOs (Vanstone & Graham, 2016), this study's findings make it possible for peak bodies and/or governments to differentiate their sector development approaches. For example, to better enable regional/remote NPOs to address local challenges, organizational development programs for these NPOs could focus on building capacity in place-based relationing, paradoxical change-making, and complexity leadership (which were found to be necessary ingredients in the 'Regional Responsiveness' recipe). For small metropolitan NPOs, on the other hand, organizational capability-building focusing on peer-to-peer collaboration within local settings may lead to an increase in their socially innovative capacity (in accordance with the capabilities outlined in the 'Place-Based Diversity' recipe).

Overall, the study's identification of a diversity of NSI pathways and influencing factors means that more flexible, non-prescriptive regulatory mechanisms on the part of government may be conducive to greater levels of NSI in the sector. However, the rollout of the NDIS has been criticized for its overly constraining rules and pricing structures and a general lack of support for NPO engagement in systemic-focused activities (Smith-Merry et al., 2018). Overcoming these issues may require that policymakers: instigate more flexible funding streams, or what is termed 'catalytic funding' (Vanstone & Graham, 2016) to support socially innovative initiatives in the sector, rather than purely funding standardized delivery of services (Baker, 2017); allow for a requisite level of risk and/or disruption within NPO program offerings, which necessitates a loss of control or power for those who have typically acted as the decision makers for people with disabilities, i.e. traditionally the state has had almost total decision-making authority in this regard (Stainton, 2002); and, improve communication between NDIS regulators and the service providers on the ground to enable knowledge about localized (and systemic) issues to filter back up (Smith-Merry et al., 2018).

STUDY LIMITATIONS AND FUTURE RESEARCH DIRECTIONS

Limitations of the Study

There are several limitations of this research that point to avenues for future research. Four of the limitations pertain to the study design, and a further three limitations stem from the difficulties in conceptualizing the social innovation phenomenon. In terms of the research design, this study's focus on a specific sector and organization type limits the transferability of findings. While the context-specific nature of this study is intentional and has been posited as a strength of the research design, it necessarily frames this study as a highly situated exploration into social innovation which may not hold direct relevance in other contexts. Second, the involvement of just two NPOs in Phase 1 as organizational case studies once again means that a narrow realm of practical experiences is captured in this study, rather than a wider ranging investigation which could be achieved by including a greater number of organizations as case studies. Limiting the number of case studies to two, however, was deemed necessary due to the extended timeframe required for the researcher to undertake data collection in Phase 1. Third, the Phase 2 survey findings are limited by the use of self-perceived outcome measures in the survey instrument. This makes it difficult to verify the accuracy of responses regarding organizational capabilities or objectively measure the benefits associated with social innovation outcomes, as the respondents were not requested to include substantiating evidence to support their claims. Fourth, the use of QCA as the data analysis technique in Phase 2 presents some methodological limitations that bear mentioning. For example, the results generated with QCA are 'sensitive to cases' such that adding or removing factors can potentially generate significantly different findings. The researcher must also subjectively select the thresholds for consistency, coverage, and the crossover point in the calibration process (Ordanini, Parasuraman, & Rubera, 2014). Despite strict adherence to the methodological guidelines in the QCA literature (Ragin, 2000, 2008), the possibility for bias stemming from the analytical assumptions of the researcher should be acknowledged. To address this, 14 semi-structured interviews (subsequent to the QCA analyses) were carried out to verify the results based on the practical insights of CEOs from the sector.

Other limitations in this study are associated with the theorizing process. Although the theory-building steps leading to the formulation of the NSI framework are detailed through the account of the Phase 1 abduction process, it is important to note that theorizing is to a large extent driven by a number of inbuilt, often unconscious, criteria (Weick, 1989). This is particularly the case when a researcher must deal with complex subject matters. It is the theorist's own judgements that act as a defining feature of the theory construction process, which according to the philosophy of Peirce (1997), means that theoretical modelling cannot be expected to recreate reality in its exactness. Moreover, an underlying assumption of the NSI framework is that the possible outcomes of social innovation may involve long-term processes of social change. This renders the specificities of the social impact achieved as inherently difficult to measure (Davies et al., 2012). Likewise, it may also be problematic to attribute a particular social innovation that was previously implemented as the initial cause of complex social change processes (Howaldt & Schwarz, 2017). A final limitation of the conceptualizing process is that the meaning and implications of what is termed 'positive social change' is not clear-cut; rather, it is subjectively determined. In real terms, it is often based on deeming certain social objectives valuable or desirable, and as a result it is important to acknowledge that some individuals or groups may experience disadvantage in terms of the outcomes of NSI (Neumeier, 2012).

Directions for Future Research

This study points to a number of valuable future directions for social innovation research. In its current state of nascent knowledge development, expanding the field's coverage of empirical knowledge and strengthening its theoretical standing requires abductive logics to be employed in further research within different fields of practice, i.e. in other organizational settings and sectors. This will deepen the current knowledge of the phenomenon's untold contextual variations and complex realities, in recognition that "meaning is context bound, but contexts are boundless" (Culler, 1983, p. 123). Eventually, by comparing a wide range of close-range phenomenological explorations and then drawing them together in holistic alignment, cohesive social innovation conceptualizations may be achieved. This can ultimately lead to better explanations of social innovation's complex patterns based on parsimonious theories. Critical for management scholars researching into social innovation is a deeper involvement with organizational contexts, as this will contribute concepts and theories with a more distinct management and organizational feel to this cross-disciplinary domain.

The study of social change is inherently problematic due to the implications of power relations at different social levels and the trajectories of long-term processes of change that will always be uncertain (Matheson et al., 2009). A potentially fruitful way forward for social innovation researchers to overcome these challenges is the use of action research that directly works with cohorts of participants belonging to a designated field or situation to test out new ideas for social action. These socially innovative interventions are trialed through an unfolding and open-ended series of evaluative and reflective exchanges occurring in real-time (Fox, 2003). Importantly, for action research processes to be truly participatory and contribute to the socio-political empowerment of marginalized groups (Moulaert, MacCallum, & Hillier, 2013), it is vital for researchers to critically contemplate who are the people with a direct stake in the research and its findings, and to directly include them in the research. For example, in the field of disability research, including people with lived experience of disability and mental health issues is essential for drawing upon their own perspectives and language in the adaptation of social innovation-related strategies and theories (Dew & Boydell, 2017).

Ultimately, the aim of social innovation research going forward should be to build new understandings that directly address the problems outlined by end-users and those involved in service delivery or policy, rather than being driving purely by 'gap-spotting' in the academic literature (Alvesson & Sandberg, 2013). This will ensure the research field can not only develop and refine theories that better respond to the variations in local situations and the nuances of social problems; it will also move social innovation scholarship into novel directions, influencing and shaping real-world possibilities within organizational domains and beyond (Starbuck, 2004). The findings of these research endeavors should be tested over long-range time periods and within multiple settings to enable researchers to better integrate the complex realities of social innovation, and the multiple and divergent steps along the pathways to social change.

CONCLUSION

This chapter summarizes the associated theory-building processes deliberately described in the authors' previous studies (i.e., Tayloret al., 2018a,b; Taylor et al., 2019a,b) to understand how Australian disability NPOs develop and implement social innovations and the societal 'system-level' impacts of such innovations, and utilizes the key findings of these studies to develop *'theories-in-practice'* in disability NPOs. Such theory-building processes involve two broad phases (Phases 1 and 2), and the culmination of these phases – which involve the close entwining of practical and theoretical realities – leads to the development of several 'theories-to-practice' that outline change-oriented solutions for managers who are working to address complex social challenges (i.e., via development of Nonprofit Social Innovation: NSI). From a complexity standpoint, it is important for managers to view their NPO-driven actions and processes, and the environments that encase them, as intrinsically linked to whole system dynamics. By viewing NSI through a complex-systems perspective, individuals and teams working 'on the ground' can come to understand how specific actions and relationships occurring at local scales are connected to broader structures, including the inequitable structures they are seeking to change. The study findings highlight the systemic potential of NPO-based social innovation; yet at the same time the findings also emphasize the role of person focused (micro) elements, along with risk-tolerance (a meso-level element), as important enabling conditions for bolstering NSI that can lead to high levels of societal benefits. From a 'whole system' perspective, these layers are inexorably linked. By understanding the intrinsic value of the human-centered interactions occurring at these micro/meso layers (i.e. in everyday situations involving both people with disabilities and non-disabled people), this study provides these disability NPO actors with a greater understanding of their own critical agency in social change processes. As Matheson (2016, p. 828) explains: "For interventions to be effective in the long-term, it is the qualitative features of real interactions – in particular the manifestation of power relationships – that matter". The authors' research findings may thus serve as a small yet meaningful contribution that revitalizes the role played by individuals and groups as they work at local scales towards social change.

ACKNOWLEDGMENT

This research has received financial support through an Australian Government Research Training Program Scholarship.

REFERENCES

Alvesson, M., & Sandberg, J. (2013). Has management studies lost its way? Ideas for more imaginative and innovative research. *Journal of Management Studies*, *50*(1), 128–152. doi:10.1111/j.1467-6486.2012.01070.x

Arena, M., & Uhl-Bien, M. (2016). Complexity leadership theory: Shifting from human capital to social capital. *People & Strategy*, *39*(2), 22–27.

Argyris, C., & Schön, D. A. (1974). *Theory in practice: Increasing professional effectiveness*. San Francisco, CA: Jossey-Bass.

Ayob, N., Teasdale, S., & Fagan, K. (2016). How social innovation 'came to be': Tracing the evolution of a contested concept. *Journal of Social Policy*, *45*(4), 635–653. doi:10.1017/S004727941600009X

Baker, K. (2017). *State of the disability sector report 2017*. Deakin West, ACT: National Disability Services.

Barraket, J., & Furneaux, C. W. (2012). Social innovation and social enterprise: Evidence from Australia. In H.-W. Franz, J. Hochgerner, & J. Howaldt (Eds.), *Challenge Social Innovation: Potentials for Business, Social Entrepreneurship, Welfare and Civil Society* (pp. 215–241). Berlin: Springer. doi:10.1007/978-3-642-32879-4_14

Battisti, S. (2014). Social innovation in living labs: The micro-level process model of publicprivate partnerships. *International Journal of Innovative Research and Development*, *5*(4-5), 328–348. doi:10.1504/IJIRD.2014.064146

Berzin, S. C., & Pitt-Catsouphes, M. (2014). A social work approach to social innovation. *International Journal of Innovation. Creativity and Change*, *1*(4), 1–14.

Brown, T. (2009). *Change by design: How design thinking transforms organisations and inspires innovation*. New York: HarperCollins.

Burnes, B. (2005). Complexity theories and organizational change. *International Journal of Management Reviews*, *7*(2), 73–90. doi:10.1111/j.1468-2370.2005.00107.x

Butkevičienė, E. (2009). Social innovations in rural communities: Methodological framework and empirical evidence. *Social Sciences*, *63*(1), 80–87.

Byrne, D. (2005). Complexity, configurations and cases. *Theory, Culture & Society*, *22*(5), 95–111. doi:10.1177/0263276405057194

Caulier-Grice, J., Davies, A., Patrick, R., & Norman, W. (2012). Defining Social Innovation (D. Research, Trans.). In E. C.-t. F. Programme (Ed.), The theoretical, empirical and policy foundations for building social innovation in Europe (TEPSIE) (pp. 1-43). Brussels: European Commission.

Chalmers, D. (2012). Social innovation: An exploration of the barriers faced by innovating organizations in the social economy. *Local Economy*, *28*(1), 17–34. doi:10.1177/0269094212463677

Chalmers, D. (2012). Social innovation: An exploration of the barriers faced by innovating organizations in the social economy. *Local Economy*, *28*(1), 17–34. doi:10.1177/0269094212463677

Chesbrough, H. (2006). Open innovation: A new paradigm for understanding industrial innovation. In H. Chesbrough, W. Vanhaverbeke & J. West (Eds.), Open Innovation: Researching a New Paradigm (pp. 1-14). Oxford, UK: Oxford University Press.

Chetkovich, C., & Frumkin, P. (2003). Balancing margin and mission: Nonprofit competition in charitable versus fee-based programs. *Administration & Society*, *35*(5), 564–596. doi:10.1177/0095399703256162

Cohen, W. M., & Levinthal, D. A. (1990). Absorptive capacity: A new perspective on learning and innovation. *Administrative Science Quarterly*, *35*(1), 128–152. doi:10.2307/2393553

Connellan, J. (2014). Big disruptive and here to stay: The impact of the National Disability Insurance Scheme on not for profit housing and homelessness agencies. *Parity*, *27*(5), 23–24.

Culler, J. (1983). *On deconstruction: Theory and criticism after structuralism*. London: Routledge.

Davies, A., Mulgan, G., Norman, W., Pulford, L., Patrick, R., & Simon, J. (2012). *Systemic Innovation*. Social Innovation Europe.

Dew, A., & Boydell, K. M. (2017). Knowledge translation: Bridging the disability researchto- practice gap. *Research and Practice in Intellectual and Developmental Disabilities*, *4*(2), 142–157.

DiMaggio, P., & Powell, W. (1983). The iron cage revisited: Institutional isomorphism and collective rationality in organizational fields. *American Sociological Review*, *48*(2), 147–160. doi:10.2307/2095101

Eppel, E., Matheson, A., & Walton, M. (2011). Applying complexity theory to New Zealand public policy: Principles for practice. *Police Quarterly*, *7*(1), 48–55.

FAHCSIA (Department of Families, Housing, Community Services and Indigenous Affairs). (2011). *2010-2020 National Disability Strategy*. Canberra: Commonwealth of Australia.

Fiss, P. (2007). A set-theoretic approach to organizational configurations. *Academy of Management Review*, *32*(4), 1180–1198. doi:10.5465/amr.2007.26586092

Fox, N. (2003). Practice-based evidence: Towards collaborative and transgressive research. *Sociology*, *37*(1), 81–102. doi:10.1177/0038038503037001388

George, G. (2014). From the editors – Rethinking management scholarship. *Academy of Management Journal*, *57*(1), 1–6. doi:10.5465/amj.2014.4001

Gertler, M. (2003). Tacit knowledge and the economic geography of context, or The undefinable tacitness of being (there). *Journal of Economic Geography*, *3*(1), 75–99. doi:10.1093/jeg/3.1.75

Goggin, G., & Wadiwel, D. (2014, Sept.). Australian disability reform and political participation. *Australian Review of Public Affairs*, 1–8.

Goldstein, J., Hazy, J. K., & Silberstang, J. (2010). A complexity science model of social innovation in social enterprise. *Journal of Social Entrepreneurship*, *1*(1), 101–125. doi:10.1080/19420671003629763

Green, J., & Mears, J. (2014). The implementation of the NDIS: Who wins, who loses? *Cosmopolitan Civil Societies Journal*, *6*(2), 25–39.

Gupta, A., Smith, K., & Shalley, C. (2006). The interplay between exploration and exploitation. *Academy of Management Journal*, *49*(4), 693–706. doi:10.5465/amj.2006.22083026

Haig, B. D. (2005). An abductive theory of scientific method. *Psychological Methods*, *10*(4), 371–388. doi:10.1037/1082-989X.10.4.371 PMID:16392993

Hallahan, L. (2013). In all its unfitness: The public's framing of the NDIS. *International Journal of Innovation. Creativity and Change*, *1*(1), 1–13.

Hammack, D. C. (1995). Accountability and nonprofit organizations: A historical perspective. *Nonprofit Management & Leadership*, *6*(2), 127–139. doi:10.1002/nml.4130060203

Hazy, J. K., & Uhl-Bien, M. (2015). Towards operationalizating complexity leadership: How generative, administrative and community-building leadership practices enact organizational outcomes. *Leadership*, *11*(1), 79–104. doi:10.1177/1742715013511483

Howaldt, J., & Schwarz, M. (2010). *Social innovation: Concepts, research fields and international trends. International Monitoring*. IMA/ ZLW.

Kabeer, N. (2005). *Inclusive Citizenship*. London: Sage.

Katz, D., & Kahn, R. L. (1978). *The Social Psychology of Organizations* (2nd ed.). New York: Wiley.

Kovács, G., & Spens, K. (2005). Abductive reasoning in logistics research. *International Journal of Physical Distribution & Logistics Management, 35*(2), 132–144. doi:10.1108/09600030510590318

Laursen, K., & Salter, A. (2006). Open for innovation: The role of openness in explaining innovation performance among UK manufacturing firms. *Strategic Management Journal, 27*(2), 131–150. doi:10.1002mj.507

Matheson, A. (2016). Reducing social inequalities in obesity: Complexity and power relationships. *Journal of Public Health, 38*(4), 826–829. PMID:28158692

Matheson, A., Dew, K., & Cumming, J. (2009). Complexity, evaluation and the effectiveness of community-based interventions to reduce health inequalities. *Health Promotion Journal of Australia, 20*(3), 221–226. doi:10.1071/HE09221 PMID:19951243

Morrison, K. (2005). Structuration theory, habitus and complexity theory: Elective affinities or old wine in new bottles? *British Journal of Sociology of Education, 26*(3), 311–326. doi:10.1080/01425690500128809

Moulaert, F. (2009). Social innovation: Institutionally embedded, territorially (re) produced. In D. MacCallum, F. Moulaert, J. Hillier, & S. V. Haddock (Eds.), *Social Innovation and Territorial Development*. London: Ashgate.

Moulaert, F., MacCallum, D., & Hillier, J. (2013). Social innovation: Intuition, precept, concept, theory and practice. In F. Moulaert, D. MacCallum, A. Mehmood, & A. Hamdouch (Eds.), *The International Handbook on Social Innovation* (pp. 13–24). Cheltenham, UK: Edward Elgar. doi:10.4337/9781849809993.00011

Moulaert, F., Martinelli, F., Swyngedouw, E., & Gonzalez, S. (2005). Towards alternative model(s) of local innovation. *Urban Studies (Edinburgh, Scotland), 42*(11), 1969–1990. doi:10.1080/00420980500279893

Mulgan, G. (2006). The process of social innovation. *Innovations, 1*(2), 145–162. doi:10.1162/itgg.2006.1.2.145

Neumeier, S. (2012). Why do social innovations in rural development matter and should they be considered more seriously in rural development research? - Proposal for a stronger focus on social innovations in rural development research. *Sociologia Ruralis*, *52*(1), 48–69. doi:10.1111/j.1467-9523.2011.00553.x

Nicholls, A., & Murdock, A. (2012). The nature of social innovation. In A. Nicholls & A. Murdock (Eds.), *Social innovation: Blurring boundaries to reconfigure markets* (pp. 1–32). Hampshire: Palgrave MacMillan. doi:10.1057/9780230367098_1

NRHA. (2013). Discussion paper: Issues relating to the NDIS in rural and remote areas of Australia. *Australian Journal of Rural Health*.

Patokorpi, E., & Ahvenainen, M. (2009). Developing an abduction-based method for futures research. *Futures*, *41*(3), 126–139. doi:10.1016/j.futures.2008.09.019

Peirce, C. S. (1955). *Philosophical writings of Peirce*. New York: Dover.

Peirce, C. S. (1997). *Pragmatism as a principle and method of right thinking: The 1903 Harvard lectures on pragmatism*. New York: State University of New York Press.

Pfeffer, J., & Salancik, G. (2003). *The External Control of Organizations: A Resource Dependence Perspective*. Stanford, CA: Stanford University Press.

Phillips, W., Lee, H., Ghobadian, A., O'Regan, N., & James, P. (2014). Social innovation and social entrepreneurship: A systematic review. *Group & Organization Management*, *40*(3), 428–461. doi:10.1177/1059601114560063

Phills, J., Deiglmeier, K., & Miller, D. (2008). Rediscovering social innovation. *Stanford Social Innovation Review*, *6*(4), 34–43.

Productivity Commission. (2010). *Contribution of the Not-for-Profit Sector* (Vol. Research Report). Canberra: Commonwealth of Australia.

Productivity Commission. (2011). *Productivity Commission Inquiry Report: Disability Care and Support* (Vol. 1). Canberra: Commonwealth of Australia.

Ragin, C. C. (1987). *The Comparative Method*. Berkeley, CA: University of California Press.

Ragin, C. C. (2000). *Fuzzy-set Social Science*. Chicago: University of Chicago Press.

Ragin, C. C. (2008). *Redesigning Social Inquiry: Fuzzy Sets and Beyond*. Chicago: University of Chicago Press. doi:10.7208/chicago/9780226702797.001.0001

Rhodes M., & Donnelly-Cox G. (2008, Sept. 30). Social entrepreneurship as a performance landscape: The case of 'Front Line'. *Emergence: Complexity and Organization*, 1–17.

Ryan, R. M., & Deci, E. L. (2000). Self-determination theory and the facilitation of intrinsic motivation, social development and well-being. *The American Psychologist*, *55*(1), 68–78. doi:10.1037/0003-066X.55.1.68 PMID:11392867

Sanzo, M. J., Álvarez, L. I., Rey, M., & García, N. (2015). Business-nonprofit partnerships: A new form of collaboration in a corporate responsibility and social innovation context. *Service Business*, *9*(4), 611–636. doi:10.100711628-014-0242-1

Schwartz, B. (2012). *Rippling: How Social Entrepreneurs Spread Innovation throughout the World*. San Francisco: Jossey-Bass.

Skelton, C. (2016, Sept.). NDIS: Is your approach to governance ready for the new environment? *Governance Directions*, 458–460.

Smith-Merry, J., Hancock, N., Bresnan, A., Yen, I., Gilroy, J., & Llewellyn, G. (2018). *Mind the Gap: The National Disability Insurance Scheme and psychosocial disability*. Final Report: Stakeholder identified gaps and solutions. University of Sydney: Lidcombe.

Sousa, M., & van Dierendonck, D. (2017). Servant Leadership and the Effect of the Interaction Between Humility, Action, and Hierarchical Power on Follower Engagement. *Journal of Business Ethics*, *141*(1), 13–25. doi:10.100710551-015-2725-y

Stacey, R. D. (2003). *Strategic Management and Organisational Dynamics: The Challenge of Complexity*. Harlow: FT/Prentice Hall.

Stainton, T. (2000). What is self-determination? *Proceedings of the First International Conference on Self Determination and Individualized Funding*.

Stainton, T. (2002). Taking rights structurally: Disability, rights and social worker responses to direct payments. *British Journal of Social Work*, *32*(6), 751–763. doi:10.1093/bjsw/32.6.751

Starbuck, W. H. (2004). Why I stopped trying to understand the real world. *Organization Studies*, *25*(7), 1233–1254. doi:10.1177/0170840604046361

Svennevig, J. (2001). Abduction as a methodological approach to the study of spoken interaction. *Norskrift*, *103*, 1–22.

Tapsell, P., & Woods, C. (2010). Social entrepreneurship and innovation: Self-organization in an indigenous context. *Entrepreneurship and Regional Development, 22*(6), 535–556. doi:10.1080/08985626.2010.488403

Taylor, R., Torugsa, N., & Arundel, A. (2018a). Leaping into real-world relevance: An 'abduction' process for nonprofit research. *Nonprofit and Voluntary Sector Quarterly, 47*(1), 206–227. doi:10.1177/0899764017718635

Taylor, R., Torugsa, N., & Arundel, A. (2018b). Thriving within the chaos: A complexity theorizing approach to social innovation by nonprofit organizations. In C. Dogru (Ed.), *Handbook of Research on Contemporary Approaches in Management and Organizational Strategy* (pp. 36–61). IGI Global.

Taylor, R., Torugsa, N., & Arundel, A. (2019a). Organizational pathways for social innovation and societal impacts in disability nonprofits. *Voluntas*. doi:10.100711266-019-00113-8

Taylor, R., Torugsa, N., & Arundel, A. (2019b). Social innovation in disability nonprofits: An abductive study of capabilities for social change. *Nonprofit and Voluntary Sector Quarterly*. doi:10.1177/0899764019873965

Toepler, S. (2003). Grassroots associations versus larger nonprofits: New evidence from a community case study in arts and culture. *Nonprofit and Voluntary Sector Quarterly, 32*(2), 236–251. doi:10.1177/0899764003032002004

Uhl-Bien, M., Marion, R., & McKelvey, B. (2007). Complexity leadership theory: Shifting leadership from the industrial age to the knowledge era. *The Leadership Quarterly, 18*(4), 298–318. doi:10.1016/j.leaqua.2007.04.002

Van de Ven, A. H. (2015). Welcome to the Academy of Management Discoveries (AMD). *Academy of Management Discoveries, 1*(1), 1–4. doi:10.5465/amd.2014.0143

Vanstone, C., & Graham, B. (2016). NDIS readiness and beyond. Adelaide: TACSI (The Australian Centre for Social Innovation).

von Hippel, E. (1986). Lead users: A source of novel product concepts. *Management Science, 32*(7), 791–805. doi:10.1287/mnsc.32.7.791

von Hippel, E. (2005). *Democratizing Innovation*. Cambridge, MA: MIT Press. doi:10.7551/mitpress/2333.001.0001

Weerawardena, J., & Mort, G. S. (2012). Competitive strategy in socially entrepreneurial nonprofit organizations: Innovation and differentiation. *Journal of Public Policy & Marketing, 31*(1), 91–101. doi:10.1509/jppm.11.034

Weick, K. E. (1989). Theory Construction as Disciplined Imagination. *Academy of Management Review*, *14*(4), 516–531. doi:10.5465/amr.1989.4308376

Westley, F., Antadze, N., Riddell, D. J., Robinson, K., & Geobey, S. (2014). Five configurations for scaling up social innovation: Case examples of nonprofit organizations from Canada. *The Journal of Applied Behavioral Science*, *50*(3), 234–260. doi:10.1177/0021886314532945

Woodside, A. G. (2014). Embrace perform model: Complexity theory, contrarian case analysis, and multiple realities. *Journal of Business Research*, *67*(12), 2495–2503. doi:10.1016/j.jbusres.2014.07.006

Zivkovic, S. (2015). A complexity based diagnostic tool for tackling wicked problems. *Emergence*, *17*(4), 1–9.

ADDITIONAL READING

Ragin, C. C. (2008). *Redesigning Social Inquiry: Fuzzy Sets and Beyond*. Chicago: University of Chicago Press. doi:10.7208/chicago/9780226702797.001.0001

Taylor, R., Torugsa, N., & Arundel, A. (2018a). Leaping into real-world relevance: An 'abduction' process for nonprofit research. *Nonprofit and Voluntary Sector Quarterly*, *47*(1), 206–227. doi:10.1177/0899764017718635

Taylor, R., Torugsa, N., & Arundel, A. (2018b). Thriving within the chaos: A complexity theorizing approach to social innovation by nonprofit organizations. In C. Dogru (Ed.), *Handbook of Research on Contemporary Approaches in Management and Organizational Strategy* (pp. 36–61). IGI Global.

Taylor, R., Torugsa, N., & Arundel, A. (2019a). Organizational pathways for social innovation and societal impacts in disability nonprofits. *Voluntas*. doi:10.100711266-019-00113-8

Taylor, R., Torugsa, N., & Arundel, A. (2019b). Social innovation in disability nonprofits: An abductive study of capabilities for social change. *Nonprofit and Voluntary Sector Quarterly*. doi:10.1177/0899764019873965

Westley, F., Antadze, N., Riddell, D. J., Robinson, K., & Geobey, S. (2014). Five configurations for scaling up social innovation: Case examples of nonprofit organizations from Canada. *The Journal of Applied Behavioral Science*, *50*(3), 234–260. doi:10.1177/0021886314532945

Woodside, A. G. (2014). Embrace perform model: Complexity theory, contrarian case analysis, and multiple realities. *Journal of Business Research*, *67*(12), 2495–2503. doi:10.1016/j.jbusres.2014.07.006

KEY TERMS AND DEFINITIONS

Abduction: An initial creative stage in scientific inquiry, where researchers begin with closely observed "surprising facts" driven by a vital yet poorly understood phenomenon, comparing these facts to the existing body of theoretical knowledge, and proposing new hypotheses that are empirically testable and practically explainable.

Australian Disability Sector: A national sector of nonprofit organizations that are predominately funded by government and/or philanthropic grants, and that provide services to people with disability or mental health issues. Services provided within this sector include accommodation support, respite, specialized skill development, and community access services.

Complexity Theory: An overarching theoretical lens to integrate multiple theories and emergent constructs, and which is concerned with the emergence of order in nonlinear systems.

Disability Nonprofit Organization (NPO): A non-profit organization that works with people who live with disability or mental health issues.

National Disability Insurance Scheme (NDIS): A major social policy reform based on a piece of legislation introduced by the Australian government, titled National Disability Insurance Scheme Act 2013. The Act sets out a funding program based on individualized support packages for people with disabilities, and commenced being rolled out nationally on July 1, 2016.

Nonprofit Processes: Combinations of practices or organizational structures in social action, social relations or human interactions within a nonprofit organization context.

Nonprofit Services: Programs, initiatives, or service products provided by a nonprofit organization.

Nonprofit Social Innovation: A new service or process enacted specifically by a disability nonprofit which promotes the broader community's inclusion of people with disabilities, and thereby has the potential to achieve systems-level (not just organization-level) impacts.

Organizational Capability: An organization's capacity to bring its resources and skills together and deploy them advantageously, thereby forming the identity of the organization by defining what strategies they are good at doing.

Qualitative Comparative Analysis (QCA): An analytical method grounded in configurational thinking and complexity theory to capture the heterogeneity, context-specificity, and casual complexity. QCA bridges 'qualitative analysis' (analyzing data by cases) and 'quantitative analysis' (producing the minimum possible number of attribute combinations). Its emphasis (i.e., finding different combinations of causal conditions capable of generating the same outcome) contrasts with the 'net effects' thinking that dominates conventional quantitative social science.

Social Innovation: A service and/or process that contributes to generating a novel solution to a social problem that is more effective, efficient, sustainable or just than existing solutions, and for which the value created accrues primarily to society as a whole rather than private individuals.

Theories-in-Practice: The theories that are instrumental in connecting localized and systemic realities. These theories provide practitioners with knowledge into the specific ways that complex problems can be addressed according to localized conditions and constraints.

Chapter 10
Explaining the Interaction Between Leader Ambidextrous Behavior, Employee Ambidexterity, and Organizational Ambidexterity

Melis Attar
https://orcid.org/0000-0002-1299-7552
Selçuk University, Turkey

Serap Kalfaoğlu
https://orcid.org/0000-0002-7248-8370
Selçuk University, Turkey

ABSTRACT

The aim of the study is to examine the relationships between ambidextrous leadership behaviors (i.e., task-oriented, relation-oriented, change-oriented, and external) in the context of leader behavior level and employee ambidextrous performance (i.e., exploitation and exploration) in the context of individual employee level by suggesting that the interaction of these two may result in organizational ambidexterity. This study aims to contribute to the immature ambidexterity literature by developing some propositions based on the previous ambidexterity researches related to ambidexterity of leaders, employees, and entire organizations. Taking into account ambidextrous leadership theory, it is proposed that while the leaders' task-oriented and relation-oriented behaviors are positively related to the employees' exploitation behaviors, the leaders' change-oriented and external behaviors are positively related to the employees' exploration behaviors. It is recommended for further studies to investigate the possible propositions with quantitative research designs.

DOI: 10.4018/978-1-7998-1108-4.ch010

INTRODUCTION

Leadership is one of the most important prophets that will enable organizations to identify and implement strategies that make sense for the future and to benefit from existing competencies and to differentiate organizations in this context. Leadership has been recognized as one of the most effective predictors of employee innovation, organizational development, and learning (Alghamdi, 2018). From this perspective, there are changes in various individual and organizational outputs with the leaders who constitute a key factor in the qualification, differentiation and final success of the business and their styles and behaviors according to their characteristics. Leadership behaviors that help organizations to make themselves ambidextrous can also be categorized mainly as task-oriented, relationship-oriented, change-oriented, and external behaviors (Yukl, 2012).

Literature review depicts that there has been an intense interest in organizational ambidexterity theory and respective researches in this particular area especially from 2000's onwards. At first sight, the main assertion thereby is the necessity to operate by interacting organizations' capacities simultaneously on exploiting existing competencies and developing exploratory research on new competencies and opportunities for long-term life and success (Alghamdi, 2018:1). In a constantly changing and dynamic environment, organizations struggle to fit in the interests of all stakeholder groups they are in for survival. Change adaptation strategies are vital in terms of transforming opportunities by observing the fluctuations in the market, the environmental threats and other stakeholders in the competition with content and methods. Organizational theory (relying on the proven facts of this fieldwork) claims that successful organizations in dynamic environments are already harmonious and effective while being adaptable to future changes; thus ambidextrous (Cao et al., 2009). In this direction, overcoming all structural, contextual, or behavioral challenges will be possible within the organizations which can both improve their current positions and adapt to change (Duncan, 1976), along with ambidextrous leaders who represent these organizations (Rosing et. al., 2011) and with ambidextrous human resources whose strategies are effective in the functional context (Alghamdi, 2018).

In particular, the conceptual examination of organizational ambidexterity and the curiosity of accessing ambidexterity capacity are still a research topic of growing interest (Yu et al., 2017: 2). When theoretical foundations of the concept are examined, it can be observed that the factors leading to the dimension of ambidexterity of an organization during the studies carried out on a day-to-day basis are mainly delineated under *three* headings. The *structural*-based factor is that the establishment of autonomous and structurally separated sub-units in relation to different people, structures and processes will benefit organizational ambidexterity (synchronicity) (Raisch and Birkinshaw, 2008; O'Reilly and Tushman, 2004). The *contextual/content-*

based factor advocates the view that ambidexterity can be achieved by establishing a set of processes or systems that encourage individuals to make their own decisions about how they divide their time between conformable and adaptable contradictory requests (Gibson and Birkinshaw, 2004: 210). It must be noted that both of these dimensions take into account organizational factors (e.g., organizational structure, culture) to facilitate synchronicity (Agostini et al., 2016). Finally, the third factor of *leadership*-oriented, in other words, the behavioral ambidexterity perspective claims that managers/entrepreneurs/leaders play an important role in the development of organizational ambidexterity, and the characteristics of managers/entrepreneurs/leaders have important influences on organizational ambidexterity (Yu et al., 2017: 3). In this context, ambidexterity's thoroughly impact on organization is of great importance in each of organizational, team and individual level. It has been also emphasized that ambidexterity requires employees to internalize and to implement strategies in order to exploit and explore in equal amounts and at the same time. At the individual level, ambidexterity is not only a possible level that can compensate for both exploration and exploitation but it is also necessary to combine synergy and create interaction between discovery and exploitation activities at a higher organizational level (Alghamdi, 2018: 2).

In the context of innovation, ambidextrous leadership theory focuses on leaders handling ambidexterity of leadership behaviors such as explicit and implicit leadership and claims to be complementary to meeting innovation needs as they encourage exploration and exploitation within the individual or a group (Zacher and Rosing 2015; Rosing et al., 2011). On this account, this study attempts to explain basic characteristics of leadership behaviors based on Yukl (2012)'s hierarchical classification, and to suggest proposals on leadership behaviors that are effective in ambidextrous leadership, accordingly. Up to present, to the best of researchers' knowledge, there has been no conceptual study that examines the hierarchical classification and the interaction of each leadership behavior with ambidextrous leadership. In furtherance, the characteristics of ambidextrous employee behaviors and their rankings of organizational impacts, as well as those leaders who lead organizations to be ambidextrous, will be addressed. The proposed ending is that ambidextrous leadership behaviors shall foster the employee ambidexterity and in turn may well increase the organizational ambidexterity.

BACKGROUND

Ambidexterity and Organizational Ambidexterity

Since the work of Tushman and O'Reilly (1996), which sets out the foundations of the concept, management and organization literature has begun to emphasize on organizational ambidexterity (synchronicity) making possible that the opposite poles, conflicts and tensions can be managed simultaneously. Two poles, one in competition (Attar, 2015) are composed of elements that create a multifaceted confusion such as adaptability rather than adoptability, the clarity of the past against the uncertainty of the future, the exploitation of existing information against the discovery of information, adopting radical changes instead of incremental innovations and so on (Agostini et al., 2016: 130). This definition requires a good relationship with two contradictory elements: efficiency and flexibility (Alghamdi, 2018, quoted from Adler et al., 1999), evolutionary and revolutionary change (Tushman and O'Reilly III, 1996) low-cost strategy and differentiation (Porter, 1996), incremental and radical innovation, and aligning available resources while adapting to a changing environment at the same time (Gibson and Birkinshaw, 2004). These conflicts, which will come from the upper hand through the access to ambidexterity power, have contributed to the widespread use and accessibility of the concept. The fact that an ambidextrous organization has the ability to explore new opportunities, as well as the ability to utilize both existing competencies and to improve the performance and competitive level of the organization is a common opinion accepted by many researchers in this field (Cao et al., 2009: 781). Therefore, it is emphasized that the ambidexterity phenomenon can be achieved by providing strategic ideas, ways and methods that help businesses to compete, while at the same time providing the synergy and harmony in the structural and contextual domain of the enterprises.

Along with Duncan (1976), the first scholar who helped to bring the concept of organizational ambidexterity to the stage of history, and the first person to present the concepts of exploring new possibilities and exploiting existing possibilities, especially in organizational learning called as the two important outputs (strategy/dimension) of a single process (organizational ambidexterity) is March (1991). In accessing to information that is among the basic inputs of organizations, redesigning this information according to the current situation and conditions, using it effectively and efficiently is called as the experiential learning process. This process necessitates the fact that organizations conduct research while making new activities and therewithal make use of existing information and in turn shows that the two basic strategies of organizational ambidexterity are applied at the same time. As a matter of fact, some studies (eg, Bakan et al., 2017) have proved that organizational learning, which includes sub-dimensions as access to information,

information separation, shared interpretation and organizational memory, has a meaningful relationship with organizational ambidexterity covering the dimensions of explorative and exploitative strategies.

Duncan (1976) argued that businesses are ambidextrous when it comes to implementing existing opportunities, seeking for new opportunities, being predictable and moving those opportunities into new activities. He also used the concept in terms of duality of organizations for innovation support. In technological innovation, which focuses on transforming new technological knowledge and ideas into new products or processes, and according to He and Wong (2004), the strategic implementation process has been expressed in two dimensions: (1) investigative technological innovation aimed at entering the domain of a new product market; and (2) utilitarian innovation aimed at improving existing product market conditions. According to researchers, while pursuing these two processes, businesses become ambidextrous of innovation strategy in two ways. First, when the enterprise scores high in both explorative and exploitative innovation strategies, the score reflected in the product as a result of the two applications will be a good criterion of ambidexterity. Second, the degree of ambidexterity can also be determined by looking at the difference between the two points. For example, if the firm has a relatively equal emphasis on both dimensions, it is considered as ambidextrous.

As He and Wong (2004) and Minister and Sezer (2017) have studied, both explorative and exploitative innovation has a major impact on firm performance. The recognition of the company improves with the increase in the influence of organizational ambidexterity. This also affects employees in terms of "*the importance of being perceived as the best in the work performed, the encouragement of the employees by the senior management to be quite confident in the task given before the start of the work*", and "*the recognition of the company's success by others and evaluations of the firm's overall performance compared to other competitors in the sector*" cause the performance target orientation of the company to increase (Bakan and Sezer, 2017: 341).

In addition, it should be stated that the changes experienced in organizations and the necessity of change in the organizational structure brought with it, serve organizations to be ambidextrous. As stated in Duncan's (1976) study, enterprises can use both mechanical and organic organizational structures in combination with these strategies, both in implementing beneficiary strategies in order to protect the current situation and initiating a different innovation (Çiftçi, 2017: 190).

According to March (1991: 71), researcher/explorer/exploration strategies include concepts such as variations (changes), risk taking, experience, flexibility, exploration and innovation and deal with the process of exploring new alternatives, or exploring opportunities. Beneficiary/Exploitation strategies include concepts such as development, production, effectiveness, selection, implementation and management are also basically related to the development of existing resources or competencies. Apart from exploitation, it is possible that the harmonization systems that deal with the exploration strategies can continue to live for short periods by ignoring many benefits with the expense of trial costs and by having many underdeveloped new ideas as well as without distinctive competence. It is clear that systems that take advantage of the exploitation strategy are in a non-optimal, stationary balance trap. The concept of ambidexterity is therefore related not only to the equilibrium of exploration and exploitation, but also to the equality of incremental and radical innovations, continuity and change, or the balancing of organic and mechanical organizational structures (Rosing et al., 2011).

It is obvious that protecting and expanding available resources can enhance a company's productivity and short-term performance (March, 1991). However, focusing solely on such exploitation processes can create proficiency traps and prevent organizational changes from being implemented (Zhang et al., 2016: 822 quoted from Levitt and March, 1988). However, in any case, the balance between exploration and exploitation strategies is inevitable given that each system has difficulties in meeting organizational demands due to the certain or limited resources. Therefore, determining an appropriate balance between exploration and exploitation strategies is crucial for the system to be sustainable and prosperous. Furthermore, although the so-called processes need different qualifications, as a result of being contradictory together with aspects of utilization of limited resources, organizational ambidexterity is suggested as one of the sources to help find the appropriate balance between the two strategies (Cao et al., 2009: 781).

Essentially, organizational ambidexterity refers to the level of skill in the process of both exploiting existing competencies and exploring new opportunities simultaneously in a system. In other words, organizational ambidexterity is the ability to be compatible and efficient in management to meet changing demands while adapting to a changing environment, as well (Zhang et al., 2017: 2). In this context, it is also described in terms of dual skills (Çiftçi, 2017: 189). It is suggested that organizational capacity is necessary in order to meet business demands and to adapt to changes in the environment at the same time via organizational ambidexterity (Gibson and Birkinshaw, 2004; Raisch and Birkinshaw, 2008). Organizational ambidexterity is also a strategic decision-making approach that entrepreneurs/leaders use to cope with the desire to meet all the anticipations surrounded by risk and uncertainty (Dai et al., 2017).

Apart from the authors who deal with two important outputs of organizational ambidexterity as exploitation and exploration strategies (March, 1991; Cao et al., 2009; Akdogan et al., 2009), several researchers who deal with adaptation and adoption strategy dimensions (Birkinshaw and Gibson, 2004; Bakan and Sezer, 2017). Adoption strategy is to determine how the activities of all sub-units of the enterprises will provide coordination and adaptation with each other in terms of purpose and functioning, and how to create the desired value (Birkinshaw and Gibson, 2004: 47; Bakan and Sezer, 2017: 329). Adaptation strategy is to be proactive towards possible opportunities, to be flexible against developments, to enter into expectations and quests for better all the time rather than to be satisfied with what is possible (Bakan and Sezer, 2017: 329). Hence, any organization that has the capacity to use contradictory exploration and exploitation strategies and manages adaptation and adoption strategies at the same time is ambidextrous (Yu et al., 2017: 2; Bakan and Sezer, 2017: 326).

Along with the above-mentioned dimensions, there are also researchers who treat duality as the balance and mixed dimension of organizational ambidexterity (Cao et al., 2009). Besides, as the balance dimension of ambidexterity, the orientation of firms to maintain close ties between exploration and exploitation activities is recognized. In the mixed dimension of mastery, the magnitudes of the respective activities are associated.

In an attempt to investigate other variables related to the concept, for example, Agostini et al. (2016) has come to the conclusion that organizational ambidexterity with its structural and contextual components has helped the innovation reach a higher level. Structural ambidexterity is achieved through structural interventions and is based on a discovery idea. It is generally associated with structures that utilize different teams or different organizational units for exploration and exploitation activities (O'Reilly and Tushman, 2004). Contextual ambidexterity requires exploitation of existing skills and exploration of a future opportunity (Alghamdi, 2018). Examining the distinction between exploitation and exploration of organizational skill in technology-oriented and market-oriented innovations, Zhang et al. (2016) concluded that investigative technology innovation, together with beneficial market innovation, had a beneficial and interactive impact on firm performance. On the other hand, Taylor and Helfat (2009: 718) have examined the wholesale transformation of a company from a business based on an existing core technology to a company based on a new technology in the context of organizational links (economic, structural, social and cognitive) and they have come to the conclusion that these links have helped them to be ambidextrous regarding their success in the transition period they experienced when adapting their existing capabilities to a new technology. In this context, managerial ambidexterity is also deemed necessary to develop organizational ties including new and old complementary assets. As a matter of fact, the literature on ambidexterity emphasizes that in a dynamic environment, organizations must trade-off between having existing jobs well organized and having new business search and exploration processes (Duncan, 1976; Tushman and O'Reilly, 1996).

Ambidextrous Leadership

The ambidexterity that emerges in the organizational context is assessed along with leadership and culture studies as to investigate how organizational problems and deadlocks are to be managed or how conflicts or contradictions can be resolved. When questions become more complicated, it is obvious that a single leadership model will be insufficient to effectively manage the organization in order to be able to flexibly adapt to the ever-changing environment (Zheng et al., 2017: 2). Ambidexterity is considered as a struggle for leadership before anything else, and to date it has not been possible to determine the best in leadership styles to sustain organizational ambidexterity (Bledow et al., 2009).

When examining the reasons that strengthen organizational conflicts and affect the performance of the organization, the current literature focuses on ambidextrous leadership with organizational capacity and ability to transform and transact (Zheng et al., 2017; Zacher and Rosing, 2015; Jansen et al., 2009). Ambidextrous leadership theory has expressed the interaction between two complementary leadership behaviors, such as explicit and implicit leadership behaviors (Alghamdi, 2018), steering and empowering leadership, and transformational and transactional leadership (Bass, 1990, Rosing et al., 2011). These complementary forces simultaneously explore new knowledge as well as values, and use existing knowledge as well as values for organizational reform (Tuan, 2017). Among the studies done in this field, Jansen et al. (2009) argue that exploratory innovation is positively associated with transformational leadership and negatively associated with transactional leadership. Rosing et al. (2011) emphasized that current researches are insufficient to explain the complex nature of the innovation processes, and that the basic requirements of innovation have the flexibility to switch between exploration and exploitation activities. They defined ambidextrous leadership as ambidexterity in innovation and tried to explain the effects of explicit leadership behavior and implicit leadership behaviors on individuals and groups. They pointed out that ambidextrous leadership is a flexible type of leadership that uses explicit and implicit leadership behaviors, meeting the ever-changing needs of the innovation process, and transitioning between the two behaviors that the situation requires. On the one hand, explicit (open) leadership behavior is a behavior that helps them to raise awareness such as encouraging subordinates to achieve different experiences, giving subordinates the opportunity to conduct their activities independently thinking and taking initiative and considering the critical approaches of your subordinates. Hence, open leadership behavior supports subordinate's exploration activities. On the other hand, implicit (close) leadership behavior is intended to reduce changes in the behavior of subordinates through corrective actions, constant rules and continuous monitoring of goals. In this context, implicit leadership behaviors are effective in achieving the exploitation activities of subordinates (Rosing et al., 2011; Zacher and Rosing, 2015; Alghamdi, 2018). It appears that the effects of different leadership styles on the ambidexterity of organizations also differ.

Being among the researchers who believe that multiple leadership behaviors will bring success to organizational effectiveness, Baskarada and colleagues (2017) suggested that increasing the effectiveness of the leaders needed a balance between the transformational and transactional leadership. The key feature of ambidexterity leadership can be put as implementing by matching multiple complementary leadership behaviors (e.g., open-close, task-relation-change-external orientation, etc.) in order to adapt to environmental dynamism (Zheng et al., 2017: 3). According to Zheng et al. (2017: 5), the most important characteristics of ambidexterity leadership are flexibility and fluidity by which leaders can switch between two leadership behaviors (transformational and transactional leadership) as the organizational task requires.

Transformational leaders, on the other hand, in combination with ambidextrous leaders, can focus on creating breakthrough innovations and experiences with an open vision; they can aim to achieve specific and clear (understandable) goals with an implicit vision, as well (Zacher and Rosing, 2015). They are more proactive and help achieve extraordinary goals by raising awareness (Antonakis, 2003). It is a leadership style that enables interactive communication between the leader and the members to help increase employee motivation. They are change-oriented, because they influence their members by glorifying their aims and providing them with confidence in working beyond the explicit and implicit expectations of the business contract. Transformational leadership features consist of four basic elements, which are also known as 4 I's: idealized influence (charisma), inspiring motivation, individualized attention (individual support), and intellectual stimulation (Bass et al., 1996). The *idealized influence* (charisma) involves creating a vision that interacts with the leaders and their members and forming a mission for that vision, and is examined in two sub-dimensions, as the behavior dimension and the attributed dimension. In terms of behavior, charisma is about giving information about the reasons for the aim and goals that the leader wants to be determined, expressing the effects and results of the decisions to be taken. The attributed influence is that the leader creates the impression that he is a strong, self-confident, holds organizational interests superior to personal interests. *Inspirational motivation* is to show a hopeful attitude about the future by creating a leader team awareness and a motivation towards the shared vision that he wants to achieve using various symbols or expressions. The *individualized interest* refers to the prioritizing needs of members and effort for this purpose. *Intellectual stimulation* is to help the leader to develop solutions for problem through different perspectives and to question traditional thought patterns of his subordinates. With this incentive, the leader helps to advance the organization in a clear, intellectual, creative and changing way (Karip, 1998).

Another factor involved in the combination of ambidextrous leadership is the transactional leadership. It includes an interaction (processor) leadership in other words, social exchange. There is a trade-off between the leader and the members. In view of the mutual interest-based approach, members will have active roles to obey the leader, to show self-sacrifice in reaching the goals, and to obey the procedures, in order to meet the basic physiological, security and social needs of the members of the leader. In addition, rewards and penalties are efficient in this trade-off. While the leader provides opportunities such as salary increase and promotion to those with high performance, they may penalize those with low performance (Bass, 1990). Transactional leadership features can be examined under three dimensions (Bass and Avolio, 1993): *1. Conditional Reward*: It reveals the prize that the leader will provide after his goals are determined. Behaviors such as determining who is responsible for achieving the desired performance goal and the nature of the performance award are accounted for example behaviors of transactional leaders. *2. Management by Exceptions*: The leader must intervene only when there are major problems that need to be resolved immediately. Intervention hereby is the criticism of errors or problems and the provision of negative feedback. The primary goal of the leader is to identify and resolve areas where problems have arisen, and where routinely worsening the performance as an exception. It is never aimed at "better or best", but fulfilling the task as it should be routinely done is enough for the leader. In active management by exceptions, the rules that the leader puts forward are aimed at preventing possible errors, while passive management by exceptions do not deal with the problems until the leading problems become inaccessible, and there are no interventional actions unless there occur mistakes or deviations. *3. Laissez-faire Leadership*: The leader must release his members and not go to any barter with his members. It is the most passive and least effective type of leadership. Delays in decisions, not being awarded, not giving importance to the motivation of members, are present in managers who lack leadership qualities (Karip, 1998: 7). Although the last feature does not actually reflect the transactional leadership, it appears to be included in some studies (Karip, 1998; Chebbi et al., 2017).

Although the style of transactional leadership does not help the employee develop his creative feature (Jung, 2001), and thus, is criticized that it does not mediate to the ambidextrous employee's application of exploration strategies, it is considered to represent an upward trend in employee performance as they focus on the exchange of transactions between leaders and their members through the combination of the transformational leadership (Luttikhuis, 2014).

Another feature of ambidextrous leaders is that they appreciate how to encourage the exploitation and exploration between subordinates when necessary (Tung, 2016). As they pursue various opportunities, purposes and goals, they may anticipate complex cognitive processes such as integrative or contradictory thinking to prevent possible conflicts. They may face conflict in connection with time horizon, risk situation, current strategy and leader's responsibilities (Alghamdi, 2018: 3). As noted by Rosing et al. (2011: 957), an ambidextrous leader is a leader who supports members (followers/subordinates) to become masters. The ambidextrous leaders adopt the open and close behavior and the transition process between them and also trying new solutions and strengthening current competencies respectively. They can also support dynamic abilities, including exploration and exploitation forces (Tuan, 2017).

In terms of bi-directionality, behavioral as well as cognitive complexity can be significant prerequisites for leaders, as Rosing et al. (2011) put it. Despite the fact that explicit and implicit leadership behaviors are not mutually exclusive, they are still very divergent behaviors. To be an ambidextrous leader, he/she must have the ability to exhibit these different behaviors. Therefore, a broad repertoire of behavior is fundamental. This repertoire articulates the degree of change between different behaviors according to situational conditions, as well as the behavioral varieties which the leader can express. For this reason, ambidextrous leadership requires both the ability to have repertoire of different leadership behaviors and the ability to switch between them (differentiation). Among the characteristics of ambidextrous leadership, having integrative thinking, emotional intelligence and prediction should also be considered. The "integrative thought" attributes to the ability to confront counter-ideas, the "emotional intelligence" attributes to the leader's understanding of his or her employees and their emotions and thus to their emotional support, and finally the awakened awareness of the environment, "forecasting and planning" functions are among the precursors of ambidextrous leadership (2011: 968-969).

Tushman et al. (2011), over a study including 112 senior management teams, have shown that employees adopt organizational effectiveness when they adopt the tension between improving the current order and exploratory research, and centralizing the idea to create a constant creativity environment. Three leadership pillars have been mentioned to help develop new facilities that will stimulate their industries and help them grow their core business in this regard: 1) Engaging the senior management team with the desire to become strategic in the future 2) Maintaining the tension between the demands of the innovation units and the basic business demands at the top of the organization 3) Embracing incoherence/inconsistency by making multiple and often conflicting strategic agendas constant. When leaders adopt this approach, they empower senior teams to engage in clear, *ongoing* and *forward-looking* discussions about the tensions that are behind the scenes in negotiating feudal interests (Tushman et al., 2011) since to balance the basic needs of enterprises and innovation efforts is a paramount leadership task.

With regard to cultural aspect of ambidextrous leadership, for example, Chebbi et al. (2017) develop five propositions including charisma, inspirational motivation, intellectual stimulation, individualized interest, and ambidextrous leadership in

India. India's seven cultural variables that characterize ambidextrous leadership (a model of nurturing spiritual interest, personal touch, expertise, simple life and high thought, loyalty, self-sacrifice and motivation) reinforce the recent successful internationalization of Indian companies.

EMPLOYEE AMBIDEXTERITY

Despite the fact that it is initially developed as an organizational skill (Gibson and Birkinshaw, 2004; March, 1991), studies of ambidexterity at the group and individual levels have also been deemed necessary to explain how relationships ambidexterity with different levels contribute to the overall performance of a firm (Schindler, 2015). Thus, this study also focuses on individual ambidexterity from a leadership perspective.

It must be noted that it is necessary for employees to be explorative and exploitative in equal amounts and concurrently in order to achieve ambidexterity. At the individual level, ambidexterity is not only a possible level through which an organization can balance both exploration and exploitation, but it is also essential to combine and achieve synergy between exploration and exploitation activities at a higher organizational level (Alghamdi, 2018: 2).

In terms of short term organizational efficacy and effectiveness, exploitation activities in individual level involve the use of existing knowledge and skills. Exploration activities are to conduct research for a new product and process innovation as well as for competitive solutions and behaviors. This requires employees to acquire new skills and adapt existing business. Both activities are related to learning. Employee ambidexterity or individual ambidexterity is a multi-dimensional structure involving all these activities. It is a behavioral competence (Kauppila and Tempelaar 2016, Alghamdi, 2018: 3-4) which involves managing to eliminate conflicts, contradictory situations and tensions rather than a psychological feature. Employee ambidexterity is to help employees to behave autonomously in managing time and distributing their energy while performing the two strategies in question.

The ability of the leader to balance and to promote various innovation-based activities depends on the roles of its members, in other words their followers, to influence their behavior and thus to implement appropriate managerial actions to achieve organizational goals (Bledow et al., 2009). Leaders have played a crucial role in supporting their employees' ambidexterity. Therefore, leadership behaviors are a leading factor in directing the behavior of subordinates. This basic doctrine belongs to the "upper echelon theory" (Rosing et al., 2010), which is based on the assumption that the underlying cause of subordinate behavior is to be found in the behavior of the leader.

The Interaction Between Leadership Behaviors and Employees' Ambidextrous Behaviors

Ambidextrous leadership theory has discussed the interaction between complementary leadership behaviors couples, such as explicit and implicit leadership behaviors, guiding and empowering leadership, transformational and transactional leadership (Zheng et al., 2017: 3). The employment of multiple leadership types or behaviors is one of the propositions that will be effective for responding to environmental changes and organizational needs. The presence of leaders in organizations ensures that all members interact in a manner to achieve organizational goals. Leaders are influential in improving performance at the team and organization level, as well as at the individual level. Although it differs by type and purpose of the organization, it is still important for future leadership researches that design and formulate theories, classifications, observable and measurable evidence of leadership behaviors (Yukl, 2012: 66).

An overview of leadership studies from the 1960's to the 1980's such as Blake and Mouton (1964)'s Managerial Grid, House (1971)'s Path-Goal Theory and Hersey and Blanchard (1977)'s Situational Leadership Theory indicate that they are mostly focused on the differences of leadership in terms of task orientation and relation orientation. From the 1980s to the recent studies, such as Conger and Kanungo (1987), Shamir, et al., (1993) found evidence of characteristics of different leadership types, e.g, transformational, transactional and charismatic leadership, and their impact on organizations. The need for meta-categorical classification, which measures the impact-response relationship with more than one measurement method, has increased due to the relatively different identification and measurement of leadership types and behavioral characteristics of research, and the researcher who fulfilled this need is Yukl (2012).

As it turns out, regarding the investigations of leaders' attitudes and behaviors towards their members, while the formerly it was divided only in task-oriented and relationship-oriented terms, as time passes and as the diversity of research methods increases, the researches are enriched with production-oriented, employee-centered, open, implicit, change-oriented classifications. It was Yukl (2012) who presented a hierarchical taxonomy of leadership behaviors that are used to influence the performance of a team, a business unit or an organization to bring these different groupings together with similar characteristics. The four main categories and sub-components of behaviors are depicted in Table 1.

As can be tracked down through the given table, each category has a different prioritized purpose, yet, all of the objectives are influential on the performance indicators. The main purpose for task-oriented behavior is to carry out the work effectively, efficiently and reliably. The main purpose of relation-oriented behaviors is to increase the quality of both human resources and also the relationships which are sometimes called "human capital". Priority objectives for change-oriented behaviors are innovation, collective learning, and adaptation to the external environment. The ultimate goal for external leadership behavior is to obtain the necessary information and resources and to develop and defend the interests of the team or organization. Each category, along with peculiar differences in its main objectives, also comprises specific behaviors to achieve the objectives. The significance of the behavior of each subcomponent depends on its aspects, and is not always positive for the effect, primary purpose, or other outcome.

Basically, this study intends to examine the interaction between exploitation and exploration behaviors of employees based on Yukl (2012)'s classification of leadership behaviors presented above. The starting point of this study is based on the theory of Leader-Member Interaction (Dansereau et al., 1975), Organization Theory and Ambidextrous Leadership Theory (Rosing et al., 2011). Besides, inspiring studies can be addressed as Zacher et al. (2015), Kauppila and Tempelaar (2016) and Alghamdi (2018).

Table 1. Hierarchical classification of leadership behaviors

Task-Oriented Efficiency and Result-Oriented	Clarification-clarity of duties and responsibilities
	Planning
	Observation-Control of Activities
	Problem solving
Relation/Employee-Oriented	Supporting
	Developing
	Recognizing
	Empowering
Change-Oriented	Defense of Change
	Creating an Inspiring Change (Vision)
	Promoting Innovation
	Facilitating Collective Learning
External	Networking
	External Observation
	Representing

Reference: Yukl (2012: 68).

The Interaction Between Task-Oriented Leadership Behaviors and Employees' Ambidextrous Behavior

Task-oriented leadership behavior includes clarifying roles and responsibilities of members, in other words; by whom, by what time, by where, by which methods the roles are planned; observing and monitoring the planned action, and endeavoring to provide a solution to the problems that arise in each sequence and result. Clarification feature of this leadership behavior type is intended to help people understand what to do and what to expect. Clarity includes identifying the responsibilities required by the job, assigning tasks, clarifying objectives, priorities and deadlines, setting performance standards, and describing related rules, principles and standard procedures. Planning process involves decisions about objectives and priorities, arranging activities, and financing of resources. Observation feature is related to concern employees to perform activities, to determine that the process is progressing as planned, and therefore to predict the availability of the targets. The last task-oriented leadership behavior, problem-solving, is handled to deal with normal operations and the disruption of illegal, destructive or unreliable member behaviors (Yukl, 2012: 70).

This study firstly has examined the relationship between Yukl (2012)'s task-oriented leadership behaviors and exploitation behaviors of employees. The supposition is that the leader's clarification, planning, monitoring activities and problem solving behaviors are related to employee exploitation activities. Leadership is not only limited to the current behavior of the leader but it is also a consequence of the behaviors of the followers/ members/ subordinates/ employees. The primary responsibility of the leaders is to direct human resources and attitudes in the direction of the fulfillment of the roles and duties undertaken. The first step of this task is to perform the right planning. It is much more effortless to commit an organization member who can meet their needs. The sharing of organizational goals clearly with employees, the equal distribution of duties and responsibilities, and the creation of platforms or networks in which employees can provide solutions to problems and criticisms at key points will contribute to sustainable internal stability so that there are consistent relationships at both the organizational and individual levels. In addition, the leader demonstrating work-centered behavior strives to work on the capacities of employees. The behavior, tendency, or attitude that the leader who has initiated the structure while performing the activities in question is due to his goal of realizing the work as planned. Therefore productivity is also high. The change in the duty-centered behavior of the leader is human qualities such as knowledge, ability, and skill, which change with duty ambidexterity. The task performance of the employee who is trained by the task-oriented behavior of the leader points out to the usefulness of the tasks and responsibilities of the tasks, so that the existing resources and competences are used effectively and efficiently. All these explanations suggest that task-oriented leadership behaviors are related to exploitation activities in the organizational context as they increase employee performance (Çekmecelioğlu, 2014; Steers, 1991; Fayol, 1917; Yukl, 1994; Ahadzie et al., 2007; Özdevecioğlu and Kanıgür, 2009).

Proposition One: Leader's task-oriented behaviors are positively related to employee exploitation behavior.

The Interaction Between Relation-Oriented Leadership Behaviors and Employees' Ambidextrous Behavior

Relation or employee-based leadership behavior includes; strengthening the link between leader and members; delegation of authority; strengthening the members in order to ensure participation in decisions; support for members; development; to be appreciated and rewarded. Supporting from this type of behavior is adopted to provide a positive perspective, to act collaboratively and to help people find solutions to the struggle. Development behaviors are displayed in order to enhance the skills and self-confidence of their members and to facilitate their career development. This leadership behavior, especially when supported by counseling and coaching, helps teach the members of the leaders' experiences. Leaders use praise and other forms of recognition to appreciate others to demonstrate effective performance, achieve significant achievements, and create significant contributions to the team or organization. Hence, recognition includes a concrete incentive recommendation, such as a prize presented at a ceremony, or an increase in leader's wage. Empowering, the type of last leadership action focused on relationships, includes activities such as allowing leaders to act autonomously in making decisions about their members, and enabling their members to self-identify in initiating and supervising their activities (Yukl, 2012: 71-72).

Secondly, the relationship between Yukl (2012)'s relation-oriented leadership behaviors and the exploitation behaviors of employees. The supposition is that the leader's support, development and empowerment behaviors are related to employee exploitation activities. The theory of Leader-Member Exchange (LME) in the literature is the theory that has been studied much and interfered with the interaction between the leader and his employees (Janssen and Van Yperen, 2004). This theory seeks to determine how leaders use the power and resources they hold by developing different relationships with their members. In addition, one of the most important factors affecting the relationship between the leader and the member is the employee's job performance. It would also be wrong to expect the leader to exhibit an average leadership behavior against his subordinates, as long as all subordinates in a working group do not exhibit a homogeneous trait in terms of job skills and achievements. What is important in assessing a job-related work is the level of resources and opportunities that the supervisor provides to accomplish these tasks, both in the conduct of work or in the taking of responsibilities beyond the job description of a co-worker. They will be able to achieve a better quality of interaction with subordinates who are influencing the leaders with their business achievements. As a consequence of this kind of interaction, he will tend to make persistent utility by going to the way of transferring the resources of his leader to these subordinates. These may be physical resources such as budget support, material, or non-physical resources such as knowledge and experience transfer (Yıldız et al., 2008; Çekmecelioğlu and Ülker, 2014)

Proposition Two: Leader's relation-oriented behaviors are positively related to employee exploitation behavior.

The Interaction Between Change-Oriented Leadership Behaviors and Employees' Ambidextrous Behavior

Change-oriented leaders have particular features such as creating innovation, pursuing opportunities and adapting to environmental changes. The most prominent feature is advocating for change. When a leader considers negative events as a unique opportunity rather than a threat, it will be easier for him/her to get support from their members and other stakeholders around them to make innovative changes. Another reason for the leaders to react positively to new strategies and initiatives is to create a clear and compelling vision of what can be achieved by the functional unit or whole organization. When the vision is aligned with organizational and individual values, the level of effectiveness of the leader will increase. Apart from being able to create innovation, leaders can encourage their members to approach the problems from different perspectives, to get out of the box when solving problems, to try new ideas and find solutions from other areas that can be added to existing problems or tasks. In this context, they encourage innovation and support their creativity and entrepreneurial activities. Finally, it is suggested that leaders, with their feature of facilitating collective learning, are not only using existing strategies but alsosupporting them with the discovery of new strategies and methods. Leaders can support in-house activities that are used to discover new information (e.g., research projects, small-scale experiments, etc.) or new information acquisition activities from external sources (Yukl, 2012: 73-74).

Thirdly, the relationship between Yukl (2012)'s change-oriented leadership behaviors and employees' exploratory ambidexterity behaviors has been examined. The supposition is that the behaviors of the leader to defend change, to create inspirational change, and to promote innovation are associated with employee exploratory activities. Thanks to change-oriented leaders, employees understand the necessity of change and feel satisfied and dedicated. Because, change-oriented leaders think they will secure their professions and their future careers. For this reason, employees are physically committing all their energy to their work, making it easier to focus on their work, and they are proud of their work and have positive feelings (Skogstad and Einarsen, 1999; Zehir and Yavuz, 2017). Change-oriented leader deals with change, creativity, new ways of achieving of old jobs within the organization. Such leaders support change and development, seek and solve new ways to solve problems, take risks and make plans for the future (Bass and Bass, 2009: 501). Employees are the most important auxiliary resources that function as a tool for leaders to realize organizational change. Many studies have concluded that especially the transformational leaders have positive effects on the creativity of employees and innovation performance (Tung, 2016; Zacher and Rosing, 2015). Creativity, innovation, creating innovation, discovering, revealing behaviors are the result of people researching and emerging on exploratory journeys. With the leaders encouraging their subordinates to create change, the employees' willingness to apply the exploration strategies rather than being content with the sufficiency of available resources.

Proposition Three: Leader's change-oriented behaviors are positively related to employee exploration behavior.

The Interaction Between External Leadership Behaviors and Employees' Ambidextrous Behavior

Leader's external environment-oriented leadership behaviors to obtain information about emerging situations in the external environment, to allocate necessary resources and to get help, contain social networking, external observation and representation. Participation in social networks include attending meetings, professional conferences and ceremonies; being a member of the relevant associations and clubs; socializing without being formal, or using tactics to communicate and build relationships with network members (e.g., finding common interests, doing good, using tolerance and so on). External observation is based on analyzing information about relevant events and changes in the external environment and determining the threats and opportunities for the organization. Representation as the last external leadership behavior refers to the representation power of the members of the leaders, the team or the whole of the organization, its peers and other stakeholders (e.g., customers, suppliers, investors and venture partners) in the periphery. Representation, lobbying for resources and assistance include advocating for the reputation of the team or organization, negotiating agreements and coordinating related activities (Yukl, 2012: 74-75). External leaders use broader support for the vision of the outward-looking organization, by accumulating external resources and keeping external connections wide, as well as using organizational change as sources of external information (Topaloglu and Dalgin, 2013: 283).

Forthly, the study has examined the relationship between Yukl (2012)'s external leadership behaviors and employees' exploratory ambidexterity behaviors. The supposition is that the leader's networking, external observation and representation behaviors are related to the employee's exploratory activities. The relevant environment here is the outer periphery, which can be described as relatively turbulent and complicated. In a predictable environment, strategy-building approaches focus on maintaining existing successful strategies rather than restructuring strategies. Mechanical and analytical approaches to such environments are more feasible. In the turbulent and complex external environment, the strategic approaches of the leaders become more aggressive in order to compete in the organizations. Decentralized strategies can be more effective than detailed analytical planning. Careful analysis, innovation, learning and creative processes are priorities in these environments (Uğurluoğlu and Celik, 2009). Leaders who understand the nature of the external environment for organizational success need the support of their employees while developing effective strategies for complex and turbulent trends. The support and implementation of innovative and creative strategies by employees is one of the indicators of their exploratory behavior. Research shows that promoting innovative thinking is an important and distinctive feature of transformational leadership (Yukl et al., 2002). Transformational leadership is also an important component of ambidextrous leadership. Leaders can lead innovative thinking by allowing employees to think outside the frame of traditional assumptions and to solve problems by generating creative ideas. External-oriented behavior is important because gathering information about competitors' activities and products, new developments in the field of technology, market trends and changes in government policies can help an organization become more competitive. By virtue of environmental monitoring, the reasons for the invalidity of existing strategies or procedures, the reasons for making quick changes and also the proposed changes can be questionable by employees who follow the leader's behavior a valid strategy, and thus the employee becomes convinced (Michel et al., 2011; Yukl, 2006). Following this leadership behavior where enthusiastic, flexible, innovative and creative environments are introduced through the adherent, outward-oriented and dynamic organizational culture, employees will feel more successful, happy and satisfied because they can take initiative, make new discoveries and support their freedoms (Cameron and Quinn, 1999; Acar, 2013). In addition, a leader is an agent and a symbol. The LMX theory also suggests that the impression of the leader is equivalent to comments and evaluations to be made. A good leader means a good organization, a good working group and ultimately organizational success.

Proposition Four: Leader's external behaviors are positively related to employee exploration behavior.

The Effect of Interaction Between Ambidextrous Leadership Behaviors and Ambidextrous Employee Behaviors on Organizational Ambidexterity

Ambidextrous leadership behaviors as discussed in previous propositions are expected to foster employees to behave ambidextrously. It is argued that in organizations top management leaders may establish the necessary energy and supportive environment by recognizing and supporting the new ideas, and that this environment shapes individual and employee behaviors, thus the leaders play a critical role in intervening in the formation of organizational ambidexterity (Gibson and Birkinshaw, 2004: 223). O'Reilly and Tushman (2004) suggest that organizations that are described as ambidextrous require ambidextrous top management teams and as Cao et al. (2010) argues, senior management team plays a significant role in achieving organizational ambidexterity. By recognizing the different needs of different business units and processes, ambidextrous leaders form an integrated and clear vision and become committed to balancing beneficiary and exploratory strategies (Attar, 2015). Leaders need to recognize opportunities and threats to develop the right vision and strategies to exploit opportunities and also to find the right balance between explorative and exploitative strategies (Güttel & Konlechner, 2009). Carmeli and Halevi (2009) emphasizes that in order to achieve organizational ambidexterity based on balancing of conflicting strategies, by switching between different leadership behaviors, top management shall have a broad repertoire of leadership behaviors and that they should be able to use different leadership practices at different times when necessary. The changing behavioral diversity in accordance with the situation shall facilitate the balancing of different strategies and provide the basis for achieving organizational ambidexterity (Attar, 2015). This evaluation is also consistent with the classification of Yukl (2012) and may shed light on the view that leaders shall engage in task-oriented, relation-oriented, change-oriented and external behaviors simultaneously or at different times as is required by the situation when necessary in order to achieve organizational ambidexterity.

Additionally, according to the classification of Yukl (2012), the leaders' ambidextrous behaviors (task-oriented, relation-oriented, change-oriented and external) will lead to changes in a number of organizational mechanisms. From the view of Baskarada et al. (2016), the focus of these mechanisms is set on how these objectives can be realized, as they are in the areas of reduction of change, strengthening institutionalized learning (*status quo*), reliability, improvement, efficiency, selection, implementation and execution. They also affect the levels of exploratory or exploitory skills of organizations, such as education, performance management, knowledge management, belonging, vision creation, mobility, risk propensity, empowerment and competitive advantage.

Organizational ambidexterity further depends on the ability of the employee to be able to smoothly switch between the two approaches (exploration and exploitation) (Rosing et al., 2011; Li et al., 2015) and the influence of the leader on the ambidexterity level (Baskarada et al., 2016: 779). Rosing et al. (2011) suggest that different leadership styles solely prove not to be sufficient and they propose an ambidextrous leadership approach; emphasizing that different leadership behavior should be integrated with each other in order to flexibly adjust leadership behaviors according to changing requirements over time and ensure total consistency (Attar, 2015). Ambidextrous leadership is defined as the ability of employees to depict behaviors consistent with the explorative and exploitative strategy and to flexibly switch between two different behavior patterns by increasing or decreasing the changes in the behaviors of the employees (Rosing et al., 2011). Thus, while ambidextrous leaders encourage their followers to push for their goals, they fulfill their expectations and create an environment in which employees support and trust each other (Gibson and Birkinshaw, 2004; Zacher and Rosing, 2015). As to clarify, ambidextrous leadership behaviors shall foster the employee ambidexterity and in turn may well increase the organizational ambidexterity.

Proposition Five: Leaders' and employees' ambidextrous behaviors are effective on organizational ambidexterity.

FUTURE RESEARCH DIRECTIONS

It can be asserted that, to the best of researchers' knowledge, this study is the first one that provides the interaction and interrelation between aforementioned variables (ambidextrous leadership behaviors and ambidextrous employee behaviors). Given the fact that the ambidextrous leadership and ambidextrous employee (or individual ambidexterity) literature is relatively new and that the constructs' definitions are not yet explicitly defined, it is recommended for further studies to investigate the possible propositions with quantitative research designs.

In addition, the quality of the sample studied can also guide the subject. The differences between employees' ambidextrous behaviors at the top management level and the lower level can be examined and compared. With a long-term work and every innovation added to the business, the degree of maintaining the current situation and accepting innovation can be checked gradually to discuss the level of ambidexterity and effectiveness of both the leader and employees.

CONCLUSION

The inclusive objective of this study is to introduce some proposals and suggestions based on the ambidexterity literature and on some significant researches (i.e., Rosing et al., 2011; Zacher et al., 2014; Kauppila and Tempelaar, 2016; Alghamdi, 2018) related to ambidexterity of leaders, employees and the entire organization. Accordingly, it can be noted that these propositions embrace the relationship between ambidextrous leadership behaviors (i.e., task-oriented, relation-oriented, change-oriented and external) in the context of leadership behavioral level that combine transformational and transactional leadership behaviors and employee ambidextrous performance (i.e., exploitation and exploration) in the context of individual employee level.

Consequently, it is firstly suggested that a leader's task oriented behaviors such as clarification of duties and responsibilities, planning, control of activities, problem solving etc., may trigger the employee's exploitation behaviors such as effective and efficient use of existing resources and competencies in organization. Due to the fact that the awareness of the clear and equal distribution of duties and responsibilities among employees, planning the inner balance and on top of everything, knowledge about the goals, the objectives and the vision of the organization and the position offered by employees on this path, result in the support and implementation of the strategies for the employee to maintain and improve the current situation, if necessary.

As a second proposal, it is suggested that a leader's relation-oriented behaviors such as supporting, developing, empowering and so on, may more be effective on the employee's exploitation behavior rather than the employee's exploration behavior. When the employee freely builds interactive and qualified communication, he/she will assist in the achievement of the goals as an indicator of doing the best he/she can and with self-commitment in the management of all resources provided to him/her.

Thirdly, various dimensions of changing environmental conditions and compliance difficulties have been addressed. Furthermore, in dynamic business environment, some change-oriented behaviors such as defensing of change, creating an inspiring change (vision), promoting innovation and so on, that help organizations survive have been reviewed. Inasmuch as leaders should adapt to change in organizations, there is also a need for employees who will adopt change and will not hesitate to perform exploratory research. The change actually brings creativity and risk taking. The change in orientation so triggers the search for "*another good, even the best*" for the employees.

It is further proposed that a leader's external behaviors such as networking, external observation, representing and so on, may well be associated with the employee's explorative behaviors rather than exploitative behaviors. The leader's external behaviors are important because to gather information about competitors' activities and products, new developments in the field of related technology, recent market trends and changes in government policies can help an organization become more competitive. With the adoption of this thought by employees, their willingness to execute exploratory research and to establish different network connections will increase.

Finally, it is suggested that ambidextrous leader behaviors may foster employees to engage in exploratory and exploitative behavior patterns and in turn shall have a positive effect on organizational ambidexterity. The changing behavioral diversity of leaders in accordance with the situation shall facilitate the balancing of different strategies and assist employees to smoothly switch between the two approaches (exploration and exploitation) and provide the basis for achieving organizational ambidexterity.

REFERENCES

Acar, A. Z. (2013). Farklı Örgüt Kültürü Tipleri ve Liderlik Tarzları ile Örgütsel Bağlılık Arasındaki İlişki: Lojistik İşletmelerinde Bir Araştırma. *İşletme Araştırmaları Dergisi, 5*(2), 5-31.

Aghamdi, F. (2018). Ambidextreous Leadership, Ambidextreous Employee, and The Interaction Between Ambidextreous Leadership and Employee Innovative Performance. *Journal of Innovation and Entrepreneurship, 7*(1), 1–14. doi:10.118613731-018-0081-8

Agostini, L., Nosella, A., & Filippini, R. (2016). Towards an Integrated View of the Ambidextrous Organization: A Second Order Factor Model. *Creativity and Innovation Management, 25*(1), 129–141. doi:10.1111/caim.12167

Ahadzie, D. K., David, G. P., & Paul, O. (2007). Towards Developing Competency Based Measures for Construction Project Managers; Should Contextual Behaviors Be Distinguished From Task Behaviours? *International Journal of Project Management*, 1–15.

Akdoğan, Ş., Akdoğan, A., & Cingöz, A. (2009). Organizational ambidexterity: An empirical examination of organizational factors as antecedents of organizational ambidexterity. *Journal of Global Strategic Management, 3*(2), 17–27. doi:10.20460/JGSM.2009318458

Antokanis, J. (2003). Context and leadership: An examination of the nine-factor full rane leadership theory using the multifactor leadership questionnaire. *The Leadership Quarterly, 14*(3), 261–295. doi:10.1016/S1048-9843(03)00030-4

Attar, M. (2015). *Üst Düzey Yöneticilerin Liderlik Özelliklerinin Örgütsel Ustalık Düzeyine Etkisi: Türk Bankacılık Sektörü Üzerine Bir Araştırma*. Ankara: Gazi Kitabevi.

Bakan, İ., Doğan, İ. F., & Sezer, B. (2017). Örgütsel Ustalık ile Örgütsel Öğrenme Arasındaki İlişki. *TURAN-SAM Uluslararası Bilimsel Hakemli Dergisi*, *9*(33), 10–17.

Bakan, İ., & Sezer, B. (2017). Örgütsel Ustalığın Performans Yönetimi Üzerindeki Etkisi: Bir Alan Çalışması. *KSÜ Sosyal Bilimler Dergisi*, *14*(2), 323–346.

Baškarada, S., Watson, J., & Cormarty, J. (2016). Leadership and Organizational Ambidexterity. *Journal of Management Development*, *35*(6), 778–788. doi:10.1108/JMD-01-2016-0004

Baškarada, S., Watson, J., & Cormarty, J. (2017). Balancing Transactional and Transformational Leadership. *The International Journal of Organizational Analysis*, *25*(3), 506–515. doi:10.1108/IJOA-02-2016-0978

Bass, B., Avolio, B. J., & Atwater, L. (1996). The Transformational and Transactional Leadership of Men and Women. *Applied Psychology*, *45*(1), 5–34. doi:10.1111/j.1464-0597.1996.tb00847.x

Bass, B. M. (1990). From Transactional to Transformational Leadership: Learning to Share The Vision. *Organizational Dynamics*, *19*(3), 19–31. doi:10.1016/0090-2616(90)90061-S

Bass, B. M., & Avolio, B. (1993). Transformational Leadership: A Response to Critiques. In Leadership Theory and Research: Perspectives and Directions. Academic Press, Inc.

Bass, B. M., & Bass, R. (2009). *The Bass Handbook of Leadership: Theory, Research, and Managerial Applications (4 b.)*. New York: Simon and Schuster. Retrieved from https://books.google.com.tr/books?id=UTZ2npL2HHgC

Bledow, R., Frese, M., Anderson, N., Erez, M., & Farr, J. (2009). A Dialectic Perspective on Innovation: Conflicting Demands, Multiple Pathways, and Ambidexterity. *Industrial and Organizational Psychology: Perspectives on Science and Practice*, *2*(3), 305–337. doi:10.1111/j.1754-9434.2009.01154.x

Cameron, K. S., & Quinn, R. E. (1999). *Diagnosing and changing organizational culture: Based on the competing values framework*. Reading, MA: Addison-Wesley.

Cao, Q., Gedajlovic, E., & Zhang, H. (2009). Unpacking Organizational Ambidexterity: Dimensions, Contingencies and Synergistic Effects. *Organization Science*, *20*(4), 781–796. doi:10.1287/orsc.1090.0426

Cao, Q., & Şimşek, Z. ve Zhang, H. (2010). Modelling the joint impact of the CEO and the TMT on organizational ambidexterity. *Journal of Management Studies*, *47*, 1272–1296.

Carmeli, A., & Halevi, M. Y. (2009). How top management team behavioral integration and behavioral complexity enable organizational ambidexterity: The moderating role of contextual ambidexterity. *The Leadership Quarterly*, *20*(2), 207–218. doi:10.1016/j.leaqua.2009.01.011

Çekmecelioğlu, H. G. (2014). Göreve ve İnsana Yönelik Liderlik Tarzlarının Örgütsel Bağlılık, İş Performansı ve İşten Ayrılma Niyeti Üzerindeki Etkileri. *Kocaeli Üniversitesi Sosyal Bilimler Dergisi*, *28*, 21–34.

Çekmecelioğlu, H. G., & Ülker, F. (2014). Lider - Üye Etkileşimi ve Çalışan Tutumları Üzerindeki Etkisi: Eğitim Sektöründe Bir Araştırma. *Kocaeli Üniversitesi Sosyal Bilimler Dergisi*, *28*, 35–58.

Chang, Y. (2015). Strategic Human Resource Management, Transformational Leadership Organizational Ambidexterity: Evidence From Taiwan. *Asia Pacific Business Review*, *21*(4), 517–533. doi:10.1080/13602381.2015.1029298

Chebbi, H., Yahiaoul, D., Vrontis, D., & Thrassou, A. (2017). The Impact of Ambidextrous Leadership on the Internationalization of Emerging-Market Firms: The Case of India. *Thunderbird International Business Review*, *59*(3), 421–436. doi:10.1002/tie.21882

Chen, Y., Tang, G., Cooke, F. L., & Jın, J. (2016). How Does Executive Strategic Human Resource Management Link To Organizational Ambidexterity? An Empirical Examination Of Manufacturing Firms In China. *Human Resource Management*, *55*(5), 919–943. doi:10.1002/hrm.21797

Çiftçi, G. E. (2017). Örgütsel Ustalık Çalışmalarının Personel Güçlendirme Üzerinde Etkisi. *International Journal of Academic Value Studies*, *3*(10), 188–202.

Conger, J. A., & Kanungo, R. N. (1987). Toward a Behavioral Theory of Charismatic Leadership in Organizational Settings. *Academy of Management Review*, *12*(4), 637–647. doi:10.5465/amr.1987.4306715

Dai, Y., Du, K., Byun, G., & Zhu, X. (2017). Ambidexterity In New Ventures:The Impact Of New Product Development Alliances And Transactive Memory Systems. *Journal of Business Research*, *75*, 77–85. doi:10.1016/j.jbusres.2017.02.009

Dansereau, F. Jr, Graen, G. B., & Haga, W. J. (1975). A Vertical Dyad Linkage Approach to Leadership Within Formal Organizations. *Organizational Behavior and Human Performance*, *13*(1), 46–78. doi:10.1016/0030-5073(75)90005-7

Diaz-Fernandez, M., Pasamar-Reyes, S., & Valle-Cabrera, R. (2017). Human Capital and Human Resource Management to Achieve Ambidextrous Learning: A Structural Perspective. *BRQ Business Research Quarterly*, *20*(1), 63–77. doi:10.1016/j.brq.2016.03.002

Duncan, R. B. (1976). The Ambidextrous Organization: Designing Dual Structures for Innovation. In R. H. Kilmann, L. R. Pondy, & D. Slevin (Eds.), *The Management of Organization*. New York: North-Holland.

Fayol, H. (1917). *Administration Industrielle et Générale*. Paris: H. Dunod Et E.Pinat Editeurs.

Gibson, C. B., & Birkinshaw, J. (2004). The Antecedents, consequences and mediating role of organizational ambidexterity. Academy of Management Journal, 47, 209-226.

Gibson, C. B., & Birkinshaw, J. (2004). The Antecedents, Consequences, And Mediating Role Of Organizational Ambidexterity. *Academy of Management Journal*, *47*(2), 209–226.

Glaister, K. W., Ahammad, M. F., & Junni, P. (2015). Organizational Ambidexterity and Human Resource Practices. *International Journal of Human Resource Management*, *26*(13), 1785–1789. doi:10.1080/09585192.2015.1028757

Gültekin, Y., Özutku, H., & Cevrioğlu, E. (2008). Lider-Üye Etkileşimine Çok Boyutlu Yaklaşım: Liden ve Maslyn'in Dört Boyutlu Lider-Üye Etkileşim Ölçeğinin Psikometrik Özelliklerine Yönelik Görgül Bir Araştırma. *Akademik İncelemeler*, *3*(1), 95–123.

Güttel, W. H., & Konlechner, S. W. (2009). Continuously hanging by a thread: Managing contextually ambidextrous organizations. *Schmalenbach Business Review*, *61*(2), 149–171. doi:10.1007/BF03396782

He, Z., & Wong, P. (2004). Exploration vs. Exploitation: An Empirical Test of the Ambidexterity Hypothesis. *Organization Science*, *15*(4), 481–494. doi:10.1287/orsc.1040.0078

Jansen, J. J. P., Vera, D., & Crossan, M. (2009). Strategic leadership for exploration and exploitation: The moderating role of environmental dynamism. *The Leadership Quarterly*, *20*(1), 5–18. doi:10.1016/j.leaqua.2008.11.008

Janssen, O., & Van Yperen, N. W. (2004). Employees' Goal Orientations, The Quality of Leader Member Exchange, and the Outcomes of Job Performance and Job Satisfaction. *Academy of Management Journal, 47*(3), 368–384.

Jung, D. I. (2001). Transformational and Transactional Leadership and Their Effects on Creativity in Groups. *Creativity Research Journal*, 13.

Karip, E. (1998). Dönüşümcü Liderlik. *Kuram ve Uygulamada Eğitim Yönetimi, 16*, 1–23.

Kauppila, O. P., & Tempelaar, M. P. (2016). The social-cognitive underpinnings of employees' ambidextrous behavior and the supportive role of group managers' leadership. *Journal of Management Studies, 53*(6), 1019–1044. doi:10.1111/joms.12192

Kostopoulos, K. C., Bozionelos, N., & Syrigos, E. (2015). Ambidexterity and Unit Performance: Intellectual Capital Antecedents and Crosslevel Moderating Effects of Human Resource Practices. *Human Resource Management, 54*(S1), S111–S132. doi:10.1002/hrm.21705

Li, C.-R., Lin, C.-J., & Tien, Y.-H. (2015). CEO transformational leadership and top manager ambidexterity: An empirical study in Taiwan SMEs. *Leadership and Organization Development Journal, 36*(8), 927–954. doi:10.1108/LODJ-03-2014-0054

Lubatkin, M. H., Şimşek, Z., Ling, Y., & Veiga, J. F. (2006). Ambidexterity and performance in small- to medium-sized firms: The pivotal role of top management team behavioral integration. *Journal of Management, 32*(5), 646–672. doi:10.1177/0149206306290712

Luttikhuis, M. O. (2014). *Influence of leadership behaviors and attitudes on innovativeness' of employees.* 1st IBA Bachelor Thesis Conference, Enschede, The Netherlands.

March, J. G. (1991). Exploration And Exploitation in Organizational Learning. *Organization Science, 2*(1), 71–87. doi:10.1287/orsc.2.1.71

Michel, J. W., Lyons, B. D., & Cho, J. (2011). Is the Full-Range Model of Leadership Really a Full-Range Model of Effective Leader Behavior? *Journal of Leadership & Organizational Studies, 18*(4), 493–507. doi:10.1177/1548051810377764

O'Reilly, C. A., & Tushman, M. L. (2004). The Ambidextrous Organization. *Harvard Business Review, 82*(4), 74–83. PMID:15077368

Özdevecioğlu, M., & Kanıgür, S. (2009). Çalışanların İlişki ve Görev Yönelimli Liderlik Algılamalarının Performansları Üzerindeki Etkileri. *KMU İİBF Dergisi, 11*(16), 53–82.

Porter, M. E. (1996). What is Strategy? *Harvard Business Review, 74*(6), 61–78. PMID:10158474

Pricto-Pastor, I., & Martin-Perez, V. (2015). Does HRM Generate Ambidextrous Employees for Ambidextrous Learning? The Moderating Role of Management Support. *XXIII Congreso Nacional de Acede.*

Raisch, S., & Birkinshaw, J. (2008). Organizational Ambidexterity: Antecedents Outcomes, and Moderators. *Journal of Management, 34*(3), 375–409. doi:10.1177/0149206308316058

Rosing, K., Frese, M., & Bausch, A. (2011). Explaining the heterogeneity of the leadership-innovation relationship: Ambidextrous leadership. *The Leadership Quarterly, 22*(5), 956–974. doi:10.1016/j.leaqua.2011.07.014

Rosing, K., Rosenbusch, N., & Frese, M. (2010). Ambidextrous Leadership in the Innovation Process. In *Innovation and International Corporate Growth* (pp. 191–204). Springer Berlin Heidelberg. doi:10.1007/978-3-642-10823-5_12

Schindler, A. (2015). *Ambidextrous Leadership* (Master Thesis). Enschede: University of Twente.

Shamir, B., House, R. J., & Arthur, M. B. (1993). The Motivational Effects of Charismatic Leadership: A Self-Concept Theory. *Organization Science, 4*(4), 1–17. doi:10.1287/orsc.4.4.577

Skogstad, A., & Einarsen, S. (1999). The İmportance of A Change-Centred Leadership Style in Four Organizational Cultures. *Scandinavian Journal of Management, 15*(3), 289–306. doi:10.1016/S0956-5221(98)00028-1

Steers, R. M. (1991). *Introduction to Organizational Behavior* (4th ed.). New York, NY: Harper Collins Publishers.

Taylor, A., & Helfat, C. E. (2009). Organizational Linkages for Surviving Technological Change: Complementary Assets, Middle Management, and Ambidexterity. *Organization Science, 20*(4), 718–739. doi:10.1287/orsc.1090.0429

Topaloğlu, C., & Dalgın, T. (2013). Algılanan Liderlik Tarzı ve Örgütsel Bağlılık İlişkisi, Marmaris'te Yer Alan Beş Yıldızlı Otel İşletmeleri Üzerinde Bir Uygulama. Balikesir University The Journal of Social Sciences Institute, 16(29), 277-301.

Tuan, L. T. (2017). Reform in Public Organizations: The Roles of Ambidextrous Leadership and Moderating Mechanisms. *Public Management Review*, *19*(4), 518–541. doi:10.1080/14719037.2016.1195438

Tung, F. C. (2016). Does Transformational, Ambidextrous, Transactional Leadership Promote Employee Creativity? Mediating Effects of Empowerment and Promotion Focus. *International Journal of Manpower*, *37*(8), 1250–1263. doi:10.1108/IJM-09-2014-0177

Tushman, M. L., & O'Reilly, C. A. III. (1996). Ambidextrous Organizations: Managing Evolutionary and Revolutionary Change. *California Management Review*, *38*(4), 1–23. doi:10.2307/41165852

Tushman, M. L., Smith, W. K. & Binns, A. (2011, June). The Ambidextrous CEO. *Harvard Business Review*, 74-80.

Uğurluoğlu, Ö., & Çelik, Y. (2009). Örgütlerde Stratejik Liderlik ve Özellikleri. *Hacettepe Sağlık İdaresi Dergisi*, *12*(2), 121–156.

Yu, X., Meng, X., Chen, Y., Chen, Y., & Nguyen, B. (2017). Work-Family Conflict, Organizational Ambidexterity and New Venture Legitimacy In Emerging Economies. *Technological Forecasting and Social Change*, 1–12.

Yukl, G. (1994). *Leadership in Organizations* (3rd ed.). Prentice Hall International Inc.

Yukl, G. (2012). Effective Leadership Behavior: What We Know and What Questions Need More Attention. *The Academy of Management Perspectives*, *26*(4), 66–85. doi:10.5465/amp.2012.0088

Yukl, G., Gordon, A., & Taber, T. (2002). A hierarchical taxonomy of leadership behavior: Integrating a half century of behavior research. *Journal of Leadership & Organizational Studies*, *9*(1), 15–32. doi:10.1177/107179190200900102

Zacher, H., & Rosing, K. (2015). Ambidextrous leadership and team innovation. *Leadership and Organization Development Journal*, *36*(1), 54–68. doi:10.1108/LODJ-11-2012-0141

Zehir, C., & Yavuz, A. (2017). Değişim Odaklı Liderlik, Örgütsel Özdeşleşme ve Örgütsel Vatandaşlık Davranışı İlişkisi. *Doğuş Üniversitesi Dergisi*, *18*(2), 97–115. doi:10.31671/dogus.2018.34

Zhang, F., Wang, Y., Li, D., & Cui, V. (2016). Configurations of Innovations across Domains: An Organizational Ambidexterity View. *Journal of Product Innovation Management*, *34*(6), 821–841. doi:10.1111/jpim.12362

Zheng, J., Wu, G., Xie, H., & Xu, H. (2017). Ambidextrous Leadership and Sustainability-Based Project Performance: The Role of Project Culture. *Sustainability*, *9*(2336), 1–24.

ADDITIONAL READING

Bresciani, S., Ferraris, A., & Giudice, M. D. (2017). The management of organizational ambidexterity through alliances in a new context of analysis: Internet of Things (IoT) smart city projects. *Technological Forecasting and Social Change*, *136*, 331–338. doi:10.1016/j.techfore.2017.03.002

Heavey, C., & Simsek, Z. (2017). Distributed Cognition in Top Management Teams and Organizational Ambidexterity: The Influence of Transactive Memory Systems. *Journal of Management*, *43*(3), 919–945. doi:10.1177/0149206314545652

Jansen, J. J. P., Vera, D., & Crossan, M. (2009). Strategic leadership for exploration and exploitation: The moderating role of environmental dynamism. *The Leadership Quarterly*, *20*(1), 5–18. doi:10.1016/j.leaqua.2008.11.008

O'Reilly, C.A. ve Tushman, M.L. (2013). Organizational ambidexterity: past, present and future. *Rock Center for Corporate Governance*, Working Paper Series No:142, 1-31.

Rao-Nicholson, R., Khan, Z., Akhtar, P., & Tarba, S. Y. (2016). The contingent role of distributed leadership in the relationship between HR practices and organizational ambidexterity in the cross-border M&As of emerging market multinationals. *International Journal of Human Resource Management*, 1–22. doi:10.1080/09585192.2016.1216882

KEY TERMS AND DEFINITIONS

Ambidexterity: A level of dexterity required to gain power.

Exploitation: A strategy that ensures benefits from the current situation.

Exploration: A strategy to conduct in-depth research to reveal a new truth.

Leader-Member Exchange (LMX) Theory: The theory proposed by Dansereau, Graen, and Haga in 1975 and formerly called Vertical Dyad Linkage Model. It examines the relationship between the leadership process and results of the process. The LMX theory also suggests that the impression of the leader is equivalent to comments and evaluations to be made. A good leader means a good organization, a good working group and ultimately organizational success.

Leadership: A qualification that pursues of ideals, inspires people, shows the right path to them, and teaches what they need to do to help them succeed.

Organizational Theory: A research theory composed of a set of propositions developed to explain how organizations construct and manage their relationship with various environmental factors while maintaining their activities.

Upper Echelon Theory: The theory proposed by Donald C. Hambrick and P. Mason in 1984 and investigates the effects of demographic characteristics of the people in the senior management level of organizations on organizational results.

Chapter 11
What Does It Take to Be a Social Entrepreneur?
Authentic Leaders and Their Effect on Innovation

Joseph Crawford
https://orcid.org/0000-0002-2191-6216
University of Tasmania, Australia

Jo-Anne Kelder
https://orcid.org/0000-0002-8618-0537
University of Tasmania, Australia

Matthew Wayne Knox
https://orcid.org/0000-0002-6092-071X
University of Tasmania, Australia

ABSTRACT

Contemporary society is characterized by the prevalence of wicked problems to which the efforts and actions of some entrepreneurs have negatively contributed to social problems. Corporate social responsibility emerged as an early response to multi-factor problems that are difficult to conceptualize or structure, but it has had limited success in engendering significant structural societal change. Social entrepreneurship is a contested construct that typically includes the social entrepreneur. A definition of social entrepreneurs is provided drawing on the literature. Social entrepreneurs offer leadership for social innovation outcomes as an antidote to prevailing social issues. This contributes to the literature on social entrepreneurs by examining how social entrepreneurs lead in the digital era. Importantly, this chapter considers the role that authentic leader behaviors (awareness, sincerity, balanced processing, positive moral perspectives, and informal influence) has in enabling social entrepreneurs to create and innovate.

DOI: 10.4018/978-1-7998-1108-4.ch011

INTRODUCTION

The construct of wicked problems has increasing currency in describing characteristics of the global world in the twenty-first century. This includes the many ethical and societal challenges facing the contemporary global citizen, in which change is an inherent part of our society. Some of our research has failed to adequately address these challenges. Some research is only just beginning to understand the complexity of unethical cases (e.g. Crawford & Newstead, 2018).

One of the greatest challenges is the role and impact of corporations and calls for greater accountability. The Global Financial Crisis (GFC) was a significant organizational challenge (Mason, 2013), with the response of growth in social entrepreneurship research and practice. The social entrepreneur offers a 'third-sector' nested between non-for-profits and for-profit entities (Aiken & Bode, 2009). This third sector proposes to do the work of the non-for-profit in a commercial private-sector setting. The results have been varied and time-lagged. European emergence, for example, is potentially the most advanced with scholars considering welfare pluralism equal to immediate economic metrics (Evers & Laville, 2004). In Australia, Governments have been slow to identify the role of the third-sector and provide a policy platform to encourage its growth (Lyons & Passey, 2006). However, to a large degree, social entrepreneurs in their infancy have created significant responses to global challenges.

A significant social entrepreneurial venture is Grameen Bank. Grameen Bank is a multibillion-dollar bank that provides microfinance loans to impoverished people without requiring collateral. The organisation still generates a profit and aims to create significant social good in the process. The Grameen Bank received a Nobel Peace Prize in 2009 for their efforts to create economic and social development from below (Nobel Prize, 2006). Other examples include: *TOMS* a company that gives a pair of shoes to needy people for every pair bought, *Seventh Generation* that produces products that reduce environmental impact, and *Fifth Estate Co.* that strives to use low impact natural ingredients for men's grooming products[1].

The social entrepreneur's mission is to improve the world. Social entrepreneurs foreground their personal values and view profit as a positive consequence of their innovations (Dees, 2017). This approach is welcome in a corporate context of challenging ethical decisions (e.g. Crawford et al., 2017). For the social entrepreneur, profit is often seen to the end of scalable social change. In this chapter, we also frame leaders as the enablers of scalable social change (Strielkowski & Chigisheva, 2018). The social entrepreneur should be distinguishable from actors like the corporate managers engaging in positive social impact strictly for lip service to their industry or meeting society's corporate social responsibility (CSR) 'quota' (Idawo, 2014). This distinction is crucial in the consideration of innovative leadership. While CSR applies to the contemporary business, the social entrepreneur regulates their mission in adherence with the perceived needs of the society in which they conduct their enterprise.

The social entrepreneur's adherence to public good is a driver for innovation; to fix a wicked problem through business-like discipline, and determination (Dorado & Ventresca, 2013). While such a mission seems rather simple in its goal, the values pertaining to social change require effective implementation and self-regulation from its leaders. Continuous innovation requires careful consideration of the kinds of behaviors that enable innovation to be both sustainable and true to its intent (Dee, 2007). Accountability, awareness, influence, and critical thinking are just some required behaviors identified (Alvord et al., 2004; Dee, 2017; Driver, 2012). This chapter proposes to explore how the encapsulation of social change goals are *more* achievable through cultivating authentic leader behaviors.

Authentic leadership was originally conceptualized as a response to evidence of corporate avarice. The behaviors of the authentic leader were articulated as an antidote to the managerial failings in corporations (Crawford et al., 2019). Authentic leadership research found that authentic leaders were more altruistic (Sosik et al., 2009), inclusive (Boekhorst, 2015), and focused on procedural justice (Hhiung, 2012). Extending the scope of the initial theory, propensity for meaningful societal change was connected to developing authentic leadership characteristics in employees and managers. We explore this possible connection further in this chapter.

In this chapter, we explore the synergies between authentic leader behaviors and their social entrepreneurial synergies. We begin with an exploration of the literature on the current state of corporate social responsibility and define social entrepreneurs. Next, we consider what it means to be innovative and to lead in a digital era, articulating the theoretical underpinnings of the authentic leader and authentic follower. As with many critical reviews (e.g. Bush et al., 2018; Carlborg et al., 2014; Civitillo et al., 2018), we synthesize the findings of the review to address our central research question: *How can authentic leadership enhance the social entrepreneurs' ability to create change and sustain innovation?* We conclude with future research opportunities.

WEAKNESSES OF CORPORATE SOCIAL RESPONSIBILITY

A systematic conceptualization of CSR emerged in the 1950s but was evident from the mid-to-late 1800s focus on employees and how to enable their productivity (Carroll, 2008). The United Kingdom, United States, Japan, and Germany are noted as prominent examples of CSR emergence in the nineteenth century (Husted, 2015). CSR is a citizenship-like mantra that seeks to enable community wellbeing and development (Pride & Ferrell, 2006). Early rhetoric asked whether such social initiatives were prompted by societal and altruistic reasons, or for corporate reasons. In 1883, for example, West Cork Railroad Company attempted to compensate for job losses resulting from a company dissolution. A judge ruled against, noting that charity and business were independent (Carroll, 2008). Bowen (1953) articulated one of the earliest cases for social responsibility on the part of the corporation, noting that businesses should pursue strategies that are in line with the objectives and values of society. From the 50s to the 80s, corporate investment in social good grew significantly (Muirhead, 1999).

Friedman (1970) argued that the social responsibility of firms is to maximize profits. Since that assumption, some firms have sought to maximize profits through CSR initiatives that have a positive effect on firm financial performance (Baron, 2007). This could take the form of carefully structured tax liability reduction or increased sales through its brand. Friedman (2005) later questioned whether the respective value of donating a stream of profit to philanthropy exceeded spending that additional amount on business growth. The metric of social responsibility through business investment only works, however, if that enterprize is inherently achieving social goals beyond economic development. Decent work and economic growth only form one of the seventeen United Nations Sustainable Development Goals (United Nation, 2015). To some degrees, good health, no poverty, zero hunger, industry and innovation, responsible consumption and production, and strong institutions could be achieved indirectly through better economic work conditions. However, the climate action, clean water, and clean energy goals would remain largely untouched by the profit maximizer. At some point, an organization would seek to clean its rivers because they require clean water, but at that point, it is perhaps a band-aid rather than an inoculation. Individuals can progress the other sixteen goals, and so can governments. Arguably, however, large-scale organizations can have one of the largest effects.

In the new millennium there is increasing concern that CSR activities from corporations come with an ulterior motive. That is, the pursuit of social goals as a metric is merely lip service, rather than arising from genuine desire to have a positive social impact (Blenkhort & MacKenzie, 2017; Wagner et al., 2009). Engagement in generally socially responsible ways, as lip service to socially expected CSR output, is not enough for organizations (Campbell, 2018). CSR communication has become an emergent field, recognizing that negative perceptions of a company can lead to negative perceptions of the motive of CSR (Shim & Yang, 2016). For example, Volkswagen's original position on the use of forced labor in World War II was that the German government had the responsibility for payment or otherwise. The latter position, in 1998, was to create a fund to compensate former employees (Schrempf-

Stirling et al., 2016). In this case, the perceptions of the past position had an impact on the effectiveness of the future CSR initiative.

The current scrutiny corporations undergo stems from a growing culture of accountability. The Global Financial Crisis, along with many other corporate malpractices, has forced business to evolve (Boddy, 2011). Societal values, and mistrust in corporations, have influenced corporate leaders to act responsibly, or at least appear to (Van Marrewijk, 2003). The greatest weakness with CSR is that some activities are political and underpinned by Machiavellian-like behavior (Joyner & Payne, 2002). Despite the growth in CSR as a form of goodwill, organizations do not have a form of insulation from corporate crises (Janssen et al., 2015). For some entities, CSR ethics is inconsistent, engaging in CSR occurs at the same temporal period as bankruptcy (Ludescher & Mahsud, 2010).

CSR has led to significant progression towards social goals. Societal pressures to consider public good not only affect public image, but business in general. CSR has reached the point wherein non-action upon social issues inhibits profit and public image (Porter & Kramer, 2006). Yet, such initiatives have not been enough in addressing significant societal challenges of sustainability (e.g. United Nations Sustainable Development Goals). For example, Australian alcohol manufacturer, Fosters Group Limited rates as a social responsibility leader, despite that adult Australians did not feel adequately informed about their alcohol consumption (Rundle-Thiele, 2009). One of the most prevailing issues, CO_2 emissions, have not seen the growth curves decreasing, nor plateauing (Peters et al., 2012). Social entrepreneurship, therefore, is a necessary response in the face of sustained corporate malfeasance, and the need to enable social change at a global scale. Political goals of fostering social growth likewise see companies competing to demonstrate their CSR through entrepreneurial innovation (Thompson et al., 2000). Economic sacrifices for the purposes of adhering to community values and needs are a requirement in ensuring financially sustainable businesses (Davis, 1973).

DEFINING SOCIAL ENTREPRENEURS

Social entrepreneurs with the right knowledge, skills, and capability can create meaningful change in a specific part of the world. Dees (1998) conceptualized a value-based understanding of the social entrepreneur paving the way for examining the behaviors of the social entrepreneur. In a summary of seven definitions, Abu-Saifan (2012) highlights that an *entrepreneur* is characterized by words such as innovator, high achiever, risk bearer, dedicated, arbitrageur, organizer, initiative taker, strategic thinker, value creator, opportunity aware, leader, holistic, persistent, and committed. In the same research, a *social entrepreneur* was defined with words such as: mission leader, persistent, emotionally charged, social value creator, change agent, highly accountable, dedicated, socially alert, opinion leader, manager, leader, innovator, initiative take, opportunity alert, visionary, and committed. Abu-Saifan defines the social entrepreneur as:

The social entrepreneur is a mission-driven individual who uses a set of entrepreneurial behavior to deliver a social value to the less privileged, all through an entrepreneurially oriented entity that is financially independent, self-sufficient, or sustainable (2012, p. 25).

We problematize this definition, in relation to the missing description of the behaviors that determine the effectiveness of a social entrepreneur's actions. To address this, we explore the existing definitions for social entrepreneur in the new millennium (See Table 1).

Social entrepreneurship theories stress motivation to achieve certain socially desirable objectives. Specifically, forming social wealth (Dees, 1998; Mort et al. 2002; Mair and Marti, 2006; Peredo and McLean, 2006; Martin and Osberg, 2007), redistributing and building wealth to promote social values (Tan et al., 2005; Abu-Saifan 2012), and to solve wicked problems (Drayton, 2002; Alford et al., 2004). Despite their similarities, an absence of consensus on a definition creates research silos, with the topic itself suffering from non-collaboration (Bruyat and Julien, 2000). Choi and Majumdar (2014) offer solution to this issue with their cluster concept, acknowledging that the concept of social entrepreneurship pertains several sub-concepts, one of five of these is the social entrepreneur. According to Choi and Majumdar (2014), the cluster concept allows for more contextual research under the wider social entrepreneurship framework. As such, consideration may be given to the traits a social entrepreneur needs to intercede with the other concepts of social innovation, social entrepreneur organization, social value creation and market orientation. Social entrepreneurial behaviors include an emphasis on relevant ethical and social motivation (Peredo and McLean, 2006; Tan et al., 2005), sincerity in presentation (Shaw 2004; Roberts & Woods, 2005), and the ability to seize opportunities while taking calculated risk (Elkington & Hartigan, 2008; Peredo & McLean, 2006; Martin & Osberg, 2007); all while effectively leading and influencing others to do the same.

The ability for the social entrepreneur to lead effectively is crucial in its conceptualization, especially in contexts of CSR. A social entrepreneur must ensure effective leadership through assessing the needs of the community beyond economic and political pressures (Martin & Osberg, 2007). As such, the ability to lead becomes the social entrepreneur's strongest asset in creating social values, along with continuous consideration, accountability, awareness, influence, and

Table 1. Definitions of social entrepreneurs

Source	Definition
Brinkerhoff (2001)	Individuals constantly looking for new ways to serve their constituencies and add value to existing services.
Mort et al. (2002)	A multidimensional construct involving the expression of entrepreneurially virtuous behaviour to achieve the social mission…the ability to recognize social value creating opportunities and key decision-making characteristics of innovation, proactiveness and risk-taking.
Drayton (2002)	A major change agent, one whose core values centre on identifying, addressing and solving societal problems.
Alford et al. (2004)	Creates innovative solutions to immediate social problems and mobilizes the ideas, capacities, resources and social arrangements required for social transformations.
Harding (2004)	Entrepreneurs motivated by social objectives to instigate some form of new activity or venture.
Shaw (2004)	The work of community, voluntary and public organizations as well as private firms working for social rather than only profit objectives.
Roberts & Woods (2005)	Social entrepreneurship is the construction, evaluation and pursuit of opportunities for transformative social change carried out by visionary, passionately dedicated individuals.
Tan et al. (2005)	Making profits by innovation in the face of risk with the involvement of a segment of society and where all or part of the benefits accrue to that same segment of society.
Mair and Marti (2006)	…a process of creating value by combining resources in new ways… intended primarily to explore and exploit opportunities to create social value by stimulating social change or meeting social needs.
Peredo and McLean (2006)	Social entrepreneurship is exercised where some person or group….aim(s) at creating social value… shows a capacity to recognize and take advantage of opportunities… employ innovation… accept an above average degree of risk… and are unusually resourceful… in pursuing their social venture.
Martin and Osberg (2007)	Social entrepreneurship is the: 1) identification a stable yet unjust equilibrium which the excludes, marginalizes or causes suffering to a group which lacks the means to transform the equilibrium; 2) identification of an opportunity and developing a new social value proposition to challenge the equilibrium, and 3) forging a new, stable equilibrium to alleviate the suffering of the targeted group through imitation and creation of a stable ecosystem around the new equilibrium to ensure a better future for the group and society.
Zahara et al., (2008)	Social entrepreneurship encompasses the activities and processes undertaken to discover, define, and exploit opportunities in order to enhance social wealth by creating new ventures or managing existing organizations in an innovative manner.
Elkington and Hartigan (2008)	These social and environmental entrepreneurs lead by example. They attack intractable problems, take huge risks, and force the rest of us to look beyond the edge of what seems possible. They seek outlandish goals, such as economic and environmental sustainability and social equity, often aiming to transform the systems whose dysfunctions help create or aggravate major socioeconomic, environmental, or political problems. In doing so they uncover new ways to disrupt established industries while creating new paths for the future.
Brinkerhoff (2009)	A social entrepreneur is someone who takes reasonable risk on behalf of the people their organization serves.
Defourny and Nyssens (2010)	Social entrepreneurs are defined as change makers as they carry out 'new combinations' in at least one the following areas: new services, new quality of services, new methods of production, new production factors, new forms of organizations or new markets. Social entrepreneurship can therefore be a question of outcomes and social impact rather than a question of incomes.
Abu-Saifan (2012)	The social entrepreneur is a mission-driven individual who uses a set of entrepreneurial behaviors to deliver a social value to the less privileged, all through an entrepreneurially oriented entity that is financially independent, self-sufficient, or sustainable.

critical thinking (Alvord et al., 2004; Dee, 2017; Driver, 2012). Mulgan et al. (2007) posits that for the social entrepreneur to be effective in their leadership they should concentrate on the four horizons of decision-making:

1. Day to day management and efficiency,
2. Effective implementation and incremental innovation over the medium term,
3. Developing more radical options – including in very different fields – that could become mainstream in 3-20 years, and
4. Taking account of generational timescales – particularly in relation to climate change and issues like pensions.

Despite the above considerations, the question remains, why does the world need such a leader? Daily news reports witness the prevalence of wicked problems embedded in contemporary society. Whether they be issues around the environment, the economy or equality, many researchers are attempting to find ways forward to solve issues. Wicked problems can be approached using social design (Thorpe & Gamman, 2011), economics (Batie, 2008), knowledge sharing (Weber & Khademian, 2008), and theoretical frameworks (Alford & Head, 2017). However, wicked problems cannot be solved by mere research, wicked problems instead require innovative and informed action from leaders informed by multiple perspectives (Rittel & Webber, 1973). Hence, the social entrepreneur, applying their business-like determination to instigate wide-scale change (Dorado & Ventresca, 2013). The social entrepreneur is differentiated through their core goal, being the pursuit to remove of wicked problems in society (Elkington and Hartigan, 2008). This requires the effective application of the values of the social entrepreneur through informed leadership. As such, we suggest the following definition:

The social entrepreneur leads social innovation in pursuit of solving wicked problems.

INNOVATION

With the consideration of society's wicked problems at the forefront of CSR and social discourse, the social entrepreneur is prompted to action (Van Marrewijk, 2003), alongside pursuing their desire to instigate change. Leadership informed by this context requires innovation and creativity in ensuring the creation and development of social values (Gumusluoglu & Ilsev, 2009). Fostering the development of innovation increases the success of social enterprises (Chell et al. 2010). Innovation's merit stems from the identification of new-business opportunities (Johne, 1999), the creation and perpetuation of social progress (Mulgan 2006). It also acts as a motivator for further innovation and productivity in motivational settings (Gumusluoglu & Ilsev, 2009). Interpretations of innovation still arguably align to Schumpeter's (1934) initial conceptualization with underlying notions of the introduction of a new product, process, method, or system (see Table 2).

Schumpeter (1934) argued that innovation was key to profit. Like social entrepreneurship, a variety of definitions of innovation have since emerged (see table 2). These interpretations can be grouped by defining innovation as a process, an outcome or a mechanism for knowledge (Quintane et al., 2011). A focus on the latter allows for innovations self-perpetuation to continue (Gumusluoglu & Ilsev, 2009; Quintane et a., 2011). As such, the application of innovation as an instigator for the creation and development of social values is required to ensure the ongoing success of the social entrepreneur. Innovation is thus framed as seeking to fix old problems through the creation of new ideas. This creation is crucial when considering innovation. Thakur et al. (2012 p.564) defines innovation 'as adoption of those best-demonstrated practices that have been proven to be successful and implementation of those practices'. We problematize such definitions in that, while they may provide

Table 2. Definitions of innovation

Source	*Definition*
Greve and Taylor (2000)	Activities are deemed innovative if they differ significantly from current or recent activities. In organizations, innovations may change the incumbent skills, standard practices, technology, services, and products of the firm.
Obstfeld (2005)	Innovation is often a process of creating new social connections between people and resources they carry, so as to produce novel combinations.
Adner (2006)	The collaborative arrangements through which firms combine their individual offerings into a coherent, customer-facing solution.
Mulgan (2006)	Social innovation refers to innovative activities and services that are motivated by the goal of meeting a social need and that are predominantly diffused through organizations whose primary purposes are social.
Gupta et al. (2007)	The production or emergence of a new idea.
Birkinshaw et al. (2008)	The invention and implementation of a management practice, project, or technique that is new to the state of the art and is intended to further organizational goals
Phills et al. (2008).	Social innovation is a novel solution to a social problem that is more effective, efficient, sustainable, or just than existing solutions and for which the value created accrues primarily to society as a whole rather than private individuals.
Carless (2012)	An attempt to bring about educational improvement by doing something which is perceived by implementers as new or different.
Garcia (2015)	A new idea, method, process, or device that creates a higher level of performance.
Herrera et al. (2015)	The innovation process in firms is a process of accumulating and creating new knowledge.
Oeij et al. (2019)	The invention, development and implementation of new ideas to solve social problems faced by individuals, groups or communities.

solution, they lack the implementation of new ideas through a creative lens. As such we present the following definition of innovation:

Innovation is the process of applying creative and original ideas to create solution(s).

Innovation in the context of Australia has been a priority for the Federal Government, to more effectively respond to the Third Sector in current policy provisions. Under Imperative Five of the government's 'Australia 2030: Prosperity through Innovation', innovation is marked as opportunistic in Australia's national development in areas of economics, science and culture. Social enterprises (or the 'Third Sector') within Australia are encapsulating the goal of development through innovation, even more so the social entrepreneurial goal of change. The social enterprise 'Who Gives A Crap?' sells sustainable toilet paper with 50 percent of profits going to the building of hygienic bathroom facilities in third world countries. 'Who Gives A Crap?' is also considered a 'Profit for a Purpose', in that they distribute their profits into different projects, having involvement in the change process of innovation (Roe & Dolton, 2019).

LEADING IN THE DIGITAL ERA

While innovation is an outlet for change, this change does not come about through passive consideration of how to be innovative. Innovations are created and implemented through the recognition of a gap in service delivery and the identification of the land, labor, and capital necessary to create change. As such, we posit that authentic leader behaviors can sustain innovation to create change. To explore this further, we examine the digital climate within which contemporary entrepreneurs are situated.

Communication in the digitalized era is far more complicated than traditional business meetings. Communication through online profiles has essentially ensured nonstop perpetuation of business trade. Whether for social, business, or pleasure, technology has encapsulated the networked communications of humanity (Tapscott, 1996). In the digital era, communications have become easy and maintaining business relationships can be aided through applying networked communications, transcending time zones and relative business hours (Larsen et al., 2006; Tapscott, 1996). For leaders to effectively influence others, their ability to engage in rapidly evolving technology-based communication methodologies is crucial.

Contemporary business is driven by networked communications, and understanding this allows a leader to ignore contextual restrictions and apply technology for their benefit. Relative understanding of a tool or an innovative creation's worth and potential can perpetuate the success of a leader in the digitalized era. Gordon Moore stated semiconductor technology would double in density of components per integrated circuit at regular intervals until 2020-2025 (Waldrop, 2016). By this metric, the rate and pace of change in the Intel microchip creates new abilities that extend human capability potentially with infinite bounds.

Oeji et al. (2018) identify twelve key components of the innovation journey including gestation, shock, plans, proliferation, setbacks, criteria shift, fluid participation of organizational personnel, investors and management, relationships, infrastructure development, adoption, and termination. These components had significant complexity prior to the emergence of technology. However, the digital era has provided a mechanism to rapidly scale enterprise under the right incubation conditions. The innovative leader is required to have the capacity to connect and collaborate with greater numbers of people, and participate in innovation, wealth creation, and social development in virtual settings (Dery et al., 2017). Those virtual settings are not limited in nature and may be immersive environments, or simply communicating via Skype-like conferencing. The challenge for a social entrepreneur is to understand the methods by which they can have meaningful impact in a busy world of people who are connected to some and disconnected from others. In this regard, a social entrepreneur needs to be authentic, and they need the characteristics that will help them make connections that sustain change. We posit these may be authentic leader behaviors.

AUTHENTIC LEADERS AND FOLLOWERS

Leader

Authentic leader behavior theory provides a rigorous theoretical concept (see Crawford et al., 2019) that explicates behaviors of a leader who understands themselves and those around them and has a strong sense of moral character in their leadership (see Table 3). Its development is a contemporary response to the growing ethical concerns in the business sector, like corporate malfeasance and mistrust (Luthans & Avolio, 2002). Underneath the inception of authentic leadership was a desire to enable a positive psychological response within organizational research. Recent studies have identified positive relationships between authentic leadership and individual creativity (Ribeiro et al., 2019), team and individual performance (Luu, 2019; Shirey et al., 2019), reduced employee turnover (Lee et al., 2019), trust (Qui et al., 2019), job satisfaction (Baquero et al., 2019), and happiness at work (Semedo et al., 2019). In an organizational performance setting, authentic leader behaviors offer a genuine response to prevailing issues. In an ethical landscape, the same argument is true. The challenge is that the field is still in its infancy.

Authentic leadership offers an antidote for the social issues the contemporary social entrepreneur seeks to tackle. The practical understanding of social values through an adherence to positive and ethical leader behaviors allows the authentic leader to likewise engage and encourage their followers in their social mission (Crawford et al., 2019). The parallel with effective social entrepreneurship becomes evident through their behaviors oriented towards enabling an ethical mission. We explore this further in the discussion.

Follower

Despite their interrelated role with authentic leaders, researchers have generally neglected the study of authentic followership (Agho, 2009), disregarding their worthiness for independent consideration (Shamir, 2007). A possible mechanism to better enable authentic followership (Lemoine et al., 2019) is by sharing core values in a psychologically safe environment (Leroy et al., 2015). To date, there is limited research that considers the authentic follower as an independent construct. The assumption often applied to authentic leadership research is that increases in authentic leadership result in increases to authentic followers (e.g. Martino, 2019; Weiss et al., 2018). For example, Avolio and Gardner (2005) briefly discuss the followers, but only after first proposing how a leader influences the followers' self-awareness, self-regulation, and clarity of self-concept. In response, Crawford (et al., 2018, p. 274) provides conceptualize the authentic follower by two behaviors: psychological capacity for authenticity (awareness, sincerity, balanced processing, and a positive moral perspective) and positive organisational engagement:

Table 3. Defining authentic leaders and their dimensions

Authentic leader	influences and motivates followers to achieve goals through their sincerity and positive moral perspective, enabled through heightened awareness and balanced processing.
Awareness	Having insight into the behaviors of one's self, other individuals, and groups.
Sincerity	Presenting one's true self to others, honestly and openly in all relationships and with consideration for the context.
Balanced processing	The tendency to consider all relevant information available, and using this to make decisions that benefit the collective (e.g. followers).
Positive moral perspectives	The commitment to one's intrinsic ethical framework, and a willingness to subdue personal interests and ego to facilitate collective interests.
Informal influence	The ability to inspire and motivate individuals to accomplish goals of their own volition, regardless of rank or position.
Definitions quoted from multiple pages in Crawford et al., (2019).	

An authentic follower is an individual who, through their capacity for authenticity and positive organisational engagement, are self-managing and follow leaders with whom they share values.

There are very little significantly tested measures for authentic followership beyond a measure developed by VanWhy (2015), limited by its sample size (n = 301) and its method (see Crawford & Kelder 2019). Some authentic followership-specific research has emerged (e.g. Sheehan, 2018), with some initial evidence to link authentic followers and entrepreneurial leadership (Imoukhuede, 2019). De Zilwa (2016) argues that holding a normative approach to leadership (where the follower is passive), results in lost opportunity and innovation potential. When seeing leadership and followership capacity by their role in organisational capacity (Herrington & Schafer, 2019), this innovation potential has more significant implications. In this case, a failure to foster the active nature of the authentic follower is likely to lead to subservience and compliance behaviors, rather than risk-taking behaviors that enable innovation (Naldi et al., 2007).

In some entrepreneurial research, an entrepreneur acts as a creative follower of a 'platform' leader (Zahra & Nambisan, 2011). In other research, the 'follower' may be another entrepreneur who may use research and development to outpace their leader (Holmes & Schmitz Jr, 2001). In this regard, a leader is argued to respond by blocking the follower from outperforming (Holmes & Schmitz Jr, 2001). With some Machiavellian-like behaviors, a leader could attempt to minimize the efficacy of their followers (e.g. Gkorezis et al., 2015). The wrong kind of leader, therefore, would have a significant impact on follower performance, and likewise affect the ability to engender change. The right kind of leadership behaviors in a social entrepreneur, therefore, would have a positive effect on enabling changes in social values.

Development

More broadly, methodological expertise and theoretical prowess is growing in organizational programs, yet various leadership styles fail to provide practical knowledge and effective innovation (Zaccaro, 2002). In many cases, leadership interventions and tools are ineffective (Crawford & Kelder, 2019). Except that there is growing interest in developing good leaders (Newstead et al., 2019), and the tools required to do so require more extensive testing (Crawford, 2015). Effective leadership development research is limited, entrepreneurial leadership is no exception. Developing entrepreneurial leadership is, arguably, developed through a social process and through the development of a leader's social capital (Leitch et al., 2013), or through service-learning pedagogy (Litzky et al., 2010). For social entrepreneurs, the need for leadership development is paramount particularly when their social mission is in constant tension with achieving commercial viability. On each side of social entrepreneurial dualism is the ability to develop skills specific to the challenges that emerge, such as assessing the adoption of abundance mentality with abundance-based grading and course assignments (Smith et al., 2012).

The development of authentic leader behaviors is arguably similar. Each of the components of the model can be broken down and attended to in part, with the assumption that an increase in one behavior likely influences an increase in another. For example, self-awareness can be influenced situationally by changing physical properties in a room (Govern & Marsch, 2001) or by constant reflection (Schussler et al., 2010). Self-awareness and self-consciousness changes situationally, with the presentation of self evolving constantly based on the environment of operation (Goffman, 1959; Rochat, 2003). However, holistic development programs seem to be far and few between. Yet, there is an opportunity to see authentic leader behavior interventions as development opportunities for social entrepreneurial outcomes.

THE SYNTHESIS

To create effective solutions to wicked problems, social entrepreneurs need positive leadership behaviors. This chapter has aimed to identify parallels in the theories of authentic leaders, authentic followers, innovation, and social entrepreneurs. To create solutions to social entrepreneurial issues, we need to focus the individual characteristics and behaviors of effective leaders (Ullrich et al., 2019). Beyond this, we also recognize the importance of positive leadership to minimize mental unhealthiness at work (Martin et al., 2017), when mental health concerns are just one reason social entrepreneurs innovate. The following section discusses how authentic leader behaviors enable the social entrepreneur. Next, we seek to understand the value proposition that authentic followers offer to a social entrepreneur and continue by exploring how social entrepreneurs lead better in the digital era with authentic leader behaviors.

Enabling Behaviors of a Social Entrepreneur

The social entrepreneur is a leader. The underlying metric that forms leadership is one person, group, or organization influencing another (e.g. Burns,1996). Social entrepreneurs influence others to achieve social goals and inspire change, and authentic leaders have relevance to social entrepreneurship (Short et al., 2009). In the cluster model of social entrepreneurship (Choi & Majumdar, 2014), the social entrepreneur is a key component of enabling social value creation. We have provided a revised definition of the social entrepreneur that builds on Choi and Majumdar's

(2014) model by defining one of its parts. We identify that the social entrepreneur leads social innovation as a way of responding to society's challenges. Social entrepreneurs are not necessarily seeking to deliver social value to the less privileged, as Abu-Saifan (2012) articulates. Some may, like TOMS giving a pair of shoes to the needy when they sell a pair. Some may not. For example, a social entrepreneur that seeks to respond to rising CO_2 emissions may do so by selling carbon negative products at a higher price than their competitors. Future generations will benefit, but arguably it is not specifically the less privileged. Instead, social entrepreneurs respond to wicked problems.

The behaviors the social entrepreneur needs to lead change could be explained as authentic leader behaviors. To further explain, we elaborate for each of the five authentic leader behaviors. Awareness in entrepreneurs enables their ability to use 'evolutionary heuristics' when they become aware of knowledge evolving through the process of variation (Breslin & Jones, 2012). De Vries (1989) articulates that entrepreneurs are constantly conduct environmental scans to heighten their awareness. The ability for an entrepreneur to identify new opportunities and see gaps in the market stem from their awareness of that market and its needs. The authentic leader is aware of the needs of those around them, and likewise is aware of their environment. Beyond the entrepreneur's awareness, is the social entrepreneur's self-awareness. Some entrepreneurs engage in unethical practice to solve problems, or their lack of self-awareness costs them (McKenna, 1996). Take Matt Harrigan, (then) CEO of PacketSled Inc, who made a threat to kill Donald Trump on social media. Harrigan founded PacketSled in 2013 and resigned as CEO in 2016 through a joint-decision from him and his board (Chasmer, 2016). In this case, the lack of self-awareness to recognize his own weakness (to speak out of turn) cost him his role in his own company.

Social entrepreneurs also require sincerity. The ability to engage and relate to others through transparent conversations, honesty, and integrity is also critical. In the pharmaceutical industry, being ethical and engaging is corporate social responsibility seems a new norm for the bigger companies (Nussbaum, 2009). Except, there is a paradox that exists. Despite their commitment to CSR, some pharmaceutical companies charge high prices for life-saving medication making it less accessible (Frederiksborg & Fort, 2014). The ability to be sincere stems from a continual commitment to present a congruent true self for others to recognize as sincere (Crawford et al., 2019). Likewise, entrepreneurs display sincerity, self-control, and perseverance to their mission (Onesimo, 2011). The toxic ingredient saga from The Honest Company where founders guaranteed there was no SLS in their products and called the Wall Street Journal 'junk science' when they argued there was SLS is just one example. Before this saga, The Honest Company was argued to be worth 1.7 billion (Ng & Terlep, 2016), and fell below $1 billion after the scandal (Korosec, 2017). The cost of the lack of transparency in this case was quantifiably up to 700 million.

The ability to think and process information in a balanced way is also critical for entrepreneurs (Prieto et al., 2012). When striving for change, there is never a complete dataset of information available (Henderson & Robertson, 2000). This means entrepreneurs need to act based on all the relevant information they can get, but failure to do so can be costly. Seeking expertise in the field is important. For example, Sean Rad, CEO of Tinder, makes regular gaffes in his media interviews and seemingly lacks a lot of key information about the way the market his company operates in works (Hennessey, 2015). The ability to innovate and grow a business is important, but sustaining it requires careful consideration and the ability to think critically.

Influence is a key element of leadership *and* entrepreneurship. Social entrepreneurs seek meaningful change to the world in which they operate. To establish their mandate for change, entrepreneurs who lack resources are required to negotiate and acquire resources through informal network ties (Stevenson & Jarillo, 1980). Their effectiveness to engage with such negotiation stems from their genuinely established influence; that is, their informal influence. In a case study by Krackhardt (1995), an employee with low informal influence was unable to affect positive change but the employee who was respected by those around them could. A social entrepreneur who goes against the grain requires informal influence to enable their pursuit. Changing behaviors is not always easy for the consumer, and changing attitudes is not always easy for stalwarts of a sector. A social entrepreneur leverages their organically generated influence, build coalitions, and enable change. When in a position of power, they may extend to also leverage their formal positional power.

Social entrepreneurs also need a moral compass, to sustain innovation and change. In some scenarios, entrepreneurs are not ethical. Parker Conrad, Founder of Zenefits, was found to be circumventing regulations through employing unlicensed health insurance salespeople. That scandal led to a 2.5 billion dollar decrease in the company's value (Tobak, 2017). In many ways, the ethical aspect of the entrepreneur is under-evidenced. An entrepreneur can be successful while conducting themselves with grey ethics. In the pursuit of new knowledge, some entrepreneurs are known for the poor way they treated others. In the short term, this may have limited consequences. In the longer term, it is not possible to maintain an unethical culture, particularly when it borders on illegality (Crawford et al., 2017). Entrepreneurs with a positive moral perspective, however, are more likely to engage in social entrepreneurial activity. Those who have a clear moral framework, like authentic leaders, remain courageous in defense of their values and champion these values in their daily enterprise.

Understanding the Authentic Follower in Social Entrepreneurship

Early research into social entrepreneurial research identified the importance of the leader-follower relationships (Prabhu, 1999). Since then, the connectivity between the follower and the social entrepreneur has been lacking. In a review of social entrepreneurship research (Short et al., 2009), there was two mentions of followers, and both were in the context of leadership research. The role of the follower, and particularly the authentic follower, in the context of social entrepreneurship is not fully understood. Social entrepreneurs often use their informal influence to form guiding coalitions. What happens when that group of people (read: *follower*) are resistant or developmentally unready? Rather than the entrepreneur creating change, they first influence those around them to change their views. Their strategy for implementing change varies. Authentic followers vary in their positive organizational engagement.

A leader creating social innovation who has a follower high in positive organizational engagement can do so with more ease. Their followers are willing to support their initiative, generate positive energy, and accept the tension that change brings. Individuals who are more aware and think critically are also able to see the merit of an idea and be willing to follow a leader with the potential to change the world. The shared ascription of values therefore enables the ability to create positive change.

In one study, entrepreneurial leadership had a positive affect on employee work outcomes in social enterprizes (Newman et al., 2017). The question that is not asked by Newman and colleagues, however, is whether the follower behaviors influence the effectiveness of such leadership on work outcomes. This same sentiment is true of most entrepreneurship research. A stubborn employee with a tenure contract will make it more difficult to create meaningful social change. A person who does not believe in the social mission is less likely to be responsive to it. The follower has a significant impact on how effective a change process will be.

CONCLUSION

Social entrepreneurs face significant challenges in understanding and navigating their pursuit to solve world problems. Their ability to have impact in their society is difficult to conceptualize, yet their ability to lead is a key element of that. Throughout this paper we articulate that social entrepreneurship stands above CSR in responding to the world's wicked problems and provides an antidote to curb injustice occurring in parts of the globe. The exploration of awareness, sincerity, balanced processing, positive moral perspectives, and informal influence create an opportunity to understand five key behaviors that a social entrepreneur adopts. Indeed, we consider the social entrepreneur to be best understood by their ability to lead social innovation for social outcomes.

We set the scene in this chapter for more research into the leadership capability of social entrepreneurs. This may include empirical experiments of differences of leadership capability across social and everyday entrepreneurs in an aim to understand what makes these two types of people different. Additionally, we begin with an exploration of the authentic follower in the context of social entrepreneurs. Our theorizing, however, was limited by scope. Far more significant research into authentic followers beyond Crawford et al.'s (2018) conceptualization is necessary to effectively understand how to measure and develop authentic follower behaviors. At that point, it will be possible to effectively apply such theoretical and empirical knowledge to the social entrepreneur. The coalescence of social entrepreneur and their authentic leader behaviors is critical to enable the sustainable change necessary to combat difficult-to-solve problems within our society. The coalescence of the social entrepreneur and authentic follower behaviors in their subordinates is also likely to further increase the efficacy and innovativeness of the entrepreneur, but it requires far more significant research to do so.

REFERENCES

Agho, A. (2009). Perspectives of senior-level executives on effective followership and leadership. *Journal of Leadership & Organizational Studies, 16*(2), 159–166. doi:10.1177/1548051809335360

Aiken, M., & Bode, I. (2009). Killing the golden goose? Third sector organizations and back-to-work programmes in Germany and the UK. *Social Policy and Administration, 43*(3), 209–225. doi:10.1111/j.1467-9515.2009.00658.x

Alford, J., & Head, B. (2017). Wicked and less wicked problems: A typology and a contingency framework. *Policy and Society, 36*(1), 397–413. doi:10.1080/14494035.2017.1361634

Alvord, S., Brown, L., & Letts, C. (2004). Social entrepreneurship and societal transformation: An exploratory study. *The Journal of Applied Behavioral Science, 40*(3), 260–282. doi:10.1177/0021886304266847

Avolio, B., Gardner, W., Walumbwa, F., Luthans, F., & May, D. (2004). Unlocking the mask: A look at the process by which authentic leaders impact follower attitudes and behaviors. *The Leadership Quarterly, 15*(6), 801–823. doi:10.1016/j.leaqua.2004.09.003

Baquero, A., Delgado, B., Escortell, R., & Sapena, J. (2019). Authentic Leadership and Job Satisfaction: A Fuzzy-Set Qualitative Comparative Analysis (fsQCA). *Sustainability*, *11*(8), 2412–2428. doi:10.3390u11082412

Batie, S. (2008). Wicked problems and applied economics. *American Journal of Agricultural Economics*, *90*(5), 1176–1191. doi:10.1111/j.1467-8276.2008.01202.x

Blenkhorn, D., & MacKenzie, H. (2017). Categorizing corporate social responsibility (CSR) initiatives in B2B markets: The why, when and how. *Journal of Business and Industrial Marketing*, *32*(8), 1172–1181. doi:10.1108/JBIM-02-2016-0036

Boddy, C. (2011). The corporate psychopaths theory of the global financial crisis. *Journal of Business Ethics*, *102*(2), 255–259. doi:10.100710551-011-0810-4

Boekhorst, J. (2015). The role of authentic leadership in fostering workplace inclusion: A social information processing perspective. *Human Resource Management*, *54*(2), 241–264. doi:10.1002/hrm.21669

Bowen, H. (1953). *Social responsibilities of the businessman*. New York, NY: Harper & Row.

Breslin, D., & Jones, C. (2012). The evolution of entrepreneurial learning. *The International Journal of Organizational Analysis*, *20*(3), 294–308. doi:10.1108/19348831211243811

Burns, J. (1996). Defining leadership: Can we see the forest for the trees? *The Journal of Leadership Studies*, *3*(2), 148–157. doi:10.1177/107179199600300212

Bush, J., LePine, J., & Newton, D. (2018). Teams in transition: An integrative review and synthesis of research on team task transitions and propositions for future research. *Human Resource Management Review*, *28*(4), 423–433. doi:10.1016/j.hrmr.2017.06.005

Campbell, J. (2018). 2017 Decade Award Invited Article Reflections on the 2017 Decade Award: Corporate Social Responsibility and the Financial Crisis. *Academy of Management Review*, *43*(4), 546–556. doi:10.5465/amr.2018.0057

Carlborg, P., Kindström, D., & Kowalkowski, C. (2014). The evolution of service innovation research: A critical review and synthesis. *Service Industries Journal*, *34*(5), 373–398. doi:10.1080/02642069.2013.780044

Carless, D. (2012). Innovation in language teaching and learning. In C. Chapelle (Ed.), *The encyclopedia of applied linguistics* (pp. 1–12). Oxford, UK: Wiley-Blackwell. doi:10.1002/9781405198431.wbeal0540

Carroll, A. (1979). A three-dimensional conceptual model of corporate performance. *Academy of Management Review*, *4*(4), 497–505. doi:10.5465/amr.1979.4498296

Carroll, A. (2008). A history of corporate social responsibility; Concepts and practices. In A. Crane, A. McWilliams, D. Matten, J. Moon, & D. Siegel (Eds.), *The Oxford handbook of corporate social responsibility* (pp. 19–46). Oxford, UK: Oxford University Press.

Chasmer, J. (2016, Nov. 15). PacketSled CEO Matt Harrigan resigns over Trump assassination threat. *The Washington Times*.

Chell, E., Nicolopoulou, K., & Karataş-Özkan, M. (2010). Social entrepreneurship and enterprise: International and innovation perspectives. *Entrepreneurship and Regional Development*, *22*(6), 485–493. doi:10.1080/08985626.2010.488396

Civitillo, S., Juang, L. P., & Schachner, M. (2018). Challenging beliefs about cultural diversity in education: A synthesis and critical review of trainings with pre-service teachers. *Educational Research Review*, *24*(1), 67–83. doi:10.1016/j.edurev.2018.01.003

Crawford, J. (2015). *Authentic leadership in student leaders: An empirical study in an Australian university.* First Class Honours Thesis. Launceston, Australia: University of Tasmania.

Crawford, J., Dawkins, S., Martin, A., & Lewis, G. (2017). Understanding the organizational climate of unethical leadership in the Australian Football League. *The Journal of Leadership Studies*, *11*(2), 52–54. doi:10.1002/jls.21525

Crawford, J., Dawkins, S., Martin, A., & Lewis, G. (2018). Conceptualising authentic followers and developing a future research agenda. In *Authentic leadership and followership* (pp. 271–293). Cham: Palgrave Macmillan. doi:10.1007/978-3-319-65307-5_11

Crawford, J., Dawkins, S., Martin, A., & Lewis, G. (2019). Putting the leader back into authentic leadership: Reconceptualising and rethinking leaders. *Australian Journal of Management*; Advanced Online Publication. doi:10.1177/0312896219836460

Crawford, J., & Kelder, J.-A. (2019). Do we measure leadership effectively? Articulating and evaluating scale development psychometrics for best practice. *The Leadership Quarterly*, *30*(1), 133–144. doi:10.1016/j.leaqua.2018.07.001

Crawford, J., & Newstead, T. (2018). The cascading effect of unethical leadership. In S. McShane & M. Von Glinow (Eds.), *Organizational behavior* (p. 384). Sydney, Australia: McGraw-Hill.

Davis, K. (1973). The case for and against business assumption of social responsibilities. *Academy of Management Journal*, *16*(2), 312–322.

De Vries, M. (1989). Can you survive an entrepreneur? In J. Kao (Ed.), *Entrepreneurship, creativity and organization* (pp. 157–165). Englewood Cliffs, NJ: Prentice-Hall.

de Zilwa, D. (2016). The strengths and capacities of Authentic Followership. *Leadership and Organization Development Journal*, *37*(3), 310–324. doi:10.1108/LODJ-01-2014-0010

Dees, J. (2007). Taking social entrepreneurship seriously. *Society*, *44*(3), 24–31. doi:10.1007/BF02819936

Dees, J. (2017). 1 The Meaning of Social Entrepreneurship. In J. Hamschmidt & M. Pirson (Eds.), *Case studies in social entrepreneurship and sustainability* (pp. 34–42). Abingdon, UK: Routledge.

Defourny, J., & Nyssens, M. (2010). Conceptions of social enterprise and social entrepreneurship in Europe and the United States: Convergences and divergences. *Journal of Social Entrepreneurship*, *1*(1), 32–53. doi:10.1080/19420670903442053

Dery, K., Sebastian, I., & van der Meulen, N. (2017). The Digital Workplace is Key to Digital Innovation. *MIS Quarterly Executive*, *16*(2), 135–152.

Dorado, S., & Ventresca, M. (2013). Crescive entrepreneurship in complex social problems: Institutional conditions for entrepreneurial engagement. *Journal of Business Venturing*, *28*(1), 69–82. doi:10.1016/j.jbusvent.2012.02.002

Driver, M. (2012). An interview with Michael Porter: Social entrepreneurship and the transformation of capitalism. *Academy of Management Learning & Education*, *11*(3), 421–431. doi:10.5465/amle.2011.0002a

Elkington, J., & Hartigan, P. (2008). *Power of Unreasonable People: How Social Entrepreneurs Create Markets that Change the World*. Cambridge, MA: Harvard Business School Press Books.

Evers, A., & Laville, J. (2004). Defining the third sector in Europe. In A. Evers & J. Laville (Eds.), *The third sector in Europe: Globalization and welfare* (pp. 11–42). Cheltenham, UK: Edward Elgar.

Frederiksborg, A., & Fort, T. (2014). The paradox of pharmaceutical CSR: The sincerity nexus. *Business Horizons*, *57*(2), 151–160. doi:10.1016/j.bushor.2013.10.006

Friedman, M. (1970). The social responsibility of business is to increase its profits. *The New York Times Magazine*, *32–33*(122), 126-132.

Friedman, M. (2005). Making Philanthropy Out of Obscenity. *Reason*. Retrieved from http://www2.pitt.edu/~woon/courses/reason.pdf

Gkorezis, P., Petridou, E., & Krouklidou, T. (2015). The detrimental effect of machiavellian leadership on employees' emotional exhaustion: Organizational cynicism as a mediator. *Europe's Journal of Psychology*, *11*(4), 619–631. doi:10.5964/ejop.v11i4.988 PMID:27247681

Goffman, E. (1959). *The presentation of self in everyday life*. New York, NY: Doubleday.

Govern, J., & Marsch, L. (2001). Development and validation of the situational self-awareness scale. *Consciousness and Cognition*, *10*(3), 366–378. doi:10.1006/ccog.2001.0506 PMID:11697870

Gumusluoglu, L., & Ilsev, A. (2009). Transformational leadership, creativity, and organizational innovation. *Journal of Business Research*, *62*(4), 461–473. doi:10.1016/j.jbusres.2007.07.032

Henderson, R., & Robertson, M. (2000). Who wants to be an entrepreneur? Young adult attitudes to entrepreneurship as a career. *Career Development International*, *5*(6), 279–287.

Hennessey, R. (2015, Nov. 18). It's time for Sean Rad to leave Tinder. For good. *Entrepreneur*.

Herrera, M. (2015). Creating competitive advantage by institutionalizing corporate social innovation. *Journal of Business Research*, *68*(7), 1468–1474. doi:10.1016/j.jbusres.2015.01.036

Herrington, V., & Schafer, J. (2019). Preparing individuals for leadership in Australasia, the United States, and the UK. *Police Practice and Research*, *20*(3), 240–258. doi:10.1080/15614263.2019.1598069

Holmes, T., & Schmitz, J. Jr. (2001). A gain from trade: From unproductive to productive entrepreneurship. *Journal of Monetary Economics*, *47*(2), 417–446. doi:10.1016/S0304-3932(01)00044-7

Hsiung, H. (2012). Authentic leadership and employee voice behavior: A multi-level psychological process. *Journal of Business Ethics*, *107*(3), 349–361. doi:10.100710551-011-1043-2

Hughes, D., Lee, A., Tian, A., Newman, A., & Legood, A. (2018). Leadership, creativity, and innovation: A critical review and practical recommendations. *The Leadership Quarterly*, *29*(5), 549–569. doi:10.1016/j.leaqua.2018.03.001

Husted, B. W. (2015). Corporate social responsibility practice from 1800–1914: Past initiatives and current debates. *Business Ethics Quarterly*, *25*(1), 125–141. doi:10.1017/beq.2014.1

Idowu, A. (2014). Corporate Social Responsibility in Nigerian Banking Industry: When Will the Lip-Service Games End? *Journal of Economics and Sustainable Development*, *5*(22), 21–26.

Imoukhuede, O. (2019). *The impact of entrepreneurial leadership on authentic followership in Nigeria and the United States* (Doctoral dissertation). Regent University.

Janssen, C., Sen, S., & Bhattacharya, C. (2015). Corporate crises in the age of corporate social responsibility. *Business Horizons*, *58*(2), 183–192. doi:10.1016/j.bushor.2014.11.002

Johne, A. (1999). Successful market innovation. *European Journal of Innovation Management*, *2*(1), 6–11. doi:10.1108/14601069910248838

Joyner, B., Payne, D., & Raiborn, C. (2002). Building values, business ethics and corporate social responsibility into the developing organization. *Journal of Developmental Entrepreneurship*, *7*(1), 113–131.

Korosec, K. (2017, Oct. 5). Jessica Alba's Honest Co. cuts its valuation in new funding round. *Fortune*.

Krackhardt, D. (1995). Entrepreneurial opportunities in an entrepreneurial firm: A structural approach. *Entrepreneurship Theory and Practice*, *19*(3), 53–69. doi:10.1177/104225879501900305

Lee, H., Chiang, H., & Kuo, H. (2019). Relationship between authentic leadership and nurses' intent to leave: The mediating role of work environment and burnout. *Journal of Nursing Management*, *27*(1), 52–65. doi:10.1111/jonm.12648 PMID:30238670

Leitch, C., McMullan, C., & Harrison, R. (2013). The development of entrepreneurial leadership: The role of human, social and institutional capital. *British Journal of Management*, *24*(3), 347–366. doi:10.1111/j.1467-8551.2011.00808.x

Lemoine, G., Hartnell, C., & Leroy, H. (2019). Taking Stock of Moral Approaches to Leadership: An Integrative Review of Ethical, Authentic, and Servant Leadership. *The Academy of Management Annals*, *13*(1), 148–187. doi:10.5465/annals.2016.0121

Leroy, H., Anseel, F., Gardner, W., & Sels, L. (2015). Authentic leadership, authentic followership, basic need satisfaction, and work role performance. *Journal of Management, 41*(6), 1677–1697. doi:10.1177/0149206312457822

Litzky, B., Godshalk, V., & Walton-Bongers, C. (2010). Social entrepreneurship and community leadership: A service-learning model for management education. *Journal of Management Education, 34*(1), 142–162. doi:10.1177/1052562909338038

Ludescher, J., & Mahsud, R. (2010). Opening Pandora's box: Corporate social responsibility exposed. *Independent Review, 15*(1), 123–131.

Luthans, F., Avolio, B., Walumbwa, F., & Li, W. (2005). The psychological capital of Chinese workers: Exploring the relationship with performance. *Management and Organization Review, 1*(2), 249–271. doi:10.1111/j.1740-8784.2005.00011.x

Luu, T. (2019). *Linking authentic leadership to salespeople's service performance: The roles of job crafting and human resource flexibility. Industrial Marketing Management*. Advanced Online Publication.

Lyons, M., & Passey, A. (2006). Need public policy ignore the third sector? Government policy in Australia and the United Kingdom. *Australian Journal of Public Administration, 65*(3), 90–102. doi:10.1111/j.1467-8500.2006.00496a.x

Martin, A., Sanderson, K., Warnecke, E., Dawkins, S., Bartlett, L., Memish, K., … Peebles, D. (2017). *An integrated approach to workplace mental health*. Hobart, Australia: University of Tasmania.

Martin, R., & Osberg, S. (2007, Spring). Social entrepreneurship: The case for definition. *Stanford Social Innovation Review*, 28–39.

Martino, J. (2019). Coaching Greatness: An Application of Authentic Leadership Development Theory to Wooden and Lombardi. *The Journal of Values Based Leadership, 12*(1), 13. doi:10.22543/0733.121.1258

Mason, C. (2013). It's not you, it's me! Breaking up social entrepreneurship identity. In S. Denny & F. Seddon (Eds.), *Social enterprise* (pp. 66–80). Abingdon, UK: Routledge.

McKenna, S. (1996). The darker side of the entrepreneur. *Leadership and Organization Development Journal, 17*(6), 41–45. doi:10.1108/01437739610130519

Muirhead, S. (1999). *Corporate contributions: The view from 50 Years*. New York: The Conference Board.

Mulgan, G. (2006). The Process of Social Innovation. *Innovations: Technology, Governance, Globalization*, *1*(2), 145–162. doi:10.1162/itgg.2006.1.2.145

Mulgan, G., Tucker, S., Ali, R., & Sanders, B. (2007). *Social innovation: what it is, why it matters and how it can be accelerated: Working paper*. Oxford, UK: Oxford Said Business School.

Naldi, L., Nordqvist, M., Sjöberg, K., & Wiklund, J. (2007). Entrepreneurial orientation, risk taking, and performance in family firms. *Family Business Review*, *20*(1), 33–47. doi:10.1111/j.1741-6248.2007.00082.x

Newman, A., Neesham, C., Manville, G., & Tse, H. H. (2018). Examining the influence of servant and entrepreneurial leadership on the work outcomes of employees in social enterprises. *International Journal of Human Resource Management*, *29*(20), 2905–2926. doi:10.1080/09585192.2017.1359792

Newstead, T., Dawkins, S., Macklin, R., & Martin, A. (2019). The virtues project: An approach to developing good leaders. *Journal of Business Ethics*; Advanced Online Publication. doi:10.100710551-019-04163-2

Ng, S., & Terlep, S. (2016, Sept. 15). Unilever is in talks to acquire Jessica Alba's Honest Co. *Wall Street Journal.*

Nobel Prize. (2006). *Grameen Bank: Facts.* The Nobel Prize. Retrieved from https://www.nobelprize.org/prizes/peace/2006/grameen/facts/

Nussbaum, A. (2009). Ethical corporate social responsibility (CSR) and the pharmaceutical industry: A happy couple? *Journal of Medical Marketing*, *9*(1), 67–76. doi:10.1057/jmm.2008.33

Oeij, P., van der Torre, W., Vaas, F., & Dhondt, S. (2019). Understanding social innovation as an innovation process: Applying the innovation journey model. *Journal of Business Research*, *101*(1), 243–254. doi:10.1016/j.jbusres.2019.04.028

Onesimo, A. (2011). What motivates entrepreneurs? A Study of the value systems of Filipino entrepreneur. *International Journal of Entrepreneurship*, *15*(1), 73–97.

Peters, G., Marland, G., Le Quéré, C., Boden, T., Canadell, J., & Raupach, M. (2012). Rapid growth in CO 2 emissions after the 2008–2009 global financial crisis. *Nature Climate Change*, *2*(1), 2–4. doi:10.1038/nclimate1332

Phills, J., Deiglmeier, K., & Miller, D. (2008). Rediscovering social innovation. *Stanford Social Innovation Review*, *6*(4), 34–43.

Porter, M., & Kramer, M. (2006). The link between competitive advantage and corporate social responsibility. *Harvard Business Review, 84*(12), 78–92. PMID:17183795

Prabhu, G. (1999). Social entrepreneurial leadership. *Career Development International, 4*(3), 140–145. doi:10.1108/13620439910262796

Pride, W., & Ferrell, O. (2006). *Marketing: Concepts and strategies.* Boston, MA: Houghton, Mifflin.

Prieto, L., Phipps, S., & Friedrich, T. (2012). Social entrepreneur development: An integration of critical pedagogy, the theory of planned behavior and the ACS model. *Academy of Entrepreneurship Journal, 18*(2), 1–82.

Qiu, S., Alizadeh, A., Dooley, L., & Zhang, R. (2019). The effects of authentic leadership on trust in leaders, organizational citizenship behavior, and service quality in the Chinese hospitality industry. *Journal of Hospitality and Tourism Management, 40*(1), 77–87. doi:10.1016/j.jhtm.2019.06.004

Ribeiro, N., Duarte, A., Filipe, R., & Torres de Oliveira, R. (2019). How authentic leadership promotes individual creativity: The mediating role of affective commitment. *Journal of Leadership & Organizational Studies*; Advanced Online Publication. doi:10.1177/1548051819842796

Rittel, H., & Webber, M. (1973). Dilemmas in a general theory of planning. *Policy Sciences, 4*(2), 155–169. doi:10.1007/BF01405730

Roberts, D., & Woods, C. (2005). Changing the world on a shoestring: The concept of social entrepreneurship. *University of Auckland Business Review, 7*(1), 45–51.

Rochat, P. (2003). Five levels of self-awareness as they unfold early in life. *Consciousness and Cognition, 12*(4), 717–731. doi:10.1016/S1053-8100(03)00081-3 PMID:14656513

Roe, R. & Dalton, P. (n.d.). *Giving Hope: The Journey of the For-Purpose Organisation and Its Quest for Success.* Gateway East, Singapore: Springer Nature.

Rundle-Thiele, S. (2009). Social gain: Is corporate social responsibility enough? *Australasian Marketing Journal, 17*(4), 204–210. doi:10.1016/j.ausmj.2009.06.006

Schrempf-Stirling, J., Palazzo, G., & Phillips, R. (2016). Historic corporate social responsibility. *Academy of Management Review, 41*(4), 700–719. doi:10.5465/amr.2014.0137

Schussler, D., Stooksberry, L., & Bercaw, L. (2010). Understanding teacher candidate dispositions: Reflecting to build self-awareness. *Journal of Teacher Education*, *61*(4), 350–363. doi:10.1177/0022487110371377

Semedo, A., Coelho, A., & Ribeiro, N. (2019). *Authentic leadership, happiness at work and affective commitment: An empirical study in Cape Verde. European Business Review*. Advanced Online Publication. doi:10.1108/EBR-01-2018-0034

Shamir, B. (2007). From passive recipients to active co-producers: The roles of followers in the leadership process. In B. Shamir, R. Pillai, M. Bligh, & M. Uhl-Bien (Eds.), *Follower-centered perspectives on leadership: A tribute to the memory of James R. Meindl*. Stanford, CT: Information Age.

Sheehan, S. (2018). *Calling and Authentic Followership* (Doctoral dissertation). Regent University.

Shim, K., & Yang, S. (2016). The effect of bad reputation: The occurrence of crisis, corporate social responsibility, and perceptions of hypocrisy and attitudes toward a company. *Public Relations Review*, *42*(1), 68–78. doi:10.1016/j.pubrev.2015.11.009

Shirey, M., White-Williams, C., & Hites, L. (2019). Integration of Authentic Leadership Lens for Building High Performing Interprofessional Collaborative Practice Teams. *Nursing Administration Quarterly*, *43*(2), 101–112. doi:10.1097/NAQ.0000000000000339 PMID:30839447

Short, J., Moss, T., & Lumpkin, G. (2009). Research in social entrepreneurship: Past contributions and future opportunities. *Strategic Entrepreneurship Journal*, *3*(2), 161–194. doi:10.1002ej.69

Smith, W., Besharov, M., Wessels, A., & Chertok, M. (2012). A paradoxical leadership model for social entrepreneurs: Challenges, leadership skills, and pedagogical tools for managing social and commercial demands. *Academy of Management Learning & Education*, *11*(3), 463–478. doi:10.5465/amle.2011.0021

Sosik, J., Jung, D., & Dinger, S. (2009). Values in authentic action: Examining the roots and rewards of altruistic leadership. *Group & Organization Management*, *34*(4), 395–431. doi:10.1177/1059601108329212

Strielkowski, W., & Chigisheva, O. (2018). Social, Economic, and Academic Leadership for Sustainable Development of Business and Education in the Future: An Introduction. In W. Strielkowski & O. Chigisheva (Eds.), *Leadership for the Future Sustainable Development of Business and Education* (pp. 3–8). Cham: Springer. doi:10.1007/978-3-319-74216-8_1

Thakur, R., Hsu, S., & Fontenot, G. (2012). Innovation in healthcare: Issues and future trends. *Journal of Business Research, 65*(4), 562–569. doi:10.1016/j.jbusres.2011.02.022

Thompson, J., Alvy, G., & Lees, A. (2000). Social entrepreneurship–a new look at the people and the potential. *Management Decision, 38*(5), 328–338. doi:10.1108/00251740010340517

Thorpe, A., & Gamman, L. (2011). Design with society: Why socially responsive design is good enough. *CoDesign, 7*(3-4), 217–230. doi:10.1080/15710882.2011.630477

Tobak, S. (2017, Jan. 4). The 10 worst entrepreneurs of 2016. *Entrepreneur.*

Ullrich, D., Koleňák, J., Ambrozová, E., Pokorný, V., & Milichovský, F. (2019). Global X-tream Index and its Partial Parameters for Identifying the Level of Potential Individual Characteristics in the Challenging Conditions of a Modern Corporate and Security Environment. *Sustainability, 11*(12), 3325. doi:10.3390u11123325

United Nations. (2015). *Resolution adopted by the General Assembly on 25 September 2015.* United Nations. Retrieved from https://www.un.org/ga/search/view_doc.asp?symbol=A/RES/70/1&Lang=E

Van Marrewijk, M. (2003). Concepts and definitions of CSR and corporate sustainability: Between agency and communion. *Journal of Business Ethics, 44*(2-3), 95–105. doi:10.1023/A:1023331212247

VanWhy, L. (2015). *Development of the Authentic Followership Profile (AFP) Test Instrument* (Doctoral dissertation). Regent University.

Wagner, T., Lutz, R., & Weitz, B. (2009). Corporate hypocrisy: Overcoming the threat of inconsistent corporate social responsibility perceptions. *Journal of Marketing, 73*(6), 77–91. doi:10.1509/jmkg.73.6.77

Waldrop, M. (2016). The chips are down for Moore's law. *NATNews, 530*(7589), 144. PMID:26863965

Weber, E., & Khademian, A. (2008). Wicked problems, knowledge challenges, and collaborative capacity builders in network settings. *Public Administration Review, 68*(2), 334–349. doi:10.1111/j.1540-6210.2007.00866.x

Weiss, M., Razinskas, S., Backmann, J., & Hoegl, M. (2018). Authentic leadership and leaders' mental well-being: An experience sampling study. *The Leadership Quarterly, 29*(2), 309–321. doi:10.1016/j.leaqua.2017.05.007

Zaccaro, S. J., & Horn, Z. N. (2003). Leadership theory and practice: Fostering an effective symbiosis. *The Leadership Quarterly*, *14*(6), 769–806. doi:10.1016/j.leaqua.2003.09.009

Zahra, S., & Nambisan, S. (2011). Entrepreneurship in global innovation ecosystems. *AMS Review*, *1*(1), 4–17. doi:10.100713162-011-0004-3

ENDNOTE

1 Disclosure: one of the authors is a co-owner of one of these companies.

Chapter 12
The Psychological Capital: A Moderator in Authentic Leadership - Organizational Commitment Relation

Nazli Berberoglu Yilmaz
Ankara Yıldırım Beyazıt University, Turkey

ABSTRACT

The psychological capital concept, which refers to developmental positive states of human beings, has proven to be important for organizations in order to gain competitive advantage. Thus, in this study, a brief discussion has been made in order to provide a better understanding of the desired organizational outcomes of this concept and its role in the authentic leadership and organizational commitment behavior of employees. It has been stated that under authentic leadership style, followers with low psychological capital are thought to be more inclined to show organizational commitment than their peers with high psychological capital.

INTRODUCTION

Today's rapidly changing environment makes both organizations and leaders to create and develop competitive advantage against their rivals. Therefore, it can be observed that organizations apply differing options in order to gain superiority. Thus, in such an environment from technological innovations to public relational strategies, every tool may make the organization one step ahead of its competitors if used correctly. However, each of these tools is traceable and repeatable resources by competitors. Organizations, on the other hand, need resources that cannot be copied and replicated by their rivals. At this point, human resources they have, are thought

DOI: 10.4018/978-1-7998-1108-4.ch012

to provide the uniqueness they need. Consequently, today's leaders realized that the real power they have is the human resources within the organization. When an employee joins the organization, s/he also brings his/her personal capabilities and strengths with himself. These strengths and abilities are also the ones that cannot be replicated and that creates a competitive advantage.

Recently, positive organizational behavior and relatedly psychological capital (PsyCap) derived from positive psychology have been attracted, organization researchers. The psychological capital is defined by Luthans and his colleagues (2007) as a positive psychological state of individuals. According to Luthans and his colleagues (2007) psychological capital of employees can enhance the uniqueness that is required for competitive advantage. With its developmental feature and its dimensions, psychological capital is argued to support desired organizational outcomes. The dimensions of the psychological capital, on the other hand, are still be developing. However, there are four dimensions, namely self-efficacy, hope, optimism, and resilience, that researchers arrived at a consensus. Self-efficacy refers to one's belief in himself/herself regard to accomplishing the given task properly(Stajkovic & Luthans, 1998). Hope, on the other hand, refers the cognitive set that composed of deciding an aim and the way that leads this aim(Snyder et al., 1991). Optimism, another dimension of psychological capital; is the general expectation that good things will come to life in a way that allows the person to continue on his/her way to reach his goal (Brissette, Scheier, & Carver, 2002). The last dimension, resiliency indicates an individual's ability to cope with all difficulties besides endurance and adaptation to change. (Luthans, 2002). It has been stated that although these dimensions affect behavioral outcomes individually, the psychological capital that they composed also have an influence on the behaviors(Luthans et al., 2007).

The psychological capital also determined to be developed through leaders' efforts that create an eligible environment to flourish in this positive state. More specifically, authentic leadership is proposed to increase the psychological capital of his/her followers via their transparency, trust, and openness. Besides, authentic leadership is known as its positive effect on other organizational behaviors such as organizational citizenship behavior, performance, trust and, finally organizational commitment. Thus, this study aims to investigate the relationship between psychological capital, authentic leadership, and organizational commitment. More specifically, the moderator effect of psychological capital in authentic leadership organizational commitment relation is the principal objective of this study.

BACKGROUND

Positive Psychology

Since there is negative psychology perspective prevailed in the disciplines such as psychology and sociology, which contribute to organizational behavior and human resource studies, the powerful and developmental positive sides of individuals are neglected in the organizational studies. The reason behind this negative approach is depended on especially the psychology studies after the second world war. Although psychology is a discipline with trivet-like aims to improve mental illnesses, to make healthy individuals happier and more effective and, to realize the potential of individuals, after the war psychologists pay more attention to the negative side of these aims. Thus, most of the studies in this field take their foci to cure the trauma and distorted parts of human psychology(Luthans et al., 2007). The richer and more positive sides of individuals and potentially higher returns of this positivity, on the other hand, was considered as risky or time-consuming, and consequently were avoided (Luthans et al., 2007).

However, at the beginning of the 21st century, the American Psychological Association's then-president Martin Seligman and his colleagues focus on the positive side of human psychology and reminded the positive psychology which is not given enough importance until that day. There were also examples of early representatives of positive psychology in the field before Seligman's studies such as Terman's studies of giftedness and mental happiness, Watson's writings on effective parenting, Jung's work on discovering and researching the meaning of life and, Maslow's self-realization and healthy individuals studies(Csikszentmihalyi & Seligman, 2000). Although there were preliminary studies on positive psychology in the field, such as Maslow's and Roger's studies in 60s, the concept re-introduced by Seligman and his colleagues by emphasizing "what is right" in humans. They stated that psychology as a discipline has explanations for the depression, racism, violence, self-esteem management, and growth under difficulties, while it does not offer much for the positive side of human psychology such as strong character traits, virtue, and conditions leading to high levels of happiness and civic engagement (Gable & Haidt, 2005). Unlike their colleagues who are interested in disease model psychology, positive psychologists are interested in the strengths, and resilience of human psychology. Furthermore, while the negative approach focus on the pathologies in human, positive approach in psychology, on the other hand, consider improving the goodness and success in human. Thus, the aim of psychology has been shifted from what is wrong in humans to what is right by positive psychologists. In other words, the focus has been replaced with human strengths, resiliency and endurance, rather than weaknessesand vulnerabilities(Csikszentmihalyi & Seligman, 2000; Linley, Joseph, Harrington, & Wood, 2006).

Overall, after the negative oriented point of view, positive psychology has changed the perspective of the psychology discipline. It focuses on to reveal the strengths and positive side of humans and build competencies to settle obstacles in life(Csikszentmihalyi & Seligman, 2000). More specifically, it considers positive human experiences and their subjective reflections on human psychology. For example, it deals with the goodness and satisfaction of past experiences, happiness and inner peace of the present time, and, hope and optimism for the future. Furthermore, it also considers the positive traits not only at the individual level but also at the group level. At the individual level, positive psychology focus on the love, sense of mission, courage, perseverance, forgiveness, originality, ability to adapt to the environment, aesthetic, foresight, spirituality, high abilities, and acumen. On the other hand at the group level it examines civic virtue, citizenship, responsibility, emotional satisfaction, sacrifice, courtesy, tolerance, and ethical values(Luthans & Youssef, 2007).

As expected, positive psychology studies have been attracted many researchers from different disciplines of social sciences. Its reflection to organizational field is positive organizational behavior which is developed by Luthans and his colleagues (2007). Since the main aim of the existing work is to investigate the effect of the psychological capital, the study will continue with positive organizational behavior, psychological capital, and its dimensions.

Positive Organizational Behavior

The effect of positive emotions on the performance of employees has been accepted as an important indicator since the time Hawthorne studies. However, considerable attention has been devoted to positively oriented studies in the organizational studies since then(Bakker & Schaufeli, 2008). As it is the case in the psychology discipline, it is seen that in the field of organizational behavior negative perspective is dominant rather than a positive approach. For example; negative emotions, stress, and burnout were studied instead of eustress. Moreover, the emphasis was placed on resistance to change, deficiencies and disorders in workplaces rather than acceptance and adaptation to change, strengths and psychological capacities of employees that have a positive effect on development and performance improvements(Wright, 2003).

However, following the positive psychology movement, organizational behavior researchers have changed their point of view and applied the positive centered approach to the field. Fred Luthans(Luthans, 2002) was the first person to introduce Positive Organizational Behavior (POB) to the field as a new focus. According to him, there was a need for developing a concept that has a positive origin and also can be developed and measured. As has been stated by him, positive organizational behavior not only a "trend" that is prevailed in today's personal development books. Instead, it is a scientific research area with its measurable and developmental features(Luthans, 2002; Luthans et al., 2007) Thus, the important criteria of positive organizational behavior as an organizational field is depended summarized as; a) being base on a theory, research, and a valid measurement technique; b) being open to development; and, c) having performance implication(Luthans & Avolio, 2009).

The first criterion was briefly developed to distinguish POB from other popular positive-origin personal development and self-healing literature such as Norman Vincent Peal's Power of Positive Thinking, Stephen Covey's Seven Habits of Effective Man(Luthans & Avolio, 2009). This important scientific criterion stems from both positive psychology and the tradition of organizational behavior studies. Without its scientific criteria, neither the sustainability nor the reliability of POB as academic research or evidence-based practice can be achieved. The second criterion of being developmental arise from the prevailed trait-like concepts in the organizational behavior field such as personality. These trait-like concepts refer to unchangeable characteristics of individual which, inevitably, are not open to development. In order to differentiate POB from other positive structures in organizational behavior, researchers have directed their efforts to define state-like psychological resources. Here, the main concern is to define the state-like psychological resources that are open and eligible for change and development. Other positive approaches and structures are based on relatively more trait-like characteristics require long-term professional interventions in which the success depends on the presence of appropriate factors and the absence of preventive factors. Besides the development of these treatments is more stable that requires almost a lifetime of a human being (Luthans & Youssef, 2007). The final criterion points out the difference of POB from both personal development concepts and positive psychology itself. In general, although it seems to narrow the research area towards the POB, the aim is to has a performance effect rather than the POB itself. Furthermore, in addition to performance, there are also antecedents, moderators, mediators and a wide variety of behavioral outcomes defined in the POB field(Luthans, Avey, Avolio, & Peterson, 2010; Luthans & Avolio, 2009).

According to the criteria mentioned above, there are four state-like psychological abilities are defined in the field. These are self-efficacy, hope, optimism, and resiliency, which are defined in detail below(Luthans et al., 2007; Muse, Harris, Giles, & Feild, 2008).

Self-Efficacy: The concept first proposed by Bandura (1977) and he demonstrated the correlation with other positive concepts such as learning, positive feedback, modelling, administration, and psychological arousal. According to the definition of self-efficacy that is harmonized with the positive psychology literature; self-efficacy is the self-confidence of an individual on his/her ability to create motivation and cognitive resources to perform his / her task successfully (Luthans, 2002). The key feature of is concept is that it can be developed for a particular task in the given situation. Self-efficacy, thus, is not only open to improvement, but also supports the individual in his/her career (Luthans, 2002). It has been reported that self-efficacy; enables individuals to embrace his/her mission and face the challenges; make him/her spend more effort and motivation to tackle the task; and make him/her be more stable in the face of challenges(Stajkovic & Luthans, 1998). Thus, it can be expected that high self-efficacy leaders or employees appears to be suitable for efficiency and high performance in today's workplaces.

Hope: Although self-efficacy is a criterion with a strong positive relationship with performance in the workplace and developed within the positive organizational behavior literature the same is not valid for the hope dimension. Different from the "hope" that is used in daily life, hope as a positive organizational behavior refers positive motivational status derived from agency (goal-directed energy) and pathways (plan that meets objectives) (Bailey, Eng, Frisch, & Snyder, 2007). The agency component of hope refers to the inner power to do (Larson & Luthans, 2006). On the other hand, the second component of the capacity of hope is defined as the capacity perceived in oneself in order to discover the path-method or the conceptual path reaching the goals of the individual (Bailey et al., 2007). Thus, hope is defined by Snyder and Feldman(2000) as "*The individual's choice to achieve the purpose of the path and the ability to move on this path.*"

Resiliency: As with other positive psychological abilities, resiliency has its roots from clinical psychology and especially child psychopathology (Luthans, 2002). There are many definitions of resiliency, however the definition that fits the best to the situational nature of positive psychology is "*The individual's ability to cope successfully with a fundamental change, failure or risk*" (Luthans, 2002). Contrary to hope and self-efficacy, resiliency appears to be a more reactive criterion rather than proactive. This ability is stated to changed and developed over time with the protective factors of the environment and the individual. Although resiliency is beyond the simple adaptation of human beings into changing conditions, it is still based the ability to adapt. Morover, it refers to process of increasing the ability and capacity of resistance against the difficulties, uncertainties, confusion, and failures. Even, it has been observed that against some positive changes indiviuals still develop resiliency(Luthans, 2002). According to Coutu (2002) resilient human profile has the following characteristics; a deep belief in the meaning of life, confidence in the ability to improvise, and adaptation to changes.

Resiliency, like self-confidence and hope, is widely used and resembles other positive abilities in daily life. Therefore, it needs to be conceptually decomposed. In the simplest way, the main difference between self-efficacy and resiliency is to have a smaller impact area and to be reactive rather than proactive (Luthans, 2002). The relation between hope and resiliency, on the other hand, is that resiliency resembles the agency component of hope, however, does not include the pathway component. Thus, it has been stated that as a dimension of POB the resiliency meets the criteria of being positive, unique and measurable while showing a feature that is open to development.

The characterization of resilient individuals is reported as social ability, problem-solving skills, autonomy, ability to make decisions and foreseeing(Luthans et al., 2007). All of these features are stated to be learned and developed through leadership implementations and human resources programs. It has been observed that individuals with high resiliency tend to be more effective in today's world of uncertainty(Walumbwa, Wang, Wang, Schaubroeck, & Avolio, 2010).

Optimism: According to Seligman's theory (Csikszentmihalyi & Seligman, 2000) and research; optimism is more consistent with positive psychology than other structures. Like hope, optimism is a commonly used term, however, Seligman's definition arises from continuity and prevalence which are two distinctive features in the individual's way of explaining good and evil (Seligman, 2004). While optimistic people evaluate bad events as temporary, it is observed that pessimistic people evaluate bad events as permanent. For the good events, on the other hand, the opposite attribution is observed, while optimists accept positive events as permanent, pessimists accept them as temporary. Thus it can be summarized that while pessimistic people make a universal reference to bad events, optimists make context-specific reference to these kinds of situations. Schulman (1999) suggests specific guidelines for optimism that could be applied for the development of psychological capital; identifying self-undermining beliefs when faced with a challenge; evaluating the accuracy of beliefs; replacing false beliefs with improved, more constructive and correct beliefs. Likewise, Schneider proposed three perspectives to develop realistic optimism in workplaces; looking at the past with tolerance, being satisfied with the present; and looking after future opportunities(Schneider, 2001).

Psychological Capital

There are many methods to measure the success of an organization, its ability to pay debt or its value. One of these methods is to measure the forms of capital. The capital of an organization can ben measured as a financial capital based on the assets or as a physical capital based on their values. Human resources managers, on the other hand, measure human capital or social capital based on their employees (Crook, Todd, Combs, Woehr, & Ketchen, 2011). Moreover, human capital can also be assessed via an implicit knowledge, constitutes personal skills, quality, talent, personal values, hopes, health, manpower, decision-making, behavior and personality of individuals. Social capital, on the other hand, is an economic and sociological concept that is constituted by the totality of ties within the social networks in a society (Adler & Kwon, 2002). However, although these approaches are appropriate and useful for the organizational research; there is limited knowledge in the literature that how individuals' psychological and behavioral strengths can be measures. Accordingly, a new type of capital has been defined in the literature as "Psychological Capital" (Luthans et al., 2007). The psychological capital concept has been composed of the aforementioned positive psychology states which are hope, resiliency, optimism, and self-efficacy. According to Luthans and his colleagues (2007) the concept can be briefly defined as a positive psychological development of individuals. Thus the psychological capital characterized as; to have confidence to overcome challenging tasks and make the necessary effort (self-sufficiency), to have a positive expectation for achieving the success today and also in the future (optimism), to lead the way to success when necessary and to be diligent for purposes (hope) and to be self-determined to continue when hindered by difficulties and problems(resiliency) (Larson & Luthans, 2006).

Psychological capital is defined as a high-level core structure composed of states that meet the criteria of POB. In other words, the whole (psychological capital) can be greater than the sum of its parts (self-efficacy, optimism, hope, and resiliency) (Luthans et al., 2007). When we consider the individually distinctive features of psychological capital, it can be assumed that each ability adds a unique variance into psychological capital with their developmental nature. Thus, as well as social capital and human capital, psychological capital can be invested and managed. Since positive psychology has a state- like background, psychological capital and its dimentions can be developed and invested by today's managers and leaders in order to gain competitive advantage. Thus, psychological capital was found to be positively associated with employee performance and satisfaction(Luthans et al., 2010).

Psychological Capital's Relation to Human and Social Capital

From the traditional point of view, economic capital has kept all the attention over many years. It is usual for enterprises with high economic capital to be considered as more successful and rich. However, today's enterprises are struggling to maintain their existence by showing high performance under intense competition due to global market conditions. In order to achieve this, they had to learn to use not only the economic resources of their enterprises but also all their tangible and intangible resources effectively in this race. Thus, the importance of intangible assets and their management is an important topic for managers and leaders.

Human capital is the leading intangible asset. According to Luthans et al. "Human" refers to people working at all levels of the organization, and "capital" refers to consumption for the expected return in the future (Luthans et al., 2007). Human capital, thus, the joint strategy that is effective of organizational performance(Larson & Luthans, 2006).

Social capital, on the other hand, refers to the networks to which the employee is linked. This type of capital, associated with whom the employee knows, takes the business one step ahead of its competitors, especially as networks that can be used to solve problems. Social capital can be used within the enterprise (to whom can I apply for a solution to this problem?), which also can be used outside of the enterprise (who gives me the best price and quality advice when making this budget decision?). It is has been stated that social capital, like human capital, recognized and invested by the managers because of its vital importance for the success and competitive advantage of firms (Larson & Luthans, 2006).

On the other hand there is a continuance, rather than divergence, among these capitals of a firm hold. Figure 1 shows this continuance between all the capital an organization may have including psychological capital.

MAIN FOCUS OF THE CHAPTER

Issues, Controversies, Problems

With the emphasis on bot positive organizational behaviours and developmental side of human resources, there has been a growing number of publications focusing on psychological capital(PsyCap) and its relations with other concepts in the organizational behaviour and human resources fields. It has been shown by researchers that application of positive psychology at the organizational level supports the

Figure 1. Psychological capital as an integrator of human and social capital
Source: (Luthans, Luthans, & Luthans, 2004)

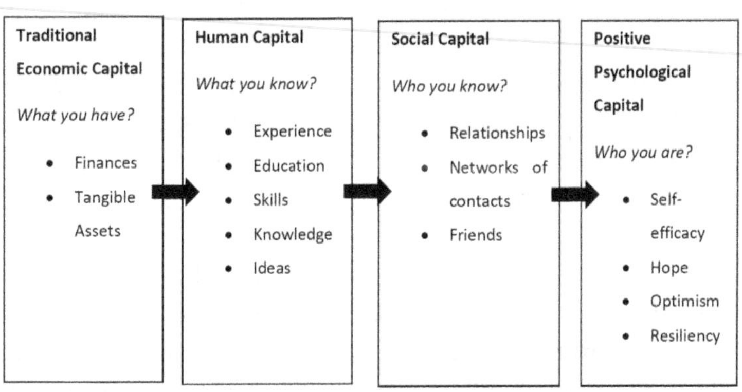

emergence of desired behaviours in organizations(Jensen & Luthans, 2006; Luthans & Youssef, 2007; Stajkovic & Luthans, 1998; Youssef, 2004). These studies are either take effects the dimensions of PsyCap individually or PsyCap concept as a whole. For either case, however, the benefits of positive organizational behavior in general and PsyCap, in particular, have been demonstrated.

The vast majority of these studies were devoted to understanding the performance impact of PsyCap. For example, according to their meta-analysis Stajkovic and Luthans(1998) have found a positive high-level correlation between self-efficacy and workplace performance of employees. Similarly, a study of Peterson and Luthans(2002) that is conducted in fast-food restaurants of the US revealed that as a psychological capital the hope level of managers is a strong indicator for explaining the workers' performance as well as their job satisfaction and turnover intention. Another study that is conducted by Luthans, Avolio, Walumbwa, and Li (2005) in China has provided more pieces of evidence for PsyCap employee performance relation. More specifically, according to the study there is a positive and strong relationship between workers' resiliency and their job performance. Furthermore, the resiliency of individuals has also been found to be positively related to individuals' job satisfaction and commitment behaviors by other researchers(Luthans & Youssef, 2007). Sweetman and his colleagues (2011) study, on the other hand, revealed that although the components of PsyCap has importance for the creative performance of employees, the overall PsyCap score of the same individuals has more impact over their performance. In other words, they have been showed that both the dimensions of PsyCap individually and the PsyCap concept in the aggregate are influential over the performance.

Besides performance studies, there are also other examinations that focus the relation of psychological capital with other concepts like leadership, organizational climate, organizational commitment, and organizational citizenship behavior etc. For example, Luthans, Norman, Avolio, and Avey (2008) have revealed that besides having positive influence on performance, satisfaction, and commitment of employees, psychological capital also mediates the relation between supportive organizational climate and job performance. They conclude that in a supportive work climate individuals' psychological capital would be increased in turn this would affect the job performance of these individuals. Moreover, Larson and Luthans (2006) found that there is a positive and strong relation between PsyCap and job satisfaction and organizational commitment. They also added to their conclusion that hope as a psychological capital of individuals may determine their academic and athletic success as well as mental and physical health. Besides its individual level implementations, it has also been showed that psychological capital can be developed at group level and this collective PsyCap may bring out desired organizational outcomes. Lopez and Snyder (2009) suggested the transmission of positive behaviors from one individual to another in a group which at the end have potential to create a positive attitude at group level. Walumbwa and his colleagues (2011), for instance, have shown that collective psychological capital and trust have a mediating role in authentic leadership and organizational citizenship behavior. Overall, it has been showed by several researchers that psychological capital has desired outcomes for organizations both at individual and organizational levels. More specifically, it has been showed that through creating a psychological base for organizational development, psychological capital creates a competitive advantage for organizations besides social and human resources.

On the other hand, there are still place for further investigations on psychological capital and its relations with organizational outcomes. In the existing study, leadership is thought to be one of the concepts that received considerable attention from PsyCap researchers. Thus, in this study the role of psychological capital in authentic leadership and organizational commitment has been investigated.

The Moderator Role of Psychological Capital in Authentic Leadership Organizational Commitment Relation

As a type of leadership, authentic leadership is defined as *"a pattern of leader behavior that draws upon and promotes both positive psychological capacities and a positive ethical climate, to foster greater self-awareness, an internalized moral perspective, balanced processing of information, and relational transparency on the part of leaders working with followers, fostering positive self- development"* by Walumbwa and his colleagues (2008). It has been stated that these leaders generally puts emphasis on the positive and developmental synergy between them and their followers(Woolley, Caza, & Levy, 2011). Thus, it can be expected that authentic leaders enhance their followers' psychological capital and foster a positive organizational climate. Moreover, authentic leaders are demonstrated as a leader that comprehends his/her own abilities, attaches importance the moral issues and has high communicative skills that give explanations to followers about the rationale behind his/her behaviors. These leaders, in this way, create a work environment in which both psychological capital and positive organizational behavior/organizational climate flourish (Avolio & Gardner, 2005). For instance, an authentic leader that restores the feeling of personal development in his/her follower would also create hope in the same follower as a psychological capital. Besides, it has been stated that authentic leaders that encourage followers through feedbacks enhance a sense of inclusion in the followers which also indicated to bring out resiliency as a psychological capital(Woolley et al., 2011).

Authentic leadership, on the other hand, has been proved to provide job performance, creativity, citizenship behavior, and commitment. Rego and his colleagues (2012), for example, showed that authentic leaders support the creativity of employees. They explain that through the characteristics of transparency, self-awareness, ethical concerns and being objective authentic leaders create trust and respect for their followers. Followers who experience trust in their leaders, in return, do not hesitate to bring creative and original ideas into the organizations. Similarly, Walumbwa et al., (2010) also suggest evidence that authentic leaders' willingness to share the information and being open to employees' ideas create a workplace where employees are aware of the importance of helping behaviors and altruism. Moreover, it has been stated by them that authentic leaders support followers to show organizational citizenship behavior and work engagement.

On the other hand, organizational commitment is an important concept in the organizational behavior field which is thought to be affected by authentic leadership. The commitment of employees is briefly defined as employees' connection to the organization and has three sub-dimensions namely; affective commitment, continuance commitment, and normative commitment. Firstly, affective commitment is defined as employees' emotional connection with the organization. This attachment is proved to arise from the perception that the organization care and respects them which is represented by the respect of the leader. Thus, these emotional motivation is thought to affect the employee to stay in the organization(Meyer & Allen, 1997). Secondly, the continuance commitment, on the other hand, refers to employees' feeling themselves to stay in the organization because of the costs of leaving. It has been stated that if employees don't have any option to stay in the job, they prefer to stay in order not to bear the costs of leaving. Thirdly, normative commitment is related to the moral obligations of employees. If employees feel responsible for the organization, they show normative commitment and don't leave the organization (Meyer & Allen, 1997).

Authentic leaders are proved to influence the organizational commitment of their followers in several studies. Walumbwa, Avolio, and Gardner (2008), for example, showed that authentic leaders are successful in reflecting their values to their followers. In return, it has been showed that followers reciprocate this behavior which consequently leads to the organizational commitment of followers. Similarly, Leroy, Palanski, and Simons (2012) showed that the openness of authentic leaders creates a trusted environment in which there is a positive social exchange between the leader and followers. Thus, in such an environment employees are reported to identify themselves with both their leader and also the organization. This identification leads them to organizational commitment, especially affective commitment. Finally, Guerrero and colleagues (2015) demonstrate that the transparency of authentic leaders creates an atmosphere in which there is a reciprocal trust, and thus, safety. In that kind of environment, it has been stated that the organizational commitment of employees is inevitable.

Although the relation between authentic leadership and organizational commitment has been demonstrated through several studies, the knowledge of the concepts that have the potential to moderate this relation is relatively limited. Accordingly, in the existing study, it has been proposed that the psychological capital concept is one of the key concepts that have capability to affect this relation. As has been discussed previously, the authentic leadership style has influence over the development of the psychological capital and organizational commitment of the employees. Especially, followers with low psychological capital is thought to be affected the positive organizational environment and reinforcement of authentic leaders which consequently make them show organizational commitment. Thus, the relation between authentic leadership an organizational commitment is expected to be higher especially when the PsyCap of the employees is low.

FUTURE DIRECTIONS

This study suggests the moderator role of psychological capital between authentic leadership and organizational commitment at the individual level. However, as it has been stated previously psychological capital can be developed at group level through the authentic leadership implementations. Thus, it is thought to be beneficial in order to provide more comprehensive understanding that this relation can be studied at group level. Furthermore, since the individual behaviors are highly determined by the context in which they live, the contextual, especially, the cultural differentiations are thought to be effective on both psychological capital and organizational commitment. Apart from the leadership style, it can be investigated how psychological capital of individuals is affected by the culture they live in. Accordingly, if there is a dominant leadership style that is promoted by the culture, how this leadership affects both the psychological capital and organizational commitment of the employees. Finally, the role of psychological capital within the leadership and desired organizational outcomes such as organizational citizenship behavior, trust, performance, etc. should be studied in order to provide better understanding of both positive organizational behaviors and its consequences in organizations.

CONCLUSION

Today, businesses no longer consist of mere tangible assets, they can create and maintain their difference with their intangible assets. For this reason, managers are trying to capture the superiority that will leave their competitors behind in an intense competition environment through recognizing the human resources that are at the forefront of abstract assets with their developmental features. At this point, psychological capital, which enables the employee to be better and to improve his / her existing features, helps the managers to keep one step ahead. Human capital gives great support to today's enterprises in order to gain a competitive advantage. In addition to human capital, making use of social capital and emerging psychological capital today will enable enterprises to gain a significant advantage in difficult competition conditions. The question "who am I" is as important as "what do I know" and "who do I know ". It is thought that going beyond the human and social capital to reach the psychological capital will give us the ability to cope with the challenges facing the organization today and in the future and to compete by recognizing the human factor better.

As an emerging concept, positive organizational studies in general and psychological capital, in particular, has been attracted researchers recently. In the existing study, first the concepts of positive organizational behavior and the psychological capital have been discussed with a comprehensive approach. Related with its potential in desired organizational outcomes and authentic leadership styles, it has been discussed that low psychological capital employees are expected to show more organizational commitment under the leading of an authentic leader. Since authentic leaders increase the psychological capital of their followers with their transparency and openness it cannot be expected to influence high PsyCap followers. On the other hand, because of their low self-efficacy, resilience, hope and optimism level, low PsyCap individuals will be affected more from their authentic leader's implementations. Thus, while their PsyCap increases it can be expected that this increase also affects their organizational commitment, due to psychological capital's influence on organizational commitment behavior.

REFERENCES

Adler, P. S., & Kwon, S. (2002). Social Capital: Prospects for a New Concept. *Academy of Management Review*, *27*(1), 17–40. doi:10.5465/amr.2002.5922314

Avolio, B. J., & Gardner, W. L. (2005). Authentic leadership development: Getting to the root of positive forms of leadership. *The Leadership Quarterly*, *16*(3), 315–338. doi:10.1016/j.leaqua.2005.03.001

Bailey, T. C., Eng, W., Frisch, M. B., & Snyder, C. R. (2007). Hope and optimism as related to life satisfaction. *The Journal of Positive Psychology*, *2*(3), 168–175. doi:10.1080/17439760701409546

Bakker, A. B., & Schaufeli, W. B. (2008). Positive organizational behavior: Engaged employees in flourishing organizations. *Journal of Organizational Behavior*, *29*(2), 147–154. doi:10.1002/job.515

Bandura, A. (1977). Self-efficacy: Toward a unifying theory of behavioural change. *Psychological Review*, *84*(2), 191–215. doi:10.1037/0033-295X.84.2.191 PMID:847061

Brissette, I., Scheier, M. F., & Carver, C. S. (2002). The role of optimism in social network development, coping, and psychological adjustment during a life transition. *Journal of Personality and Social Psychology*, *82*(1), 102–111. doi:10.1037/0022-3514.82.1.102 PMID:11811628

Coutu, D. L. (2002). How resilience works. *Harvard Business Review*, *80*(5), 46. PMID:12024758

Crook, T. R., Todd, S. Y., Combs, J. G., Woehr, D. J., & Ketchen, D. J. (2011). Does human capital matter? a meta-analysis of the relationship between human capital and firm performance. *The Journal of Applied Psychology*, *96*(3), 443–456. doi:10.1037/a0022147 PMID:21244126

Csikszentmihalyi, M., & Seligman, M. (2000). Positive psychology: An introduction. *The American Psychologist*, *55*(1), 5–14. doi:10.1037/0003-066X.55.1.5 PMID:11392865

Gable, S. L., & Haidt, J. (2005). What (and why) is positive psychology? *Review of General Psychology*, *9*(2), 103–110. doi:10.1037/1089-2680.9.2.103

Guerrero, S., Lapalme, M. È., & Séguin, M. (2015). Board Chair Authentic Leadership and Nonexecutives' Motivation and Commitment. *Journal of Leadership & Organizational Studies*, *22*(1), 88–101. doi:10.1177/1548051814531825

Jensen, S., & Luthans, F. (2006). Relationship between Entrepreneurs ' Psychological Capital and Their Authentic Leadership. *Journal of Managerial Issues*, *18*(2), 254–273.

Larson, M., & Luthans, F. (2006). Potential Added Value of Psychological Capital in Predicting Work Attitudes. *Journal of Leadership & Organizational Studies*, *13*(2), 75–92. doi:10.1177/10717919070130020601

Leroy, H., Palanski, M. E., & Simons, T. (2012). Authentic Leadership and Behavioral Integrity as Drivers of Follower Commitment and Performance. *Journal of Business Ethics*, *107*(3), 255–264. doi:10.100710551-011-1036-1

Linley, A. P., Joseph, S., Harrington, S., & Wood, A. M. (2006). Positive psychology: Past, present, and (possible) future. *The Journal of Positive Psychology*, *1*(1), 3–16. doi:10.1080/17439760500372796

Lopez, S. J., & Snyder, C. (2009). *Oxford handbook of positive psychology*. New York: Oxford University Press. doi:10.1093/oxfordhb/9780195187243.001.0001

Luthans, F. (2002). The need for and meaning of positive organizational behavior. *Journal of Organizational Behavior*, *23*(6), 695–706. doi:10.1002/job.165

Luthans, F., Avey, J. B., Avolio, B. J., & Peterson, S. J. (2010). The development and resulting performance impact of positive psychological capital. *Human Resource Development Quarterly*, *21*(1), 41–67. doi:10.1002/hrdq.20034

Luthans, F., & Avolio, B. J. (2009). The "point" of positive organizational behavior. *Journal of Organizational Behavior*, *30*(2), 291–307. doi:10.1002/job.589

Luthans, F., Avolio, B. J., Walumbwa, F. O., & Li, W. (2005). The Psychological Capital of Chinese Workers: Exploring the Relationship with Performance. *Management and Organization Review*, *1*(2), 249–271. doi:10.1111/j.1740-8784.2005.00011.x

Luthans, F., Luthans, K. W., & Luthans, B. C. (2004). Positive psychological capital: Beyond human and social capital. *Business Horizons*, *47*(1), 45–50. doi:10.1016/j.bushor.2003.11.007

Luthans, F., Norman, S. M., Avolio, B. J., & Avey, J. B. (2008). The mediating role of psychological capital in the supportive organizational climate—Employee performance relationship. *Journal of Organizational Behavior*, *29*(2), 219–238. doi:10.1002/job.507

Luthans, F., & Youssef, C. M. (2007). Emerging positive organizational behavior. *Journal of Management*, *33*(3), 321–349. doi:10.1177/0149206307300814

Luthans, F., Youssef, C. M., & Avolio, B. J. (2007). Psychological Capital: Developing the Human Competitive Edge. In Psychological Capital: Developing the Human Competitive Edge. Academic Press.

Meyer, J. P., & Allen, N. J. (1997). *Commitment in the Workplace: Theory, Research and Application*. SAGE Publication Ltd.

Muse, L., Harris, S. G., Giles, W. F., & Feild, H. S. (2008). Work-life benefits and positive organizational behavior: Is there a connection? *Journal of Organizational Behavior*, *29*(2), 171–192. doi:10.1002/job.506

Peterson, S. J., & Luthans, F. (2002). Does the manager's level of hope matter? Preliminary research evidence of a positive impact. *Proceedings of the Midwest Academy of Management.*

Rego, A., Sousa, F., Marques, C., & Cunha, M. P. e. (2012). Authentic leadership promoting employees' psychological capital and creativity. *Journal of Business Research, 65*(3), 429–437. doi:10.1016/j.jbusres.2011.10.003

Schneider, S. (2001). In search of realistic optimism: Meaning, knowledge, and warm fuzziness. *The American Psychologist, 56*(3), 250–263. doi:10.1037/0003-066X.56.3.250 PMID:11315251

Schulman, P. (1999). Applying learned optimism to increase sales productivity. *Journal of Personal Selling & Sales Management, 19*(1), 31–37.

Seligman, M. (2004). *Authentic Happiness: Using the new positive psychology to realize your potential for lasting fulfillment.* Simon and Schuster.

Shahnawaz, M., & Hassan, M. (2009). Psychological Capital as Predictors of Organizational Commitment and Organizational Citizenship Behaviour. *Journal of the Indian Academy of Applied Psychology, 35*(October), 78–84.

Snyder, C. R., Feldman, D. B., Taylor, J. D., Schroeder, L. L., & Adams, V. H. III. (2000). The roles of hopeful thinking in preventing problems and enhancing strengths. *Applied & Preventive Psychology, 9*(4), 249–269. doi:10.1016/S0962-1849(00)80003-7

Snyder, C. R., Harris, C., Anderson, J. R., Holleran, S., Irving, L., Sigmon, S., & Harney, P. (1991). The Will and the Ways: Development and Validation of an Individual-Differences Measure of Hope. *Journal of Personality and Social Psychology, 60*(4), 570–585. doi:10.1037/0022-3514.60.4.570 PMID:2037968

Stajkovic, A. D., & Luthans, F. (1998). Self-Efficacy and Work-Related Performance: A Meta-Analysis. *Psychological Bulletin, 124*(2), 240–261. doi:10.1037/0033-2909.124.2.240

Sweetman, D., Luthans, F., Avey, J. B., & Luthans, B. C. (2011). Relationship between positive psychological capital and creative performance. *Canadian Journal of Administrative Sciences, 28*(1), 4–13. doi:10.1002/cjas.175

Walumbwa, F. O., Avolio, B. J., Gardner, W. L., Wernsing, T. S., & Peterson, S. J. (2008). Authentic leadership: Development and validation of a theory-based measure. *Journal of Management, 34*(1), 89–126. doi:10.1177/0149206307308913

Walumbwa, F. O., Luthans, F., Avey, J. B., & Oke, A. (2011). Retracted : Authentically leading groups: The mediating role of collective psychological capital and trust. *Journal of Organizational Behavior, 32*(1), 4–24. doi:10.1002/job.653

Walumbwa, F. O., Wang, P., Wang, H., Schaubroeck, J., & Avolio, B. J. (2010). Psychological processes linking authentic leadership to follower behaviors. *The Leadership Quarterly, 21*(5), 901–914. doi:10.1016/j.leaqua.2010.07.015

Woolley, L., Caza, A., & Levy, L. (2011). Authentic Leadership and Follower Development. *Journal of Leadership & Organizational Studies, 18*(4), 438–448. doi:10.1177/1548051810382013

Wright, T. A. (2003). Positive organizational behavior: An idea whose time has truly come. *Journal of Organizational Behavior, 24*(4), 437–442. doi:10.1002/job.197

Youssef, C. M. (2004). *Resiliency Development of Organizations, Leaders and Employees: Multulevel Theory Building and Individual- Level, Path-Analytical Emprical Testing*. University of Nebraska.

KEY TERMS AND DEFINITIONS

Authentic Leadership: A pattern of leader behavior that draws upon and promotes both positive psychological capacities and a positive ethical climate, to foster greater self-awareness, an internalized moral perspective, balanced processing of information, and relational transparency on the part of leaders working with followers, fostering positive self-development.

Organizational Commitment: Attachment to, identification with, and involvement in the organization.

Psychological Capital: An individual positive psychological state.

Chapter 13
Authentic Leadership, Leader Optimism, and Follower Affective Commitment:
An Experimental Study

Nevra Baker
https://orcid.org/0000-0002-5557-8235
Altinbas University, Turkey

ABSTRACT

The aim of this study is to investigate the effect of authentic leadership together with the leader emotional expressivity of optimism on follower affective commitment towards the organization. This study was undertaken in the format of an experiment that is followed by a survey in which four groups of different fictional leader types were created, and respondents were asked to answer questions about their self-rated potential affective commitment towards the organization with the assumption that the respondents were actually working with the described leader. The findings of this study reveal that both authentic and inauthentic leaders who display optimism arouse higher levels of affective comitment by their followers as compared to their non-optimistic counterparts.

INTRODUCTION

Authentic leadership is one of the most widely researched theories in leadership. The antecedents and the consequences of authentic leadership have been analyzed by many researchers. For example, according to Bennis (2003) and Eriksen (2009),

DOI: 10.4018/978-1-7998-1108-4.ch013

self-knowledge is a prerequisite for authentic leadership. Second, Shamir and Eilam (2005) cite that leaders who have a high level of self-knowledge are clear about their values and convictions. Another antecedent for authentic leadership is self-consistency (Peus, Wesche, Streicher, Braun, and Frey, 2012). In addition, Walumbwa, Avolio, Gardner, Wernsing, and Peterson (2008) argue that it is of utmost importance for leaders to show consistency between their values, beliefs, and actions in order to be perceived as authentic. On the other hand, as cited in Peus et al. (2012), followers' intrinsic motivation, self-esteem, and creativity (Ilies, Morgeson, and Nahrgang, 2005), trust, engagement and well-being (Gardner, Avolio, Luthans, May, and Walumbwa, 2005), as well as organizational citizenship and performance (Walumbwa et al., 2008) are among the consequences of authentic leadership. Also, Baker (2018) found in her recent study that authentic leadership has a positive affect on follower job satisfaction.

Emotions are omnipresent in leader-follower interactions, originating from and also affecting them (Rubin, Munz, and Bommer, 2005; Sy, Côté, and Saavedra, 2005). Because leaders have a deep influence on the activity of organizations and their insiders (Yukl, 2005), leader emotional expositions have solid capacity to affect how their subordinates feel, think, and act (George, 2000).

Former studies have investigated the effect of authentic leadership on follower affective commitment (e.g., Peus et al., 2012; Leroy, Palanski, and Simons, 2012; Guerrero, Lapalme, Séguin, 2014), and the effect of leader optimism on leadership outcomes (e.g., Chemers, Watson, and May, 2000; Murphy and Ensher, 1999; Avey, Avolio, and Luthans, 2011; Wang, Zhuang, Yang, and Sheng, 2014). Nevertheless, a gap exists in the leadership and emotions literatures in terms of what kind of an effect leader emotional expressivity of discrete emotions together with leader authenticity might have on follower job outcomes. The aim of this study is to fill this gap by analyzing the effect of authentic leadership together with the leader emotional expressivity of a discrete positive emotion, which is optimism, on follower affective commitment towards the organization. In order to do so, an experimental study is undertaken by describing four different fictional leader types to the respondents and asking them about their possible affective commitment with the assumption that they were actually working with the described leader.

BACKGROUND

Authenticity can be retraced to ancient Greek philosophy and is expressed by the Greek saying "Know Thyself" which was inscribed in the Temple of Apollo at Delphi (Parke and Wormell, 1956). In fact, the word authentic can be traced back to the Greek word, authento, "to have full power" (Trilling, 1972), meaning that an individual is "the master of his or her own domain" (Kernis and Goldman, 2006). An early reference to authentic functioning is Socrates' focus on self-inquiry as he debated that an "unexamined" life is not worth living. Aristotle pursued with a view of ethics that focused on one's chasing after the "higher good" attained via self-realization when the activity of the soul is aligned with virtue to generate a whole life (Hutchinson, 1995). This self-realization is interconnected with one's well-being or "eudaimonia", a form of happiness that stems from victoriously accomplishing activities that are associated with one's genuine function, contrary to hedonism which aims at happiness and delight as aspired end states (Kernis and Goldman, 2006). Harter (2002) depicts authenticity as accepting one's individual experiences, together with one's thoughts, emotions, needs, desires, or beliefs. Therefore, it includes being self-aware and acting harmoniously with one's true self by stating what one honestly thinks and believes (Luthans and Avolio, 2003). Although the accession of thorough authenticity is an optimum, Erickson (1995) warns that authenticity should not be thought of as an either/or condition, due to the fact that people are never absolutely authentic or inauthentic. Hence, it is more proper to portray a person as more or less authentic (Gardner, Cogliser, Davis, and Dickens, 2011).

A number of studies have declared advantageous influences of positive emotional displays on assessments of leader effectiveness (Gaddis, Connelly, and Mumford, 2004), ratings of charisma and attraction to the leader (Bono and Ilies, 2006), high quality leader-member exchanges (Engle and Lord, 1997) and improved team climate (Pirola-Merlo, Härtel, Mann, and Hirst, 2002). Bono and Ilies (2006) found support for mood contagion effect, namely the spreading of positive affect from leaders to followers. Also, Engle and Lord (1997) put forth that positive leader affect is not only contagious, but also builds trust.

Shamir and Eilam (2005) argue that authentic leaders can be distinguished from inauthentic leaders in four aspects. These are the degree of person-role merger, the level of self-concept clarity, the extent to which their goals are self-concordant, and the degree to which their behavior is consistent with their self-concept. According to Bass and Steidlmeier (1999), a characteristic of inauthentic leadership is self-interest that motivates leaders to treat followers as means to their own ends. In addition, they argue that inauthentic leaders are characterized as self-concerned, exploitative, and power oriented, and they believe in distorted utilitarian moral principles.

THE EFFECTS OF AUTHENTIC LEADERSHIP AND LEADER OPTIMISM ON FOLLOWER AFFECTIVE COMMITMENT

Organizational commitment by followers has been ascertained in a number of studies as a result of authentic leadership (Avolio, Gardner, Walumbwa, Luthans, and May, 2004; Jensen and Luthans, 2006; Walumbwa et al., 2008). Affective organizational commitment can be defined as the employee's positive sentimental adherence to and identification with the organization (Allen and Meyer, 1990). Avolio and Gardner (2005) put forth that we can comprehend the correlation between authentic leadership and follower affective organizational commitment by means of the hypothetical mechanisms of favorable social exchanges and individual and social identification of the follower with the leader. Authentic leaders communicate in an open and non-defensive way—and therefore exhibit themselves to followers as defenseless (Leroy et al., 2012). This mutual, reliant relationship among leaders and followersalso lay outindividual and social identification between followers and leaders (Walumbwa, Christensen, and Hailey, 2011). Followers will recognize, appreciate, and cherish their leader's personality, desires, and demands (personal identification), along with their role–position as a leader and thereof as an ambassador for the entire organization (social identification).

Avolio et al. (2004) assert that one procedure authentic leaders can have an effect on their followers' optimism is to raise follower identification with the leader by setting an example by manifesting desirable positive emotions, paving the way to realistic optimism. As an empirical evidence, Youssef and Luthans (2007) reveal that optimism is positively related to organizational commitment, which affective commitment is a constituent of. The authors inform that an optimistic explanatory style can guide the leader to individually shoulder less guiltiness and widen his or her point of view to more thoroughly evaluate the outer, momentary, and conditional experimentations. Therefore, Youssef and Luthans (2007) denote that considering new actualities positively can give rise to more personal assessments of contentment and prosperity, both in the new job role and with regard to quality of life altogether. This kind of positive evaluations, widened point of view, and heading towards future can induce greater overall commitment to the organization.

In the light of the arguments listed above, the author of this study claims that in case of authentic leaders, the display of optimism by leaders will augment the positive effect of the high level of leader authenticity and will therefore result in higher levels of foreseen follower affective commitment as compared to followers who imagine having an authentic leader who does not display optimism. On the other hand, in case of inauthentic leaders, the author believes that leader optimism will make up for the potential negative effect of the lower levels of leader authenticity and will therefore result in higher levels of foreseen follower affective commitment. Therefore, the author puts forth the following hypotheses:

Hypothesis One: Optimistic authentic leaders will arouse higher affective commitment by their followers as compared to non-optimistic authentic leaders.

Hypothesis Two: Optimistic inauthentic will arouse higher affective commitment by their followers as compared to non-optimistic inauthentic leaders.

Research Model

The aim of this study is to test the effects authentic leadership and leader optimism on follower affective commitment. The model depicting the hypothetical relationships is presented in Figure 1.

Research Design and Methodology

This research was undertaken in the format of an experiment. Four 2*2 experiments were carried out in separate groups consisting of at least 10 persons. Each experiment was composed of two paragraphs. The first paragraph described a fictional authentic leader or a fictional inauthentic leader. After the first paragraph, a negative situation about a project that the fictional leader is leading is depicted. The second paragraph described the same leader expressing optimism or being non-optimistic about the negative situation. In order to describe authentic leadership, the items of the Organizational Leader Authenticity Scale (Henderson & Brookhart, 1996), and in order to describe optimism, items of the Life Orientation-Optimism Scale by Carver et al. (2010) were utilized, respectively. Example items from the Organizational Leader Authenticity Scale by Henderson and Brookhart (1996) used in the experiment for the description of the authentic leader are: "[The leader] often pays attention to team members during the project but does not behave like a know-it-all", "[The leader] listens to the ideas and suggestions of team members and is open to criticism", and "[The leader's] beliefs and actions are consistent and does not manipulate team

Figure 1. Conceptual model of the study

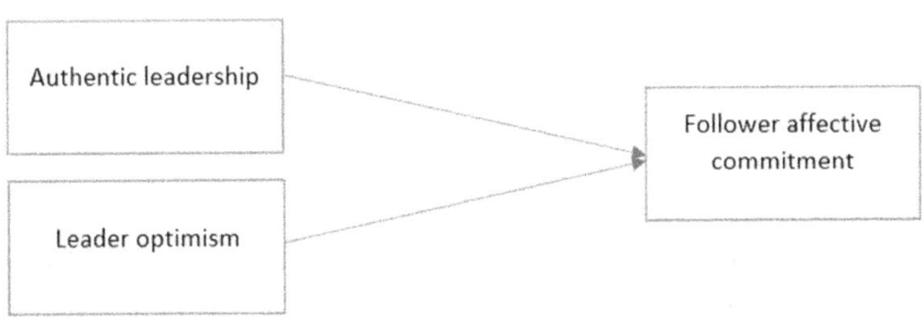

members". Example items used for the description of the inauthentic leader are: "If [the leader] makes mistakes, [the leader] does not accept and learn from them", "If something is wrong in the organization, [the leader] certainly puts the blame on one of the team members", and "[The leader] does not support that authority is delegated to team members". Example items from the Life Orientation-Optimism Scale by Carver et al. (2010) used in the experiment for the description of the optimistic leader are: "[The leader] looked at the bright side of the situation, like [the leader] always did in negative situations", "[The leader] said that this situation would motivate them to work more efficiently as a team, and let staff members know that they could easily overcome the situation", "In addition, [the leader] denoted that he was optimistic about the future of the project." Example items for the description of the non-optimistic leader are: "This situation destroyed [the leader's] morale, and he made staff members feel that", "[The leader] especially denoted that as a team, they would not easily overcome the situation and would get exhausted", "[The leader] added that he was not very optimistic about the future of the project". Because the respondents were Turkish speakers, the scale items of related scales were translated to Turkish. After reading the two paragraphs, the participants were given a survey so that they would be able to rate their own level of potential affective commitment to the organization, assuming that the fictional leader was their actual leader. This survey employed the Affective Commitment Scale by Meyer et al. (1993) which consists of eight items. Example items are "I would be very happy to spend the rest of my career with this organization", "I would feel like 'part of the family' at my organization", and "I would feel a strong sense of belonging to my organization". The answers of respondents were coded on a 5-point Likert scale.

Sample

A total of 64 employees working in the services industry were contacted, making up 4 different groups. The average age of the employees is 28.52, ranging from 20 to 50. 37 (57.8%) of the contacted employees are male. 8 (12.5%) of the contacted employees attended only elementary school, 37 (57.8%) are high school graduates, 18 (28.1%) attended university, and 1 (1.6%) completed higher education. The average working years add up to 8.34, ranging from 2 to 30. The average tenure of the respondents is 3.55 years, ranging from a minimum of 1 to a maximum of 23 years. 12 (18.8%) of the total of 64 respondents are from the retail industry, 11 (17.2%) work in the customer services industry, 7 (10.9%) come from the finance industry, again 7 (10.9%) work in the health sector, 5 (7.8%) are from the textile industry, 4 (6.3%) work in the printing industry, again 4 (6.3%) come from the food industry, and the remaining 14 respondents (21.9%) are from other services industries such as education, electronics, transportation, and communications.

SOLUTIONS AND RECOMMENDATIONS

For this study, 4 groups of different fictional leader types were created, and respondents were asked to answer questions about these leaders assuming that they work with the described leader.

Because there are two independent samples in this study, and the measurement level is ratio, independent samples t-test is used to analyze the differences between groups in terms of follower affective commitment towards the organization.

Table 1 and Table 2 below show the group statistics and the independent samples t-test results for Group 1 (optimistic authentic leader) and Group 2 (non-optimistic authentic leader):

As the above tables suggest, Group 1 consists of 13 subjects, and Group 2 is composed of 16 participants. According to the results of the Levene's Test for Equality of Variances, equal variances are not assumed for affective commitment ($p < .05$).

Table 1. Group statistics for Group 1 (optimistic authentic leader) and Group 2 (non-optimistic authentic leader)

Dependent variable		Groups	N	Mean	Std. Deviation	Std. Error Mean
Affective commitment		Optimistic authentic leader	13	3.63	.46	.13
		Non-optimistic authentic leader	16	2.57	.22	.06

Table 2. Independent samples t-test results for Group 1 (optimistic authentic leader) and Group 2 (non-optimistic authentic leader)

		Levene's Test for Equality of Variances		t-test for Equality of Means						
		F	Sig.	t	df	Sig. (2-tailed)	Mean Diff.	Std. Error Diff.	95% Confidence Interval of the Difference	
									Lower	Upper
Affective commitment	Equal variances assumed	5.53	.03	8.12	27	.00	1.06	.13	.80	1.33
	Equal variances not assumed			**7.59**	**16.49**	**.00**	**1.06**	**.14**	**.77**	**1.36**

Independent samples t-test revealed that there is a significant difference between the two groups in terms of affective commitment ($t = 7.59$, $p < .05$). Group 1 has a significantly higher mean value (3.63) than Group 2 (2.57). Thus, Hypothesis One (Optimistic authentic leaders will arouse higher affective commitment by their followers as compared to non-optimistic authentic leaders) is supported.

Table 3 and Table 4 below show the group statistics and the independent samples t-test results for Group 3 (optimistic inauthentic leader) and Group 4 (non-optimistic inauthentic leader):

As seen from the tables above, Group 3 consists of 13 subjects, and Group 4 is made up of 22 participants. According to the results of the Levene's Test for Equality of Variances, equal variances are not assumed for affective commitment ($p < .05$). Independent samples t-test revealed that there is a significant difference between the two groups in terms of affective commitment ($t = 3.03$, $p < .05$). The mean value for Group 3 (2.05) is significantly higher than the mean value of Group

Table 3. Group statistics for Group 3 (optimistic inauthentic leader) and Group 4 (non-optimistic inauthentic leader)

Dependent variable		Groups	N	Mean	Std. Deviation	Std. Error Mean
Affective commitment		Optimistic inauthentic leader	13	2.05	.48	.13
		Non-optimistic inauthentic leader	22	1.63	.21	.05

Table 4. Independent samples t-test results for Group 3 (optimistic inauthentic leader) and Group 4 (non-optimistic inauthentic leader)

<table>
<tr><th colspan="2" rowspan="3"></th><th colspan="2">Levene's Test for Equality of Variances</th><th colspan="7">t-test for Equality of Means</th></tr>
<tr><th rowspan="2">F</th><th rowspan="2">Sig.</th><th rowspan="2">t</th><th rowspan="2">df</th><th rowspan="2">Sig. (2-tailed)</th><th rowspan="2">Mean Diff.</th><th rowspan="2">Std. Error Diff.</th><th colspan="2">95% Confidence Interval of the Difference</th></tr>
<tr><th>Lower</th><th>Upper</th></tr>
<tr><td rowspan="2">Affective commitment</td><td>Equal variances assumed</td><td>9.64</td><td>.00</td><td>3.63</td><td>33</td><td>.00</td><td>.42</td><td>.12</td><td>.19</td><td>.66</td></tr>
<tr><td>Equal variances not assumed</td><td></td><td></td><td>3.03</td><td>14.83</td><td>.01</td><td>.42</td><td>.14</td><td>.12</td><td>.72</td></tr>
</table>

4 (1.63). Thus, Hypothesis Two (Optimistic inauthentic leaders will arouse higher affective commitment by their followers as compared to non-optimistic inauthentic leaders) is supported.

FUTURE RESEARCH DIRECTIONS

This research contributes to the leadership and emotions literatures by explaining the combined effect of authentic leadership and leader optimism on follower affective commitment. The results of this study suggest that in case of authentic leaders, the expression of optimism will add to the positive image of an authentic leader. On the other hand, in case of inauthentic leaders, the expression of optimism will make up for the possible negative image of an inauthentic leader from the perspective of the followers. Therefore, during their interactions with their followers, leaders should not forget that authentic leaders, who express a positive emotion such as optimism, will contribute to the positive job outcomes of their followers.

Previous studies suggested positive effects of leader optimism on leadership outcomes. For instance, research by Chemers, Watson, and May (2000) suggests that optimism was associated with rated leadership potential. Another study by Murphy and Ensher (1999) found that leader optimism predicted subordinates' ratings of LMX quality for female supervisors. Moreover, Avey, Avolio, and Luthans (2011) found a positive relationship between leader positivity and follower positivity and performance. Also, Wang, Zhuang, Yang, and Sheng (2014) argue that a team benefits more by having an optimistic agent as its leader.

In line with the previous studies, the results of the two hypotheses tested in this study reveal the enhancing effect of the expressivity of a positive discrete emotion by leaders, whether they are perceived as an authentic or an inauthentic leader in the eyes of their followers, in terms of their expected affective commitment. As a result, the author concludes that leaders should not ignore the power of positive emotions in terms of ameliorating their already established positive image in the eyes of their followers and in terms of compensating for their negative image in other aspects, such as being considered as an inauthentic leader. However, it is advisable that leaders should not rely too much on the expressivity of positive emotions but also check their image in the eyes of their followers in terms of positive leadership aspects such as leader-member exchange relationship, authenticity, and the like.

This research is, to the author's knowledge, the first to bring the concepts of authentic leadership, follower affective commitment towards the organization, and leader emotional expressivity of optimism together, and therefore shall contribute to the progress of leadership research.

As to the limitations of this research, it is a cross-sectional study, therefore data pertaining to the long term of the hypothesized relationships are not present. Secondly, all the data is collected from Istanbul, Turkey. Therefore, findings of this research might bear an effect of the Turkish culture and therefore limit the generalizability of the findings. Accordingly, the author suggests that the same study can be undertaken in different cultural contexts. Other than this, further research examining the moderating effect of the leader emotional expressivity of optimism on the relationship between authentic leadership and other follower job outcomes can be undertaken. Moreover, the moderating effect of the leader emotional expressivity of other discrete leader emotions on the relationship between authentic leadership and follower affective commitment towards the organization can be studied.

CONCLUSION

Intrinsic motivation and creativity, which are among the consequences of authentic leadership according to Ilies, Morgeson, and Nahrgang (2005), are among the most important hallmarks of innovative organizations in the era of digitalization. Thus, it is no surprise that the world's most innovative companies such as Apple, Dyson, Herman Miller are led by authentic leaders. One can assume that one of the reasons of the long-term success of these innovative firms it might be the high level of the affective commitment that their creative employees feel towards the organization they are working for.

The aim of this study was to uncover what kind of an effect an additional leader characteristic besides authentic leadership, which is leader optimism, would have on the followers' self-projected level affective commitment towards the organization, with the assumption that they were working with the described leader. As hypothesized and found in Hypothesis One, authentic leaders who display optimism aroused higher affective commitment by their followers as compared to authentic leaders who do not. Optimism is a positive emotion that goes hand in hand with authentic leadership. Hence, in line with the author's expectations, the expression of optimism by authentic leaders strengthened the positive relationship between authentic leadership and follower affective commitment. Optimism expressed by the immediate supervisors will add to their perception as a likeable leader in the eye of their follower, which will add to the affective commitment felt by them. As hypothesized and found in Hypothesis Two, inauthentic leaders who display optimism aroused higher affective commitment by their followers as compared to inauthentic leaders who do not. Hence, the mere expression of a positive emotion by an inauthentic leader might have alleviated his/her negative image in the eyes of the followers, making him/her be perceived as more positive in comparison to an inauthentic leader who acts neutral.

The results of the two hypotheses tested in this chapter reveal the enhancing effect of the expressivity of a positive discrete emotion by leaders, whether they are perceived as authentic or inauthentic in the eyes of their followers, in terms of their affective commitment towards the organization they are working in. As a result, the author concludes that leaders should not ignore the power of positive emotions in terms of ameliorating their already established positive image in the eyes of their followers and in terms of compensating for their negative image in other aspects, such as being considered as inauthentic. However, it is advisable that leaders should not rely too much on the expressivity of positive emotions but also check their image in the eyes of their followers in terms of positive leadership aspects such as authenticity, leader-member exchange relationship, and the like.

ACKNOWLEDGMENT

I would like to thank Prof. Hayat Kabasakal and Prof. Hakan Ozcelik for their detailed comments and encouragement. This research was supported by Bogazici University Research Fund (BAP) [grant number 11183].

REFERENCES

Allen, N. J., & Meyer, J. P. (1990). The measurement and antecedents of affective, continuance and normative commitment to the organization. *Journal of Occupational Psychology*, *63*(1), 1–18. doi:10.1111/j.2044-8325.1990.tb00506.x

Avey, J. B., Avolio, B. J., & Luthans, F. (2011). Experimentally analyzing the impact of leader positivity on follower positivity and performance. *The Leadership Quarterly*, *22*(2), 282–294. doi:10.1016/j.leaqua.2011.02.004

Avolio, B. J., & Gardner, W. L. (2005). Authentic leadership development: Getting to the root of positive forms of leadership. *The Leadership Quarterly*, *16*(3), 315–338. doi:10.1016/j.leaqua.2005.03.001

Avolio, B. J., Gardner, W. L., Walumbwa, F. O., Luthans, F., & May, D. R. (2004). Unlocking the mask: A look at the process by which authentic leaders impact follower attitudes and behaviors. *The Leadership Quarterly*, *15*(6), 801–823. doi:10.1016/j.leaqua.2004.09.003

Baker, N. (2018). Authentic leadership and follower job satisfaction: The moderating effect of leader emotional expressivity. *International Journal of Social Sciences and Education Research*, *4*(4), 689–697. doi:10.24289/ijsser.472863

Bass, B. M., & Steidlmeier, P. (1999). Ethics, character, and authentic transformational leadership behavior. *The Leadership Quarterly*, *10*(2), 181–217. doi:10.1016/S1048-9843(99)00016-8

Bennis, W. (2003). *On Becoming a Leader*. Cambridge, MA: Perseus Publishing.

Bono, J. E., & Ilies, R. (2006). Charisma, positive emotions and mood contagion. *The Leadership Quarterly*, *17*(4), 317–334. doi:10.1016/j.leaqua.2006.04.008

Carver, C. S., Scheier, M. F., & Segerstrom, S. C. (2010). Optimism. *Clinical Psychology Review*, *30*(7), 879–889. doi:10.1016/j.cpr.2010.01.006 PMID:20170998

Chemers, M. M., Watson, C. B., & May, S. T. (2000). Dispositional affect and leadership effectiveness: A comparison of self-esteem, optimism, and efficacy. *Personality and Social Psychology Bulletin*, *26*(3), 267–277. doi:10.1177/0146167200265001

Engle, E. M., & Lord, R. G. (1997). Implicit theories, self-schemas, and leader-member exchange. *Academy of Management Journal*, *40*(4), 988–1010.

Erickson, R. J. (1995). The importance of authenticity for self and society. *Symbolic Interaction*, *18*(2), 121–144. doi:10.1525i.1995.18.2.121

Eriksen, M. (2009). Authentic leadership: Practical reflexivity, self awareness, and self-authorship. *Journal of Management Education*, *33*(6), 747–771. doi:10.1177/1052562909339307

Gaddis, B., Connelly, S., & Mumford, M. D. (2004). Failure feedback as an affective event: Influences of leader affect on subordinate attitudes and performance. *The Leadership Quarterly*, *15*(5), 663–686. doi:10.1016/j.leaqua.2004.05.011

Gardner, W. L., Avolio, B. J., Luthans, F., May, D. R., & Walumbwa, F. (2005). Can you see the real me? A self based model of authentic leader and follower development. *The Leadership Quarterly*, *16*(3), 343–372. doi:10.1016/j.leaqua.2005.03.003

Gardner, W. L., Cogliser, C. C., Davis, K. M., & Dickens, M. P. (2011). Authentic leadership: A review of the literature and research agenda. *The Leadership Quarterly*, *22*(6), 1120–1145. doi:10.1016/j.leaqua.2011.09.007

George, J. M. (2000). Emotions and leadership: The role of emotional intelligence. *Human Relations*, *53*(8), 1027–1055. doi:10.1177/0018726700538001

Guerrero, S., Lapalme, M. È., & Séguin, M. (2014). Board chair authentic leadership and nonexecutives' motivation and commitment. *Journal of Leadership & Organizational Studies*, *22*(1), 88–101. doi:10.1177/1548051814531825

Harter, S. (2002). Authenticity. In C. S. Snyder & S. J. Lopez (Eds.), *Handbook of positive psychology* (pp. 382–394). Oxford, UK: Oxford University Press.

Henderson, J. E., & Brookhart, S. M. (1996). *Organizational Leader Authenticity Scale*. OLAS.

Hutchinson, D. S. (1995). Ethics. In J. Barnes (Ed.), *The Cambridge companion to Aristotle* (pp. 195–232). Cambridge, UK: Cambridge University Press.

Ilies, R., Morgeson, F. P., & Nahrgang, J. D. (2005). Authentic leadership and eudaemonic well-being: Understanding leader- follower outcomes. *The Leadership Quarterly*, *76*(3), 373–394. doi:10.1016/j.leaqua.2005.03.002

Jensen, S. M., & Luthans, F. (2006). Relationship between entrepreneurs' psychological capital and their authentic leadership. *Journal of Managerial Issues*, *18*(2), 254–273.

Kernis, M. H., & Goldman, B. M. (2006). A multicomponent conceptualization of authenticity: Theory and research. In M. P. Zanna (Ed.), Advances in experimental social psychology (Vol. 38, pp. 283–357). San Diego, CA: Academic Press. doi:10.1016/S0065-2601(06)38006-9

Leroy, H., Palanski, M. E., & Simons, T. (2012). Authentic Leadership and Behavioral Integrity as Drivers of Follower Commitment and Performance. *Journal of Business Ethics*, *107*(3), 255–264. doi:10.100710551-011-1036-1

Luthans, F., & Avolio, B. J. (2003). Authentic leadership development. In K. S. Cameron, J. E. Dutton, & R. E. Quinn (Eds.), *Positive organizational scholarship: Foundations of a new discipline* (pp. 241–261). San Francisco: Barrett-Koehler.

Meyer, J. P., Allen, N. J., & Smith, C. A. (1993). Commitment to organizations and occupations: Extension and test of a three-component conceptualization. *The Journal of Applied Psychology*, *78*(4), 538–551. doi:10.1037/0021-9010.78.4.538

Murphy, S. E., & Ensher, E. A. (1999). The effects of leader and subordinate characteristics in the development of leader–member exchange quality. *Journal of Applied Social Psychology*, *29*(7), 1371–1394. doi:10.1111/j.1559-1816.1999.tb00144.x

Parke, H., & Wormell, D. (1956). *The Delphic Oracle*. Oxford, UK: Blackwell.

Peus, C., Wesche, J. S., Streicher, B., Braun, S., & Frey, D. (2012). Authentic Leadership: An Empirical Test of Its Antecedents, Consequences, and Mediating Mechanisms. *Journal of Business Ethics*, *107*(3), 331–348. doi:10.100710551-011-1042-3

Pirola-Merlo, A., Härtel, C., Mann, L., & Hirst, G. (2002). How leaders influence the impact of affective events on team climate and performance in R&D teams. *The Leadership Quarterly*, *13*(5), 561–581. doi:10.1016/S1048-9843(02)00144-3

Rubin, R. S., Munz, D. C., & Bommer, W. H. (2005). Leading from within: The effects of emotion recognition and personality on transformational leadership behavior. *Academy of Management Journal*, *48*(5), 845–858. doi:10.5465/amj.2005.18803926

Shamir, B., & Eilam, G. (2005). What's your story? A life-stories approach to authentic leadership development. *The Leadership Quarterly*, *16*(3), 395–417. doi:10.1016/j.leaqua.2005.03.005

Sy, T., Côté, S., & Saavedra, R. (2005). The contagious leader: Impact of the leader's mood on the mood of group members, group affective tone, and group processes. *The Journal of Applied Psychology*, *90*(2), 295–305. doi:10.1037/0021-9010.90.2.295 PMID:15769239

Trilling, L. (1972). *Sincerity and authenticity*. Cambridge, MA: Harvard University Press.

Walumbwa, F. O., Avolio, B. J., Gardner, W. L., Wernsing, T. S., & Peterson, S. J. (2008). Authentic leadership: Development and validation of a theory-based measure. *Journal of Management*, *34*(1), 89–126. doi:10.1177/0149206307308913

Walumbwa, F. O., Christensen, A. L., & Hailey, F. (2011). Authentic leadership and the knowledge economy. *Organizational Dynamics*, *40*(2), 110–118. doi:10.1016/j.orgdyn.2011.01.005

Wang, J., Zhuang, X., Yang, J., & Sheng, J. (2014). The effects of optimism bias in teams. *Applied Economics*, *46*(32), 3980–3994. doi:10.1080/00036846.2014.948678

Youssef, C. M., & Luthans, F. (2007). Positive Organizational Behavior in the Workplace: The Impact of Hope, Optimism, and Resilience. *Journal of Management*, *33*(5), 774–800. doi:10.1177/0149206307305562

Yukl, G. A. (2005). *Leadership in organizations* (6th ed.). New York: Prentice Hall.

ADDITIONAL READING

Brief, A. P., & Weiss, H. M. (2002). Organizational behavior: Affect in the work place. *Annual Review of Psychology, 53*(1), 279–307. doi:10.1146/annurev.psych.53.100901.135156 PMID:11752487

Cooper, C. D., Scandura, T. A., & Schriesheim, C. A. (2005). Looking forward but learning from our past: Potential challenges to developing authentic leadership theory and authentic leaders. *The Leadership Quarterly, 16*(3), 475–493. doi:10.1016/j.leaqua.2005.03.008

Fisher, C. D., & Ashkanasy, N. M. (2000). The emerging role of emotions in work life: An introduction. *Journal of Organizational Behavior, 21*(2), 123–129. doi:10.1002/(SICI)1099-1379(200003)21:2<123::AID-JOB33>3.0.CO;2-8

Fredrickson, B. L., & Joiner, T. (2002). Positive emotions trigger upward spirals toward emotional well-being. *Psychological Science, 13*(2), 172–175. doi:10.1111/1467-9280.00431 PMID:11934003

Gardner, W. L., Rozell, E. J., & Walumbwa, F. O. (2004). Positive and negative affect and explanatory style as predictors of work attitudes. In M. J. Martinko (Ed.), *Attribution theory in the organizational sciences: Theoretical and empirical contributions* (pp. 49–81). Greenwich, CT: Information Age Publishing.

Grossman, R. J. (2000). Emotions at work. *Health Forum Journal, 43*, 18–22. PMID:11066984

May, D. R., Chan, A. Y. L., Hodges, T. D., & Avolio, B. J. (2003). Developing the moral component of authentic leadership. *Organizational Dynamics, 32*(3), 247–260. doi:10.1016/S0090-2616(03)00032-9

Sparrowe, R. T. (2005). Authentic leadership and the narrative self. *The Leadership Quarterly, 16*(3), 419–439. doi:10.1016/j.leaqua.2005.03.004

Tate, B. (2008). A longitudinal study of the relationships among self-monitoring, authentic leadership, and perceptions of leadership. *Journal of Leadership & Organizational Studies, 15*(1), 16–29. doi:10.1177/1548051808318002

KEY TERMS AND DEFINITIONS

Affective Commitment: An employee's positive emotional attachment to the organization.

Authentic Leader: A leader who is self-conscious and acts in harmony with his/her true ideas and emotions.

Fictional Leader: Not a real leader, a leader described in an experiment.

Follower: Employees who are operating under the guidance of a leader.

Leader Emotional Expressivity: The degree two which leaders are expressing their emotions.

Moderating Effect: The interacting effect of a variable on the relationship between the independent and the dependent variables.

Optimism: The positive emotion that makes a person able to focus more on the the positive sides of life events rather than the bad sides.

Chapter 14
E-Participating Decision-Making Mechanism in the Public Administration System:
Example of E-Government Application

Konur Alp Demir
Tekirdağ Namik Kemal University, Turkey

ABSTRACT

In this chapter, an analysis of the electronic decision making system, which is thought to benefit from the heavy bureaucratic system which does not take into account the expectations of the citizen in the public administration system, will be used to make a more flexible structure. The focus of this chapter is on the need to design the decision-making mechanisms of the state according to the expectations of the citizen. For this purpose, requests and complaints from the citizens through the electronic environment should be taken into consideration in the decision-making process. In fact, this situation is reflected in the application of electronic participation management model. The application of this management model in the public administration system is the citizen participation complaint and demand system which is carried out under the name of electronic government. The examination of this system, which is an example of the application of participatory democracy, is important for the reflection of democratic values on the administration system.

DOI: 10.4018/978-1-7998-1108-4.ch014

1. INTRODUCTION

The result of the movement of technology with information is the integration of the government and the citizens. The government has succeeded in removing the obstacles between itself and the citizen by using information technologies. For this reason, the government has started to provide public services in an electronic environment. Electronicisation of government public service delivery methods has also led to a transformation in the infrastructure of public institutions (Sharma & Pokharel, 2016: 19).

The realization of management activities within an organization in a digital environment converges the goal that management wants to achieve. It is known that digital transformation forms used in the management process increase productivity. The process is shortened thanks to the use of technology in the planning and implementation stages of an organization. This saves time and energy (Bredmar, 2017: 115).

The electronic government is a technology that facilitates access to public services provided by the state over the internet. The basic philosophy of this technology is the fulfillment of public services that can be transferred to the virtual environment without person to person communication and away from the complexity of printed documents (Jaeger, 2003: 323).

This manuscript mainly consists of four basic sections.

In the first chapter, the concepts of public administration, public services, citizen oriented management, democracy, participation of the people in administration and electronic democracy (will be discussed. At the same time, it will be tried to form the conceptual framework necessary to ensure the integrity of the subject. In the second chapter, the concept of e-government and the issue of electronic political participation will be discussed. In order to ensure the integrity of the subject, the subject of public services in electronic environment will be examined. In this chapter, a comparative analysis of classical state administration and e-government administration will also be carried out. In the third and last chapter, the subject will be completed by giving examples of the e-government applications of different governments with a universal perspective.

2. CONCEPTUAL FRAMEWORK

The government provides public services that can affect the daily life of the citizen. These public services are aimed not only to meet the needs of individuals but also to provide the public interest (Holzer & Schwester, 2011: 19). At this point, public administration emerges as a practical tool used by a neutral administrator (Holzer & Schwester, 2011: 31).

Public administration is a technical concept which defines the administrative and financial orientation of a group of people towards a specific target. In general, public administration is the fulfillment of government affairs. In other words, public administration is the general name of the system that mediates the performance of the government's activities. In this respect, it can be stated that the main task of the public administration is to implement the decisions of the political nature determined by the decision makers (Khan, 2008: 1-3).

Public administration is the executive authority of the state. The state performs its basic functions through public administration (Öztekin, 2012: 240). Public administration is responsible for the implementation of public policies determined by political authorities. Therefore, public administration is a management system that is responsible for reflecting the issues that concern the citizens (Tortop et al., 2017: 267).

Decision making for public administration is an important topic. Because decision-making is a form of behavior that is consistent with public administration. In other words, the decision-making and the implementation of the decision are in the nature of the public administration system (Çevik, 2012: 125).

Public administration is at the intersection of scientific fields such as sociology, business, political science and psychology. For this reason, public administration can be interpreted as a social science. In this context, public administration maintains its existence as an effective discipline in politics production, participation and leadership. The most important benefit of public administration is to integrate the government and its environment. In other words, it is the public administration that allows the government to participate in decision-making mechanisms externally. Therefore, public administration acts as a bridge between the government and various actors and institutions (Khan, 2008: 5).

All of the public policies determined for the administration of the state were fictionalised decision making (Çevik, 2012: 131). The basis of this fiction is the desire to increase the functionality of democratic values. As democracy rises on the idea of self-governance of people, every citizen must be directly involved in the management and take part in decision-making mechanisms. However, it is impossible to implement this system in the current situation (Çevik, 2012: 136).

Decision-making is a conscious and optional act of choosing between at least two options. When the public administrators cannot make a decision, the society is alone with the problems it faces. In such a situation, the problem of deterioration of the social order may arise. Decision-making is basically an action carried out for two purposes. The first is the desire to bring any situation to a better position and the second one to achieve a projected goal. The decision-maker acts consciously in accordance with these two objectives (Akdoğan, 2013: 212).

The emergence of a contradictory situation by the public service within the democratic administration systems is a central problem that continues to be solved in the public administration. Democracies do not acquire assets that are not supported by a technical management, lacking an effective, efficient, strong and flexible public service (Hamilton, 2007: 3).

The public service is a concept that describes the distribution of taxes from the public to the public in a way that is beneficial to the public, and then to the public. The government offers public services in different areas, particularly in the areas of education, health, diplomacy and security. In this context, the government offers a public service to meet the needs of the citizen in terms of housing, income and social life to the family institution. In the final stage, it can be stated that the pulic service is the whole of the actions taken by the government for the welfare of its citizens (Spicker, 2009: 970).

Citizen's point of view towards public services is at the key point of citizen-oriented management. Because the duties, structure and managerial practices of public institutions are affected by the value judgments of the society and are shaped in the final stage. In a liberal capitalist system, public services are in a state of transformation that is transformative and at the stage of fulfillment. Because society wants to live under an effective and efficient government. At the same time, the society approaches the public sector with suspicion and wants to control and limit the public services carried out by public administrators (Box, 2007: viii).

In the circumstances of the day, it is possible to call democracy as a control system designed to prevent the administrator from exercising his powers in a way that is against the interests of the citizen. In the democratic system, decision makers must take into account the wishes and complaints of the society. In other words, the main fiction of decision makers to have such a necessity is at the core of the democratic system (Young, 2002: 17).

The purpose of founding a government is not to implement democracy. Democracy, in the process after establishment of the state to avoid social disruption and is a tool used to create a platform for people to live in a free environment. The ideal of democracy can be defined as "the declaration of independence" (Box, 2007: viii-ix).

In the classical democratic order, the participation of the citizen in administration system is limited. For example, there are limited ways of participation, such as voting, participation in political campaigns and participation in social demonstrations. Using these democratic methods, citizens are trying to put pressure on government decisions. In this way, they aim to direct the decisions of the government in accordance with their wishes. However, participating in political decision-making processes by voting only does not satisfy social expectations. At the same time, this situation cannot be called absolute political participation (Verba, Nie & Kim, 1987: 46-47).

Participation in the decision-making process is like a two-edged sword. Public administrators should be prepared to work in partnership with the citizen and to take advantage of their ideas. On the other hand, citizens should be both active and competent in matters of interest and participation (Fagence, 1977: 4).

The name of the system that makes the participation of citizens in the process of public administration without time and space limitation is electronic participation. Because the innovations that emerged as a result of social transformation and technology development are two types of variables that support each other. The industrial society, which emerged with the transformation of the social structure following the Industrial Revolution, evolved towards the information society with the contribution of the development of technology at the point reached. At this stage, the public administration system has naturally directed its own process in an environment equipped with information and technology to the new practices supported by classical systems from electronic systems. Thus, states tried to increase their functionality by benefiting from the information and digital innovations created by the developing technology (Parlak, 2011: 217).

Electronic democracy is a concept that refers to the involvement of citizens in the process of political participation through new information technologies (Vintar et al., 2001: 924). This concept creates a convenience without the limitation of time and space, whereby citizens can explain ideas, take part in discussion groups and participate in public information meetings (Hague & Loader, 1999: 3).

The application of electronic government is an issue that both public administrators and theoreticians emphasize. Thus, a new perspective is taken against the way in which the public administration system is implemented. For this reason, it is important to examine the electronic government applications (West, 2007: 2).

This review needs to be made under the four basic headings of the electronic government. In this context, related topics can be listed as follows (Parlak, 2011: 218):

- "Electronic Public Service (E-Public Service)"
- "Electronic Democracy (E-Democracy)"
- "Electronic Commerce (E-Commerce)"
- "Electronic Management (E-Management)"

In this chapter, Electronic Public Service (E-Public Service), Electronic Democracy (E-Democracy), Electronic Management (E-Management) and Electronic Government (E-Government) topics will be examined in order to achieve the targeted subject integrity.

The electronic service aims to reach the state-owned information and documents electronically in order to make public administration transparent and to facilitate the delivery of public services. However, electronic services aim to participate in public services without time and space limitations. With electronic democracy, it is aimed to enable citizens to participate in decision making processes via electronic mail, internet and other electronic communication devices. Electronic management has a fiction that aims to integrate information and documents among public institutions, integrate interrelated processes that operate under the control of different public institutions, and harmonize the public administration system to the digital age (Parlak, 2011: 219).

The electronic government application has 8 objectives. These objectives can be listed as follows (Parlak, 2011: 220-221):

- Transparency of public administration system,
- Resolving the work of citizens with the state without encountering the problem of bureaucracy and paperwork,
- Increasing the efficiency level of public administration service units,
- Ensuring integrity by creating an information and document sharing environment between public institutions,
- Ensuring the access of citizens to public services without time and space restrictions,
- Ensuring that public services are delivered to all segments of the society in an easy way,
- Ensuring that the requests and complaints of the citizens are taken into consideration in the process of creating public policies,
- Ensuring appropriate conditions for the efficient and efficient operation of public institutions.

Through electronic government applications, the government economizes many public expenditures. This situation also has a similar character for the citizens. Because public services are maintained through an online system, there is no financial loss. In addition, the application of electronic government has introduced the concept of new citizens. The new citizen is a citizen model that is more knowledgeable, more involved in the decision-making process of the public administration. Thus, the citizens can interact easily with their government (Garson, 2004: 3-4).

3. ELECTRONIC GOVERNMENT (E- GOVERNMENT)

The aims of e-government applications are to gain speed in public service delivery and to move away from bureaucratic process. At this point, related objectives can be listed as follows (Altınok, 2005: 138):

- Increasing the speed and quality of public service offerings,
- Removal of public institutions from the heavy working process of the bureaucratic system,
- Transparency in the state of law practice,
- Accelerating the implementation of laws,
- Prioritizing development in the field of economy,
- Ensuring that every part of the society reaches public services without limitation of time and space,
- Realization of multi-player management model and direct participation ideal.

In order to understand the functions of the e-government in the management process, it should be examined under three sub-headings. These headings can be listed as "centralized", "decentralized" and "hybrid" systems. The application of state electronic systems in these three different areas constitutes one of the responsibilities of the e-government application (Heeks, 2006: 18).

- *Centralized*: Decisions are taken in a central system (Heeks, 2006: 18).
- *Decentralized*: Decisions are taken at a lower level than a central system. Decentralization is a concept that defines the end user to have a say in administration. To evaluate this situation for state administration, citizens are involved in the decisions taken by the central government in the idea of decentralization (Heeks, 2006: 18).
- *Hybrid*: Decisions are taken by both the central government and the lower level decision-making authorities. Such a system is involved in a federal or unitary administration system (Heeks, 2006: 18).

The information society that emerged as a result of technological developments basically focuses on seven strategic points. These points can be listed as follows (Kaya Bensghir, 2013: 117):

- *Social Evolution*: Creation of a social evolution where every citizen can benefit from technological opportunities,
- *Information Communication Technology*: Use of information and communication technologies in competitive environments,

- *Citizen Oriented Administration*: Continue the management process in a way that is citizen-oriented,
- *Modernization of Public Administration Process*: Transportation of public administration operation process to an electronic environment,
- *Information Technology Management Sector Within the Scope of Global Competitiveness*: Maintaining an information technology management process with a competitive approach in the international arena,
- *Competitive, Easy to Access and Cheap Communication Systems*: Ensuring that citizens have easy and cheap access to communication facilities,
- *Innovations in Research and Development Activities*: The discovery of needed innovation in the global arena.

One of the most important results to be reached with e-government applications is to bring the citizen and the state together in a virtual environment and access the strategic use of information in the state administration. Thus, the state will be restructured on the basis of information systems. Therefore, the e-government is, in fact, a state restructuring project (Bozkurt, 2017: 100). This project is not the unexpected result of a consciously produced action by the state, but rather the expected positive outcome of a planned movement that comes up in the context of the need to adapt to the emergence of the necessity of technology development (Bozkurt, 2017: 101).

The reason for the emergence of the e-government applications is not just the right approach to the development of technology. For this, it is necessary to look at the global picture. Therefore, linking the starting point of electronic government applications with environmental, managerial and political factors will be the most realistic approach. Environmental factors are becoming an important source of globalization. With globalization, the universalization of knowledge brought along the necessity of sharing. Thus, communication technologies have started to gain importance. Managerial factors emerged as a result of a comparative analysis of the private sector and the public sector. As a result of the demands of the citizens, it is focused on the necessity of maintaining the functioning of public institutions within the framework of the private sector logic. For example, the benefits of uninterrupted operation of banking services through online systems have revealed the necessity of integrating public institutions into the same process. Because, in a globalizing world, information does not remain at a fixed point. Political factors are a topic that questions the position of transparent and accountable state management in the political process. The main purpose of this title is the integration of citizens into the participatory management process (Yıldız, 2003: 308).

As a result, the state is trying to reach the objectives of informing its citizens through the e-government application developed as a result of technological developments, expanding and accelerating the communication process between the state and the citizens, and realizing the majority of the public service delivery on online systems (İnce, 2001: 9).

3.1. Expected Results From E-Government Implementation

The main goal of the e-government fiction is to produce, distribute and make use of information in the technological environment. Therefore, the expected result of the application of e-government is formed under these three headings (İnce, 2001: 22).

E-government application has formed the information society. Emerging technological tools and ways of obtaining information have led to social transformation. In an environment where the society is rapidly transformed, the elimination of the state's traditional methods constitutes a positive reflection of the expected outcome (European Commission, 1999: 8).

The e-government, in terms of the services it offers, tries to reach the conclusion under three headings. The first is the establishment of information services within the institutional integrity of the state. The second is the strengthening of the communication infrastructure with an internet-supported environment. The third is the shaping of public service units within an online order (European Commission, 1999: 8).

The concept of the ordinary citizen constituted by the conception of the traditional state rose to the position of citizen (customer) equipped with information as a result of the state's electronicization. The concept of ordinary citizens, conceived by the conception of the traditional state, has risen to the position of citizen equipped with knowledge as a result of the state's electronicization. This transformation in the understanding of state management has a priority that takes care of customer satisfaction. Prioritization of citizen satisfaction in public service delivery constitutes one of the basic rules of participatory democracy. Thus, an environment in which public service offerings are provided is fast, transparent, lack bureaucratic obstacles and citizen satisfaction is ensured (Bozkurt, 2017: 101).

In order to reach the target under the three headings supported by the above statement, the results expected from the e-government application should be classified. According to this, e-government is evaluated under the three categories of "Daily Life", "Remote Management" and "Political Participation". Daily life describes how citizens are informed about basic requirements such as housing, education, work life and transportation by using information technologies and services. For example, citizens can buy their tickets through online systems in order to use the means of transportation. For example, citizens can buy their tickets via online systems in order to travel by means of transportation. Remote management aims to support electronic systems in accessing public services and delivering requests and complaints to the management. For this purpose, citizens can communicate with the public officials via e-mail and can fill the documents online and send them to the desired service unit without printed papers. Political participation describes the electronic inclusion of citizens in participation mechanisms such as referendums, elections and surveys. With this system, citizens can communicate with the politicians who form the administration policies of the state by electronic mail, vote at the decision stages and communicate their opinions, requests and complaints to the concerned people (European Commission, 1999: 8).

The transfer of a significant portion of public services to the e-government environment saves time and money issues that are very important for the state. Thanks to online systems, paperwork applications (red tape) were eliminated and the problem of time loss caused by the transfer of printed papers between authorized personnel was eliminated (Demirhan & Türkoğlu, 2014: 249-250).

It is important to know in which areas the state can provide public service by using information and communication technologies and to learn how to facilitate the electronicization of public institutions to the citizens. Accordingly, the government can provide the following activities to its citizens in electronic environment (İnce, 2001: 2):

- "Social Welfares"
- "Open Public Administration Applications"
- "Government and Citizen Interaction"
- "Tax Services"
- "Public Goods Purchase and Sale"
- "International Trade"
- "Implementation of Laws"
- "Education and Health Services"

3.2. Comparison of Traditional State Administration and E-Government Management

The most important issue to be expressed in this section, in which a comparative analysis of traditional state administration and e-government administration is tried to be carried out, is that these two differences of understanding and application are not rival. A faster, more accessible, easy-to-use and efficient management process is established by avoiding traditional management techniques in e-government administration. In other words, e-government administration is a new form of application in which the provision of public services is changed (İnce, 2001: 24).

Citizen in the traditional state administration is not the most valuable element of management. If considered in this context, the citizen is a service recipient in a passive position. The communication between the state and the citizen is carried out through printed paper documents in a compressed form. In this system, hierarchical structuring is very important. There is no improvement in traditional public service delivery. In other words, it is not allowed to experience new methods for public service delivery within the traditional public administration system. As a result, the state continues to exist as a closed system within its own internal system. In the management system formed by the e-government system, the citizen is removed from the passive character and becomes an active customer. The communication between the state and the citizen is carried out electronically in a manner devoid of time and space by stripping from the strict rules of the bureaucracy. As a result of the rise of the state on an electronic platform, there is a notable improvement in the public service delivery. Because the state has the opportunity to try new methods in public service delivery thanks to the new possibilities brought by technology. In this way, the state can operate in a public-oriented management system with open society and transparent management principles (Parlak, 2011: 220).

Thanks to e-government applications, the workloads of civil servants have decreased. The system, in which a staff member is assigned to each job in the traditional state administration, has been transformed with the e-government. At this point, the citizen has been able to carry out his own business in the internet environment. Therefore, the concept of senior citizen in public services has emerged (Bozkurt, 2017: 100).

As a result of the new possibilities created by the e-government, the citizen can send the documents requested by the state to the units related to online systems. This situation was carried out by means of a civil servant. Therefore, such activities were quite time-consuming and a delayed process (Bozkurt, 2017: 100).

As a result of the widespread use of information technologies, security services have gained a new dimension. With the electronicization of the state, the information needed for internal and external security issues can be accessed through online services, the records of the wanted criminals can be accessed instantaneously and the late processes in the legal system can be accelerated. The maintenance of such services in the traditional state structure causes time, money and energy loss (İnce, 2001: 12).

In the traditional state administration, centralization practices are generally carried out through institutional centralization. In e-government administration centralization focuses on the citizen by changing its shape. In other words, the citizen stands at the center of the e-government. However, the difference between the center of the country and the country is also eliminated. A city located at a central point in the country and a city standing at its distant point can access public services instantly via electronic systems. Therefore, injustices caused by spatial differences can be solved (Odabaş, 2009: 12).

Each document recorded in the traditional state administration should be stored in a separate area. This situation also causes space occupation. However, in the e-government application, millions of files can be stored in a very small computer (Odabaş, 2009: 17).

The most important benefit of e-government administration is that a citizen who wants to benefit from public services can perform their transactions without going to the public institution and writing a petition. Thus, there is no need for a civil servant performing public service. A significant emphasis should be placed at this point. The e-government is not only a virtual environment that offers public services to the citizen through the internet, but also a public management system that integrates citizens and the state and synchronizes public services (Uça Güneş, 2017: 34).

4. EXAMPLES OF ELECTRONIC PUBLIC SERVICES

In this section, which public services are provided by the state in electronic form is to be examined. Each service provided by the government in electronic environment will be evaluated under different headings. This assessment will be carried out in the Republic of Turkey State. The reason for choosing the study area as an example of Turkey at this point was to create a focal point of comparative analysis. In other words, the basic point of comparison of selected state applications is Turkey and in the Turkish public service electronic government applications. In this way, an important reference point for electronic public services will be created.

4.1. Central Population Administration System (CPAS)

CPAS is an electronic based public service. This service has an infrastructure that allows the registration of events such as birth, death, marriage, divorce, population registration change and adoption related to the legal assets of each citizen from an individual center via an online system and instant access to them if necessary. CPAS system can be accessed instantaneously from 970 centers in Turkey. The most important feature of CPAS is that it is a first in the world in terms of services it fulfills and sets an example for European states. Thanks to CPAS, different identification numbers of public institutions for each citizen have been unified (MERNİS, Accessed: 10.02.2019).

4.2. Accountancy Automation System

The first initiatives for the establishment of the Accounting Office of the Ministry of Finance started in 1985. The Finance Department of the Ministry of Finance was terminated in 2000 as a project and the accountancy automation system project was introduced. Thanks to the accountancy automation system project, the accountants who made the revenue and expenditure accounts of the state switched to the automation system. In this way, each accounting is linked to each other through centralized management and the online system. In this way, each accountancy is connected to each other through centralized administration and the online system. Thanks to the accountancy automation system project, the salary and corporate information of all civil servant has been kept in a central database (Seferoğlu et al., 2011: 286-287).

4.3. Tax Office Automation Project

It is aimed to reduce the workload of tax offices by electronic declaration which is a result of the e-government application. Thus, taxpayers are able to carry out tax transactions via an online system. (Arslan et al., 2007: 46). With a shorter narrative, The aim of Tax Office Automation Project is to carry out tax office transactions through online systems. The activities of the tax office such as paper entry, accrual, collection, debt inquiry, accounting and foreclosure transactions can be carried out through Tax Office Automation Project (Kuran, 2005: 27-28).

4.4. Public Expenditure and Accounting Information System

The main purpose of the Public Expenditure and Accounting Information System is to meet the need for rapid access to information due to the necessity of effective and efficient use of public resources. This system provides a fast, economical and safe information flow between the unit making the payment and the unit that will make the payment. In this way, the transfer of documents between the spending units and accounting units are avoided by minimizing the problems raised by the bureaucratic system (Kamu Harcama ve Muhasebe Bilişim Sistemi (KBS) Nedir?, Accessed: 12.02.2019).

4.5. National Real Estate Automation Project

The main purpose of the National Real Estate Automation Project is to register private property and state owned immovable properties. This project is carried out from a single center. It is used in an integrated manner with Geographic Information System (GIS) (Uygulama Şube Müdürlüğü, Accessed: 11.02.2019).

4.6. Police Information System (POLNET)

The application of POLNET, which is put into service of the General Directorate of Security, is a system where the police can access all kinds of information they need during the task. In this way, the bureaucracy decreases and public service costs are minimized (Seferoğlu et al., 2011: 287). Thanks to POLNET, the police can make inquiries at national and international levels regarding their duties. However, the police can provide access to services such as DNA mapping, fingerprint inquiry, ballistic comparisons, access to information resulting from examination of blood and tissue samples, through the use of POLNET (Kuran, 2005: 28).

4.7. The Modernization of Customs Administration of the Undersecretariat of Customs Project

With the Modernization of Customs Administration of Customs Undersecretariat, the information requested by the customs administrations in an instantaneous manner was transferred to online systems. Thus, bureaucratic procedures were reduced, transaction processes were accelerated and service quality increased (Arslan et al., 2007: 49).

4.8. Ministry of National Education Internet Access Project

With this project, it is aimed to provide access to education channels of teachers and students via internet in schools. However, this project aims to provide support services for both teachers and students in the field of education (Kuran, 2005: 29).

4.9. Ministry of National Education Information Systems

With this system, the appointment of teachers, registration and promotion processes, budget and higher education subjects are carried out (Seferoğlu et al., 2011: 288). Teachers, administrative staff and students can access all services provided by the Ministry of National Education through this system (MEBBİS Nedir?, Accessed: 11.02.2019).

4.10. Land Registry and Cadastre Information System

Property information of citizens can be accessed through the Land Registry and Cadastral Information System. The purpose of this point is to transfer the records of land registry and cadastral records to the computer environment and to make them inquire immediately. Thus, it is easier to follow the records of real estates owned by the public and private individuals (TAKBİS Nedir?, Accessed: 11.02.2019).

4.11. National Judicial Network Project

The National Judicial Network Project envisages that the functioning of the Ministry of Justice is transferred to the electronic environment and the activities of the judicial system are monitored through online systems. With this project, paperwork such as file tracking in the judicial system has disappeared and information can be accessed via the automation system (Arslan et al., 2007: 43). Thanks to this system, judicial authorities can communicate among themselves through the automation system, transfer information and access the files instantly (Fırat, 2017: 56).

5. ELECTRONIC DEMOCRACY (E-DEMOCRACY)

As a result of the Internet entering into daily life, the ideal of direct democracy in the representation, where the debates on existence is experienced, has been found to be reflected on the application. Because the Internet has brought the convenience of time and space limitation has disappeared. The projection of this unrestricted world into the process of the state can be interpreted as a great revolution. This interpretation may be clarified by the fact that the application of e-government focuses on participation (Altınok, 2005: 134-135).

The model of democracy in the electronic environment creates an abstract representation system. With this concept of virtual reality, citizens can participate in the process of political decision making. In this context, the way in which the politician and the citizen interact with the information and communication technologies in the virtual processes of the political processes is opened. Thus, electronic voting and transparent state administration ideals can be more easily implemented (Akrivopoulou, 2013: xvi).

The reason for the existence of internet technology in the field of democracy is that it creates an autonomous space for individuals. Individuals can share their own political ideas by using this autonomous space. At the same time, individuals can experience the ease of direct participation in management. This type of democracy is called virtual democracy. Virtual democracy constitutes an open channel for citizens and citizens. Thus, the idea of classical citizen abandons its place to the online or electronic citizen model (Erdoğan Tosun, 2004: 413). Electronic citizens can communicate in a virtual environment, obtain information, use the information, reproduce information in accordance with different purposes, share information, conduct education and training activities in a virtual environment and exchange in the same environment and the person who knows the rights and obligations (Fırat, 2017: 54).

The most important problem encountered in the field of democracy is that it is gradually beginning to lose its influence. It is possible to come across the basic starting point of this idea in the fiction of theoretical studies in the field of democracy. The reason for this is that the level of participation of the citizens in the administration begins to decrease. This problem also causes the deterioration of the concept of citizens. The most important solution is the implementation of the participatory citizen model by creating virtual public spaces on the internet (Erdoğan Tosun, 2004: 415).

Citizens are able to read newspapers that share different views over the Internet, participate in discussions, share their views and share them with voice, participate in campaigns carried out in the political process and follow the activities of civil society organizations. Thus, it takes place in the decision-making process as both a listener and a participant. This is the electronic democracy itself (Erdoğan Tosun, 2004: 416).

Political activities called electronic democracy can be listed as follows (Erdoğan Tosun, 2004: 420):

- "Online political campaigns"
- "Online lobbying activities"
- "E-government applications"
- "Civil society activities"
- "Electronic citizen applications"

6. SAMPLE E-GOVERNMENT APPLICATIONS FROM THE WORLD

Globalization has transformed the world to a smaller place. Nation states have encountered many challenges that exceed their capacities. Cross-border partnerships have become an important solution to address these challenges (Barrenechea & Jenkins, 2014: 93).

6.1. United States of America (USD)

On May 23, 2012, President Obama issued a guideline called Establishing the Digital Government of the 21st Century. In this context, the Obama administration has launched a comprehensive digital government strategy to better serve the American people. The strategy is based on different initiatives. In this context, the Decree No.

13571 on "Improving Service Delivery and Improving Customer Service" and the Decree No. 13576 an "Efficient, Effective and Accountable Government Service" were published in order to provide fast and high quality public service (Barrenechea & Jenkins, 2014: 117).

The relationship between states in the USA is provided through state portals. The states operate through these portals in different areas, from renewing the driver license to paying traffic tax. State administrations and local administrations fulfill their tax obligations through online systems. Similarly, citizens who live within the borders of the state can pay their taxes via credit cards online. Each state in the USA has a separate e-government application (Paşaoğlu, 2017: 75).

E-government service areas in the USA are listed as follows (West, 2008: 1):

- "Department of Agriculture"
- "General Services Administration"
- "Postal Service"
- "Internal Revenue Service"
- "Department of Education"
- "Small Business Administration"
- "Business Administration"
- "Library of Congress"
- "Department of Treasury"
- "Federal Reserve Board"

In the spring of 2004, the US Government and Budget Office identified five key collaborative initiatives to redesign the government under a new model, to redefine the public services which is being offered to citizens, and to make significant savings, particularly, in financial areas. These decision-making powers were chosen because they represent the main areas of work common to many departments and institutions that belong to the state and have the potential to produce financially effective results. Relevant cooperation initiatives and main objectives are listed below (Seifert & Chung, 2009: 11):

6.1.1. Financial Management

The object needed to produce a government-wide financial management solution is efficiency. Productivity, while it is ensuring integrity in accountability, financial audits and task activity, also increases business performance (Seifert & Chung, 2009: 11).

6.1.2. Human Resources Management

Human resources management undertakes important tasks to promote the strategic management of human capital. It is important to develop modern, inexpensive, specific and rigorous and harmonized human resources solutions that provide common basic functionality for the government (Seifert & Chung, 2009: 11).

6.1.3. Grant Management

Grant management is a system that promotes customer service, citizen access, financial management and technical management of public institutions. This system has been developed in order to support the grant management activities and to offer government-specific solutions (Seifert & Chung, 2009: 11).

6.1.4. Status Management

Situation management has been developed as a systemic integrity tool to facilitate information management and information sharing between federal and regional law enforcement institutions and citizens. This system accomplishes its task by creating common solutions and using predetermined data patterns (Seifert & Chung, 2009: 11).

6.1.5. Federal Health System

It is a system developed to improve the health and safety of citizens by accessing reliable information and services related to health and to increase the interoperability of medical service providers and health information and technology (Seifert & Chung, 2009: 11).

6.2. Canada

E-government applications began in Canada in the late 1980s. Towards the end of 1999, internet networks were widely established in state-owned libraries and schools (Paşaoğlu, 2017: 77).

The Canadian state has taken the views of its citizens and private enterprises when establishing its electronic service infrastructure. Thus, the most beneficiaries of the realization of government services in electronic environment have been citizens. Canadian citizens benefit greatly from the services that the state has provided to them electronically (Kırçova, 2003: 153).

Canada provides its citizens with electronic government services in the following areas (Kırçova, 2003: 154):

- "Student Scholarship and Credit Services"
- "Public Pension Plans"
- "Social Insurance Services"
- "Unemployment Insurance"
- "Housing Assistance"
- "Social Help"
- "Social Aid for Disadvantaged Groups"

As of 2004, Canada has become known as the state that communicates directly with its citizens on the internet. Canadian citizens are able to access every service provided by the state in an electronic environment without having to limit the time and space to the moment they wish (Fraser, 2009: 2).

The most important reason why the Canadian state has begun to benefit from online systems in public services is to ensure that the resources of the state are used more effectively and efficiently by ensuring that citizens can access the information, documents and services they want in a cost-free manner (Fraser, 2009: 6).

6.3. Singapore

Singapore has a public policy in terms of state administration that is open to the electronic world and integrates with the digital world. In this context, Singapore has carried out a series of national programs over the past two decades, with targets, policies and projects, each designed and based on the success of the previous year (Yoke Sin, 2019: 2).

Singapore recognizes the existence of an effective public service sector as an important value for the growth of the economy. Therefore, a series of reforms were made in the 1960s to prevent corruption or illegality and to improve public service. With the status of an advanced nation, Singapore strives to improve its public service delivery habits to adapt to a rapidly changing and globalizing world and to equip public officials in a dynamic, fluid environment with different minds, behaviors and skills (Chua, 2012: 48).

Thanks to the Civil Service Computerization Program (CSCP) that was introduced in 1982, all public services were transformed into a worldwide information technology user and the efficiency of the public administration system was increased. Applications focus on transaction processing, data modelling and database management, designed to increase productivity in the public sector (Yoke Sin, 2019: 2).

Historically, responsibility for developing sectoral policies of Information and Communication Technologies in Singapore has been supported by separate government institutions. Information Technologies, telecommunications and broadcasting industries are managed by three private sector institutions under the supervision of three different ministries (Chua, 2012: 45):

- It is managed by the National Computer Board (NCB), which is organized under the Ministry of Information Communication, Trade and Industry (MTI), including electronic commerce.
- Telecommunication is managed by the Telecommunication Authority of Singapore (TAS), which was previously affiliated to the Ministry of Communication.
- The national broadcasting system is managed by the Singapore Broadcasting Authority (SBA). The SBA also regulated Internet related issues because it was accepted as a form of broadcasting of the Internet.

In Singapore, electronic government practices are carried out simultaneously with the National Information Technologies plans to undertake radical changes in the work, interaction and public services of the Government of Singapore (Chua, 2012: 49).

The focus of e-government activities in Singapore has been transformed in every phase of planning and implementation issues below (Chua, 2012: 49):

- "Civil Service Computerisation Programme (1980-1999)"
- "e-Government Action Plan I) (2000–2003)"
- "e-Government Action Plan II) (2003–2006)"
- "Integrated Government) (iGov2010) (2006–2010)"
- "E-government 2015 (eGov2015) (2011 (2015)"

6.4. United Kingdom

Since 2010, the United Kingdom has been recognized internationally as a leading state in its efforts to develop e-government applications using digital technology. In 2016, the United Kingdom was ranked first in the e-government survey (Thornton & Campbell, 2017: 2).

In its 2017 declaration, the Conservative Party promised that the new government would take pioneering initiatives in digital use. In this context, in all systems used by state institutions, technology was seen as an effective tool to provide better public services to citizens and to adapt the concept of digital citizens to daily life. Thus, it was aimed that the citizen would get more public service in electronic environment (Thornton & Campbell, 2017: 2).

The United Kingdom government considers electronic government services as an opportunity for citizens and businesses to get better services and to make more efficient use of overnment resources (Paris, 2005: 307). In this context, studies of electronic government applications in the UK are not considered as an agent to automate existing processes for public services such as tax payments and qualification (license) applications. In contrast, electronic government applications are seen as a way of modernizing public services (Jain & Kesar, 2007: 4).

Although the United Kingdom is governed by the central government, local governments develop strategies and implement plans in line with the objectives set by the central government in providing services to the public. For this reason, while the central government offers a limited level of public service, local administrations carry out public works in the field of service in 100 areas, ranging from graveyards and burials to dog feeding licenses, which are directly related to citizens (Paris, 2005: 308-309).

In the United Kingdom, there is the Electronic Government Unit (The e-Government Unit), which operates in the Cabinet Office for the management of electronic government applications. The e-government unit is the largest service unit connected to the Cabinet. The main task of this service unit is to support the commercial transformation of the state through Information Processing Technologies. The functions of the e-government unit can be listed as follows (Bennett, 2009: 2):

- To formulate the information technology strategy and policy,
- To develop common information technology components for government use,
- To promote the best practice or service across the government,
- To provide citizen-centered online services.

7. FUTURE RESEARCH DIRECTIONS

Electronic government is a modern public administration system based on information and communication tools over the internet. With electronic government applications, a significant majority of public services are provided electronically. In this way, public services can be provided faster, better quality and cheaper. Within the electronic public management system, the only means a citizen needs to access public services is a computer that provides access to the Internet or an electronic device that can do this job. Moving the public administration system to the electronic environment is a very accessible and useful solution especially for disabled citizens (Sztop-Rutkowska et al., 2019: 9).

The electronic state has assumed the role of an intermediary institution for the social transformation of society. As the internet took its place in daily life rapidly and spread to the whole society effectively and used in all areas of social life, the institutional structure of the state perceived the transformation as an obligatory task and entered into the need to act on an electronic plane. To this end, states have begun to develop strategies to establish their own public administration systems on an electronic level (Alshehri & Drew, 2010: 79).

The electronicization of the public administration system also increases accountability. Because the margin of error in the provision and reporting of public services is minimized. While the electronic government and the public workloads are reduced, it is easier to establish a transparent management system (Turnip et al., 2018: 1310).

8. CONCLUSION

The greatest benefit of the electronicisation of a system is to avoid being dependent on human beings. At this point, the human being is not the one who does the work, but the director. Thus, human errors can be minimized. Concurrently, as a result of e-government applications, the time that one should use for himself will increase.

There has been a significant change in the state administration with globalization. The concept of human-oriented management has become the key word for a democratic system. At this point, remote participation in the management process has now moved away from an acceptable reality. The rapid increase in the human population has made the ideal of direct participation in management difficult. However, the difficulties that have emerged should not be allowed to become invisible. At this point, the concept of e-government emerged with a liberating identity.

Thanks to the e-government, citizens can access public services without time and space restrictions. In fact, this situation benefits both the citizen receiving the public service and the public institution that serves the public service. From this point of view, it can be stated that e-government is an exemplary model of self-management practice.

REFERENCES

Akdoğan, A. A. (2013). Kamuda Karar Alma. In Kamu Yönetiminde Çağdaş Yaklaşımlar. Eskişehir: Anadolu Üniversitesi Yayını No: 2997.

Akrivopoulou, C. M. (2013). Preface. In *Digital Democracy and the Impact of Technology on Governance and Politics: New Globalized Practices* (pp. xiv–xx). Information Science Reference (an imprint of IGI Global). doi:10.4018/978-1-4666-3637-8

Alshehri, M., & Drew, S. (2010). Implementation of E-Government: Advantages and Challenges. *International Association For Scientific Knowledge (IASK) E-ALT Conference Proceedings*, 79-86.

Altınok, A. R. (2005). *E-Dönüşüm Yolundan Türkiye: E-Devlet Stratejisi Hazırlama Süreci ve Yaşanan Sorunlar. In Bilgi Çağında Türk Kamu Yönetiminin Yeniden Yapılandırılması-I* (pp. 133–167). İstanbul: Beta Yayınevi.

Arslan, M., Kılıç Akıncı, S., & Karapınar, P. B. (2007). e-İş, e-Devlet, eTik. Ankara: Siyasal Kitabevi.

Barrenechea, M. J., & Jenkins, T. (2014). *E-Government or Out of Government.* Open Text Corporation.

Bennett, O. (2009). *Elektronic Government (e-Government), Standard Note: SN/SC/1202*. Retrieved from http://researchbriefings.files.parliament.uk/documents/SN01202/SN01202.pdf

Box, R. C. (2007). Introduction: The Importance of Democracy for Public Administration. In Democracy and Public Administration. M.E. Sharpe (Routledge).

Bozkurt, A. (2017). Türkiye'de e-devlet Uygulamaları. In Bilgi Toplumu ve E-Devlet. Eskişehir: Anadolu Üniversitesi Yayını No: 3559.

Bredmar, K. (2017). Digitalisation of Enterprises Brings New Opportunities to Traditional Management Control. *Business Systems Research*, *8*(2), 115–125. doi:10.1515/bsrj-2017-0020

Çevik, H. H. (2012). *Kamu Yönetimi Kavramlar – Sorunlar – Tartışmalar, 2. Baskı.* Ankara: Seçkin Yayıncılık.

Chadwick, A. (n.d.a). *E-democracy*. Retrieved from https://www.britannica.com/topic/e-democracy

Chadwick, A. (n.d.b). *E-government*. Retrieved from https://www.britannica.com/topic/e-government

Chapman, B., Edward, C., Page, E. C., & Mosher, F. C. (n.d.). *Public Administration*. Retrieved from https://www.britannica.com/topic/public-administration

Chua, J. (2012). The e-Transformation Journey of Singapore. In National Strategies to Harness Information Technology, Seeking Transformation in Singapore, Finland, The Philippines, and South Africa. New York: Springer Science Business Media, LLC. doi:10.1007/978-1-4614-2086-6_2

Demirhan, Y., & Türkoğlu, İ. (2014). Türkiye'de E-Devlet Uygulamalarının Bazı Yönetim Süreçlerine Etkisinin Örnek Projeler Bağlamında Değerlendirilmesi. *Uluslararası Yönetim İktisat ve İşletme Dergisi, 10*(22), 235–256.

Democracy. (n.d.). In *Macmillan Dictionary*. Retrieved from https://www.macmillandictionary.com/dictionary/british/democracy

Erdoğan Tosun, G. (2004). *Elektronik Demokrasi. In Çağdaş Kamu Yönetimi II (Konular-Kuramlar-Kavramlar)* (pp. 413–440). Ankara: Nobel Yayınevi.

European, C. (1999). Public Sector Information: A Key Resource For Europe, Green Paper On Public Sector Information. In *The Information Society, COM(1998) 585*. Belgium: Office for Official Publications of the European Communities.

Fagence, M. (1977). *Citizen Participation in Planning*. Pergamon Press.

Fırat, M. (2017). e-Devlet Altyapı ve Teknolojileri. In Bilgi Toplumu ve E-Devlet. Eskişehir: Anadolu Üniversitesi Yayını No: 3559.

Fraser, C. (2009). E-Government: The Canadian Experience. *Dalhousie Journal of Interdisciplinary Management, 5*, 1–14.

Garson, G. D. (2004). The Promise of Digital Government. In A. Pavlichev & G. David Garson (Eds.), *Digital Government: Principles and Best Practices* (pp. 1–15). Idea Group Publishing. doi:10.4018/978-1-59140-122-3.ch001

Hague, B. N., & Loader, B. D. (1999). Digital Democracy: An Introduction. In B. N. Hague & B. D. Loader (Eds.), *Digital Democracy: Discourse and Decision Making in the Information Age* (pp. 3–23). New York: Routledge.

Hamilton, M. R. (2007). Democracy and Public Service. In Democracy and Public Administration. M.E. Sharpe (Routledge).

Heeks, R. (2006). *Implementing and Managing eGovernment: An International Text*. SAGE Publications.

Holzer, M., & Schwester, R. W. (2011). *Public Administration: An Introduction*. Routledge.

İnce, M. N. (2001). *Elektronik Devlet Kamu Hizmetlerinin Sunulmasında Yeni İmkanlar*. Ankara: TC Başbakanlık Devlet Planlama Teşkilatı.

Jaeger, P. T. (2003). The Endless Wire: E-Government As Global Phenomenon. *Government Information Quarterly, 20*(4), 323–331. doi:10.1016/j.giq.2003.08.003

Jain, V., & Kesar, S. (2007). E-Government Implementation Challenges in UK: A Case Study of Trading Standards Department. *Americas' Conference on Information Systems (AMCIS) 2007 Proceedings*, 135. Retrieved from http://aisel.aisnet.org/amcis2007/135.

Kamu Harcama ve Muhasebe Bilişim Sistemi (KBS) Nedir? (n.d.). Retrieved from https://www.muhasebat.gov.tr/kbs/kbs-nedir

Kaya Bensghir, T. (2013). E-devlet ve Türkiye'de Uygulamalar. In Kamu Yönetiminde Çağdaş Yaklaşımlar. Eskişehir: Anadolu Üniversitesi Yayını No: 2997.

Khan, H. A. (2008). *An Introduction to Public Administration*. University Press of America.

Kırçova, İ. (2003). *E-Devlet Uygulamaları ve Ekonomiye Etkileri, Yayın No: 2003-38*. İstanbul: İstanbul Ticaret Odası Yayınları.

Kuran, N. H. (2005). *Türkiye İçin E-Devlet Modeli, Devlet Baba'dan E-Devlet'e, Analiz ve Model Önerisi*. İstanbul: İstanbul Bilgi Üniversitesi Yayınları.

MEBBİS Nedir? (n.d.). Retrieved from http://www.hurriyet.com.tr/egitim/mebbis-nedir-nasil-kullanilir-40033473

Merkezi Nüfus İdaresi Sistemi (MERNİS). (n.d.). Retrieved from https://www.icisleri.gov.tr/merkez-nufus-idare-sistemi-mernis

Odabaş, H. (2009). *E-Devlet Sürecinde Elektronik Belge Yönetimi, Yayın No: 35*. İstanbul: Türk Kütüphaneciler Derneği İstanbul Şubesi Yayınları.

Öztekin, A. (2012). *Yönetim Bilimi, Gözden Geçirilmiş ve Genişletilmiş 5. Baskı*. Ankara: Siyasal Kitabevi.

Paris, M. (2005). Local E-Government and Devolution: Electronic Service Delivery in Northern Ireland. *Local Government Studies*, *31*(3), 307–319. doi:10.1080/03003930500095137

Parlak, B. (2011). *Yönetim Bilimi ve Çağdaş Yönetim Teknikleri*. İstanbul: Beta Yayınevi.

Paşaoğlu, D. (2017). Dünyada e-Devlet Uygulamaları. In Bilgi Toplumu ve E-Devlet. Eskişehir: Anadolu Üniversitesi Yayını No: 3559.

Seferoğlu, S. S., Çelen, F. K., & Çelik, A. (2011). Türkiye'de E-Devlet Uygulamaları: Sorunlar ve Çözüm Önerileri Üstüne Bir Analiz. In Türkiye'de E-Öğrenme: Gelişmeler ve Uygulamalar II. Eskişehir: Anadolu Üniversitesi İletişim Bilimleri Fakültesi Yayınları, Yayın No:77.

Seifert, J. W., & Chung, J. (2009). Using E-Government to Reinforce Government–Citizen Relationships Comparing Government Reform in the United States and China. *Social Science Computer Review*, *27*(1), 3–23. doi:10.1177/0894439308316404

Sharma, G. & Pokharel, M. (2016). The Strategic Role of Information Technology In E-Government: A Case Of Nepal. *International Journal Of eBusiness And eGovernment Studies*, *8*(1), 19-39.

Spicker, P. (2009). The Nature of a Public Service. *International Journal of Public Administration*, *32*(11), 970–991. doi:10.1080/01900690903050927

Sztop-Rutkowska, K., Zagulski, S., Maszkowska, A., Ignatowska, B., Holat, R., Oraloglu, Z., . . . Ballesta, A. G. (2019). E-Government Closer To The People, E-Citizen- Effective Inclusion Through E-Governance Services. Bialystok.

TAKBİS Nedir? (n.d.). Retrieved from https://www.tkgm.gov.tr/tr/sayfa/tapu-ve-kadastro-bilgi-sistemi-takbis

Thornton, D., & Campbell, L. (2017). Improving The Management of Digital Government. Institute for Government 2017.

Tortop, N., İsbir, E. G., Aykaç, B., Yayman, H., & Özer, M. A. (2017). Yönetim Bilimi, 11. Ankara: Nobel Yayınevi.

Turnip, K.; Lubis, A. H., & Lubis, M. S. (2018). A Review of ICT in Government Bureaucracy: Psychological and Technology Skill Perspectives. *International Journal of Civil Engineering and Technology, 9*(9), 1309–1319.

Uça Güneş, E. P. (2017). e-Devlet Tanımı ve Özellikleri. In Bilgi Toplumu ve E-Devlet. Eskişehir: Anadolu Üniversitesi Yayını No: 3559.

Uygulama Şube Müdürlüğü. (n.d.). Retrieved from http://kurumsal.milliemlak.gov.tr/Sayfalar/islemlerimiz/bim/uygulama.aspx

Verba, S., Nie, N. H., & Kim, J. (1987). *Participation and Political Equality: A Seven-Nation Comparison*. Chicago: The University of Chicago Press.

Vintar, M., Dečman, M., & Kunstelj, M. (2001). The Use Of Ict In The Public Sector And Its Influence On Communication With Citizens In Slovenia. *Global Co-Operation in the New Millennium, The 9th European Conference on Information Systems*, 923-934.

West, D. M. (2007). *Digital Government: Technology and Public Sector Performance*. Princeton University Press.

West, D. M. (2008). *State and Federal Electronic Government in the United States, 2008*. Governance Studies at Brookings.

What Is Political Participation? (n.d.). Retrieved from https://www.igi-global.com/dictionary/political-participation/22952

Yıldız, M. (2003). *Elektronik (E)-Devlet Kuram ve Uygulamasına Genel Bir Bakış ve Değerlendirme. In Çağdaş Kamu Yönetimi I (Konular-Kuramlar-Kavramlar)* (pp. 305–327). Ankara: Nobel Yayınevi.

Yoke Sin, C. (2019). *Foundation of Effective E-Government-The Singapore Experience*. Belfer Center for Science and International Affairs, Harvard Kennedy School.

Young, I. M. (2002). *Inclusion and Democracy*. Oxford University Press. doi:10.1093/0198297556.001.0001

ADDITIONAL READING

Bhatnagar, S. (2004). *E-Government: From Vision to Implementation-A Practical Guide with Case Studies*. New Delhi: Sage Publication India Pvt. Ltd.

Homburg, V. (2008). *Understanding E-Government: Information Systems in Public Administration*. USA: Routledge Taylor & Francis Group. doi:10.4324/9780203885642

Reddick, C. G. (2010). *Citizens and E-Government: Evaluating Policy and Management: Evaluating Policy and Management* (C. Reddick, Ed.). New York: Information Science Reference (an imprint of IGI Global). doi:10.4018/978-1-61520-931-6

Singh, H., Das, A., & Joseph, D. (2007). Country-Level Determinants of E-Government Maturity. *Communications of the Association for Information Systems, Volume, 20*, 631–649.

United Nations. (2014). *E-Government Survey: E-Government for The Future We Want*. New York: United Nations.

KEY TERMS AND DEFINITIONS

Democracy: Democracy is "a system of government in which people vote in elections to choose the people who will govern them." (Macmillan Dictionary, *Democracy*, Accessed: 20.08.2019).

Electronic Democracy (E-Democracy): E-democracy is a system that replaces representative democracy in the development of information and communication technologies and in some accounts (Chadwick, *E-democracy*, Accessed: 20.08.2019).

Electronic Government (E-Government): E-government is a system where information and communication technologies, especially the internet, are used extensively during the state's provision of public services (Chadwick, *E-government*, Accessed: 20.08.2019).

Public Administration: Public administration is a state science that involves the implementation of government policies. Public administration generally has a role in determining the policies and programs of governments. The main tasks of public administration include planning, organizing, managing, coordinating and controlling government activities. Public administration, regardless of the governmental system, is a feature of all nations (Chapman et al., Accessed: 20.08.2019).

Political Participation: The concept of political participation can be defined by direct citation as follows: "Taking part in the process of the decision making and conducting the public policies." (What is Political Participation?, Accessed: 20.08.2019).

Compilation of References

Aaron, S. (1993). Keeping management's door open: How to establish an open-door policy that works? *Leadership and Organization Development Journal, 14*(2), 8–12. doi:10.1108/01437739310032665

Aarum Andersen, J. (2009). When a servant-leader comes knocking. *Leadership and Organization Development Journal, 30*(1), 4–15. doi:10.1108/01437730910927070

Acar, A. Z. (2013). Farklı Örgüt Kültürü Tipleri ve Liderlik Tarzları ile Örgütsel Bağlılık Arasındaki İlişki: Lojistik İşletmelerinde Bir Araştırma. *İşletme Araştırmaları Dergisi, 5*(2), 5-31.

Adair, J. (2003). *Effective Leadership. How to Develop Skill, How to Book.* London: Macmillan.

Adams, C. A. (2016). *CSR and Accountability in an Age of Digitization.* Retrieved from: https://drcaroladams.net/csr-and-accountability-in-an-age-of-digitization/

Adams, C. A. (2002). Internal organizational factors influencing corporate social and ethical reporting: Beyond current theorizing. *Accounting, Auditing & Accountability Journal, 15*(7), 223–250. doi:10.1108/09513570210418905

Adler, P. S., & Kwon, S. (2002). Social Capital: Prospects for a New Concept. *Academy of Management Review, 27*(1), 17–40. doi:10.5465/amr.2002.5922314

Agarwal, R., Sarkar, M. B., & Echambadi, R. (2002). The conditioning effect of time on firm survival: An industry life cycle approach. *Academy of Management Journal, 45*(5), 971–994.

Agboola, A. A., & Salawu, R. O. (2011). Managing deviant behaviour and resistance to change. *Journal of Business and Management, 6*(1), 235–243.

Aghamdi, F. (2018). Ambidextreous Leadership, Ambidextreous Employee, and The Interaction Between Ambidextreous Leadership and Employee Innovative Performance. *Journal of Innovation and Entrepreneurship, 7*(1), 1–14. doi:10.118613731-018-0081-8

Agho, A. (2009). Perspectives of senior-level executives on effective followership and leadership. *Journal of Leadership & Organizational Studies, 16*(2), 159–166. doi:10.1177/1548051809335360

Agostini, L., Nosella, A., & Filippini, R. (2016). Towards an Integrated View of the Ambidextrous Organization: A Second Order Factor Model. *Creativity and Innovation Management*, *25*(1), 129–141. doi:10.1111/caim.12167

Aguinis, H. (2011). Organizational responsibility: Doing good and doing well. In S. Zedeck (Ed.), *APA handbook of industrial and organizational psychology* (pp. 855–879). Washington, DC: American Psychological Association. doi:10.1037/12171-024

Aguinis, H., & Glavas, A. (2012). What we know and don't know about corporate social responsibility: A review and research agenda. *Journal of Management*, *38*(4), 932–968. doi:10.1177/0149206311436079

Ahadzie, D. K., David, G. P., & Paul, O. (2007). Towards Developing Competency Based Measures for Construction Project Managers; Should Contextual Behaviors Be Distinguished From Task Behaviours? *International Journal of Project Management*, 1–15.

Aiken, M., & Bode, I. (2009). Killing the golden goose? Third sector organizations and back-to-work programmes in Germany and the UK. *Social Policy and Administration*, *43*(3), 209–225. doi:10.1111/j.1467-9515.2009.00658.x

Akdoğan, A. A. (2013). Kamuda Karar Alma. In Kamu Yönetiminde Çağdaş Yaklaşimlar. Eskişehir: Anadolu Üniversitesi Yayini No: 2997.

Akdoğan, Ş., Akdoğan, A., & Cingöz, A. (2009). Organizational ambidexterity: An empirical examination of organizational factors as antecedents of organizational ambidexterity. *Journal of Global Strategic Management*, *3*(2), 17–27. doi:10.20460/JGSM.2009318458

Akella, D., & Eid, N. (2018). Social enterprises in Palestine: A critical analysis. *Journal of Enterprising Communities: People and Places in Global Economy*, *12*(4), 510–544. doi:10.1108/JEC-01-2017-0010

Akrivopoulou, C. M. (2013). Preface. In *Digital Democracy and the Impact of Technology on Governance and Politics: New Globalized Practices* (pp. xiv–xx). Information Science Reference (an imprint of IGI Global). doi:10.4018/978-1-4666-3637-8

Alford, J., & Head, B. (2017). Wicked and less wicked problems: A typology and a contingency framework. *Policy and Society*, *36*(1), 397–413. doi:10.1080/14494035.2017.1361634

Allen, N. J., & Meyer, J. P. (1990). The measurement and antecedents of affective, continuance and normative commitment to the organization. *Journal of Occupational Psychology*, *63*(1), 1–18. doi:10.1111/j.2044-8325.1990.tb00506.x

Alshehri, M., & Drew, S. (2010). Implementation of E-Government: Advantages and Challenges. *International Association For Scientific Knowledge (IASK) E-ALT Conference Proceedings*, 79-86.

Altair, S. R. (n.d.). *Generative Design and Topology Optimisation by Altair*. Retrieved from https://cdn2.hubspot.net/hubfs/47251/Altair_Generative_Design_Report.pdf

Altinok, A. R. (2005). *E-Dönüşüm Yolundan Türkiye: E-Devlet Stratejisi Hazirlama Süreci ve Yaşanan Sorunlar. In Bilgi Çağinda Türk Kamu Yönetiminin Yeniden Yapilandirilmasi-I* (pp. 133–167). İstanbul: Beta Yayinevi.

Alvesson, M., & Sandberg, J. (2013). Has management studies lost its way? Ideas for more imaginative and innovative research. *Journal of Management Studies*, *50*(1), 128–152. doi:10.1111/j.1467-6486.2012.01070.x

Alvord, S. B., Brown, L. D., Christine, W. D., & Letts, C. (2002). *Social entrepreneurship and social transformation: an exploratory study.* Hauser Center for Nonprofit Organizations and The Kennedy School of Government, Harvard University, Working Paper 15, 1-26.

Alvord, S. H., Brown, L. D., & Letts, C. W. (2004). Social Entrepreneurship: Leadership that Facilitates Societal Transformation-An Exploratory Study. *The Journal of Applied Behavioral Science*, *40*(3), 260–282. doi:10.1177/0021886304266847

Amabile, T. M. (1983). The social psychology of creativity: A componential conceptualization. *Journal of Personality and Social Psychology*, *45*(2), 357–376. doi:10.1037/0022-3514.45.2.357

Amah, O. E. (2017). Leadership styles & relational energy in high quality mentoring relationship. *Indian Journal of Industrial Relations*, *53*(1), 59.

Amanchukwu, R. N., Stanley, G. J., & Ololube, N. P. (2015). A Review of Leadership Theories, Principles and Styles and Their Relevance to Educational Management. *Management*, *5*(1), 6–14.

Anderson, J. (2008). *The writings of Robert, K. Greenleaf: An interpretative analysis and the future of servant leadership.* Paper presented at servant leadership research round table, Virginia Beach, VA.

Antokanis, J. (2003). Context and leadership: An examination of the nine-factor full rane leadership theory using the multifactor leadership questionnaire. *The Leadership Quarterly*, *14*(3), 261–295. doi:10.1016/S1048-9843(03)00030-4

Archer, B. (1995). The nature of research. *Co-Design Journal*, *2*(11), 6–13.

Archibugi, D., & Iammarino, S. (2002). The globalization of technological innovation: Definition and evidence. *Review of International Political Economy*, *9*(1), 98–122. doi:10.1080/09692290110101126

Arena, M., & Uhl-Bien, M. (2016). Complexity leadership theory: Shifting from human capital to social capital. *People & Strategy*, *39*(2), 22–27.

Argyris, C., & Schön, D. A. (1974). *Theory in practice: Increasing professional effectiveness.* San Francisco, CA: Jossey-Bass.

Arslan, M., Kiliç Akinci, S., & Karapinar, P. B. (2007). e-İş, e-Devlet, eTik. Ankara: Siyasal Kitabevi.

Attar, M. (2015). *Üst Düzey Yöneticilerin Liderlik Özelliklerinin Örgütsel Ustalık Düzeyine Etkisi: Türk Bankacılık Sektörü Üzerine Bir Araştırma*. Ankara: Gazi Kitabevi.

Austin, J. E., & Seitanidi, M. M. (2012a). Collaborative value creation: A review of partnering between nonprofits and businesses: Part I. Value creation spectrum and collaboration stages. *Nonprofit and Voluntary Sector Quarterly*, *41*(5), 726–758. doi:10.1177/0899764012450777

Austin, J. E., & Seitanidi, M. M. (2012b). Collaborative value creation: A review of partnering between nonprofits and businesses. Part 2: Partnership processes and outcomes. *Nonprofit and Voluntary Sector Quarterly*, *41*(6), 929–968. doi:10.1177/0899764012454685

Austin, J., Stevenson, H., & Wei-Skillern, J. (2006). Social and commercial entrepreneurship: Same, different, or both? *Entrepreneurship Theory and Practice*, *30*(1), 1–22. doi:10.1111/j.1540-6520.2006.00107.x

Avey, J. B., Avolio, B. J., & Luthans, F. (2011). Experimentally analyzing the impact of leader positivity on follower positivity and performance. *The Leadership Quarterly*, *22*(2), 282–294. doi:10.1016/j.leaqua.2011.02.004

Avolio, B., Luthans, F., & Walumbwa, F. O. (2004). *Authentic leadership: Theory building for veritable sustained performance*. Working paper. Lincoln, NE: Gallup Leadership Institute, University of Nebraska.

Avolio, B. J., & Gardner, W. L. (2005). Authentic leadership development: Getting to the root of positive forms of leadership. *The Leadership Quarterly*, *16*(3), 315–338. doi:10.1016/j.leaqua.2005.03.001

Avolio, B. J., Gardner, W. L., Walumbwa, F. O., Luthans, F., & May, D. R. (2004). Unlocking the mask: A look at the process by which authentic leaders impact follower attitudes and behaviors. *The Leadership Quarterly*, *15*(6), 801–823. doi:10.1016/j.leaqua.2004.09.003

Ayob, N., Teasdale, S., & Fagan, K. (2016). How social innovation 'came to be': Tracing the evolution of a contested concept. *Journal of Social Policy*, *45*(4), 635–653. doi:10.1017/S004727941600009X

Bachrach, P., & Baratz, M. S. (1962). Two faces of power. *The American Political Science Review*, *56*(4), 947–952. doi:10.2307/1952796

Back, P. F. (2006). Principles or Rules? *Public Money & Management*, *26*(1), 5–9. doi:10.1111/j.1467-9302.2005.00493.x

Bailey, T. C., Eng, W., Frisch, M. B., & Snyder, C. R. (2007). Hope and optimism as related to life satisfaction. *The Journal of Positive Psychology*, *2*(3), 168–175. doi:10.1080/17439760701409546

Bakan, İ., Doğan, İ. F., & Sezer, B. (2017). Örgütsel Ustalık ile Örgütsel Öğrenme Arasındaki İlişki. *TURAN-SAM Uluslararası Bilimsel Hakemli Dergisi*, *9*(33), 10–17.

Bakan, İ., & Sezer, B. (2017). Örgütsel Ustalığın Performans Yönetimi Üzerindeki Etkisi: Bir Alan Çalışması. *KSÜ Sosyal Bilimler Dergisi*, *14*(2), 323–346.

Baker, K. (2017). *State of the disability sector report 2017*. Deakin West, ACT: National Disability Services.

Baker, N. (2018). Authentic leadership and follower job satisfaction: The moderating effect of leader emotional expressivity. *International Journal of Social Sciences and Education Research*, *4*(4), 689–697. doi:10.24289/ijsser.472863

Bakker, A. B., & Schaufeli, W. B. (2008). Positive organizational behavior: Engaged employees in flourishing organizations. *Journal of Organizational Behavior*, *29*(2), 147–154. doi:10.1002/job.515

Bandura, A. (1977). Self-efficacy: Toward a unifying theory of behavioural change. *Psychological Review*, *84*(2), 191–215. doi:10.1037/0033-295X.84.2.191 PMID:847061

Bandura, A. (1977). *Social learning theory*. Eaglewood Cliffs, NJ: Prentice Hall.

Baquero, A., Delgado, B., Escortell, R., & Sapena, J. (2019). Authentic Leadership and Job Satisfaction: A Fuzzy-Set Qualitative Comparative Analysis (fsQCA). *Sustainability*, *11*(8), 2412–2428. doi:10.3390u11082412

Barbuto, J. E. Jr, & Wheeler, D. W. (2006). Scale development and construct clarification of servant leadership. *Group & Organization Management*, *31*(3), 300–326. doi:10.1177/1059601106287091

Baregheh, A., Rowley, J., & Sambrook, S. (2009). Towards a multidisciplinary definition of innovation. *Management Decision*, *47*(8), 1323–1339. doi:10.1108/00251740910984578

Barker, R. A. (2001). The nature of leadership. *Human Relations*, *54*(4), 469–494. doi:10.1177/0018726701544004

Barney, J. (1991). Firm resources and sustained competitive advantage. *Journal of Management*, *17*(1), 99–120. doi:10.1177/014920639101700108

Barney, J. B. (1986). Strategic factor markets: Expectations, luck, and business strategy. *Management Science*, *32*(10), 1231–1241. doi:10.1287/mnsc.32.10.1231

Barney, J. B. (2001). Resource-based theories of competitive advantage: A ten-year retrospective on the resource-based view. *Journal of Management*, *27*(6), 643–650. doi:10.1177/014920630102700602

Barraket, J., & Furneaux, C. W. (2012). Social innovation and social enterprise: Evidence from Australia. In H.-W. Franz, J. Hochgerner, & J. Howaldt (Eds.), *Challenge Social Innovation: Potentials for Business, Social Entrepreneurship, Welfare and Civil Society* (pp. 215–241). Berlin: Springer. doi:10.1007/978-3-642-32879-4_14

Barrenechea, M. J., & Jenkins, T. (2014). *E-Government or Out of Government*. Open Text Corporation.

Barrs, M. Y., Muller, M. J., Galhofer, B., & Netter, P. (2011). Depressive and aggressive responses to frustration: Development of a questionnaire and its validation in a sample of male alcoholics. *Depression Research and Treatment, 2011*, 352048. doi:10.1153/2011/352048 PMID:21461367

Baškarada, S., Watson, J., & Cormarty, J. (2016). Leadership and Organizational Ambidexterity. *Journal of Management Development, 35*(6), 778–788. doi:10.1108/JMD-01-2016-0004

Baškarada, S., Watson, J., & Cormarty, J. (2017). Balancing Transactional and Transformational Leadership. *The International Journal of Organizational Analysis, 25*(3), 506–515. doi:10.1108/IJOA-02-2016-0978

Bass, B. M., & Avolio, B. (1993). Transformational Leadership: A Response to Critiques. In Leadership Theory and Research: Perspectives and Directions. Academic Press, Inc.

Bass, B. M. (1985). *Leadership and performance beyond expectations*. New York, NY: Free Press.

Bass, B. M. (1990). From transactional to transformational leadership: Learning to share the vision. *Organizational Dynamics, 18*(3), 19–31. doi:10.1016/0090-2616(90)90061-S

Bass, B. M. (1997). Does the transactional-transformational leadership paradigm transcend organizational and national boundaries? *The American Psychologist, 52*(2), 130–139. doi:10.1037/0003-066X.52.2.130

Bass, B. M. (1999). Two decades of research and development in transformational leadership. *European Journal of Work and Organizational Psychology, 8*(1), 9–32. doi:10.1080/135943299398410

Bass, B. M., & Avolio, B. J. (Eds.). (1994). *Improving organizational effectiveness through transformational leadership*. Sage Publications.

Bass, B. M., Avolio, B. J., Jung, D. I., & Berson, Y. (2003). Predicting unit performance by assessing transformational and transactional leadership. *The Journal of Applied Psychology, 88*(2), 207–218. doi:10.1037/0021-9010.88.2.207 PMID:12731705

Bass, B. M., & Bass, R. (2009). *The Bass Handbook of Leadership: Theory, Research, and Managerial Applications (4 b.)*. New York: Simon and Schuster. Retrieved from https://books.google.com.tr/books?id=UTZ2npL2HHgC

Bass, B. M., & Steidlmeier, P. (1999). Ethics, character, and authentic transformational leadership behavior. *The Leadership Quarterly, 10*(2), 181–217. doi:10.1016/S1048-9843(99)00016-8

Bass, B., Avolio, B. J., & Atwater, L. (1996). The Transformational and Transactional Leadership of Men and Women. *Applied Psychology, 45*(1), 5–34. doi:10.1111/j.1464-0597.1996.tb00847.x

Bass, B., & Bass, R. (2008). *The bass handbook of leadership: Theory, research, and managerial applications* (4th ed.). New York: The Free Press.

Batie, S. (2008). Wicked problems and applied economics. *American Journal of Agricultural Economics, 90*(5), 1176–1191. doi:10.1111/j.1467-8276.2008.01202.x

Battisti, S. (2014). Social innovation in living labs: The micro-level process model of publicprivate partnerships. *International Journal of Innovative Research and Development, 5*(4-5), 328–348. doi:10.1504/IJIRD.2014.064146

Baumol, W. J. (1996). Entrepreneurship: Productive, unproductive, and destructive. *Journal of Business Venturing, 11*(1), 3–22. doi:10.1016/0883-9026(94)00014-X

Baumol, W. J. (2002). *The free-market innovation machine: Analyzing the growth miracle of capitalism*. Princeton University Press. doi:10.1515/9781400851638

Benner, M. J., & Tushman, M. L. (2003). Exploitation, exploration, and process management: The productivity dilemma revisited. *Academy of Management Review, 28*(2), 238–256. doi:10.5465/amr.2003.9416096

Bennett, O. (2009). *Elektronic Government (e-Government), Standard Note: SN/SC/1202*. Retrieved from http://researchbriefings.files.parliament.uk/documents/SN01202/SN01202.pdf

Bennis, W. (2003). *On Becoming a Leader*. Cambridge, MA: Perseus Publishing.

Benni, W., & Nanus, B. (2007). *Leaders*. New York: Collins Business Essentials.

Berger, I. E., Cunningham, P. H., & Drumwright, M. E. (2004). Social alliances: Company/nonprofit collaboration. *California Management Review, 47*(1), 58–90. doi:10.2307/41166287

Berzin, S. C., & Pitt-Catsouphes, M. (2014). A social work approach to social innovation. *International Journal of Innovation. Creativity and Change, 1*(4), 1–14.

Bettencourt, L. A., Lusch, R. F., & Vargo, S. L. (2014). A service lens on value creation: Marketing's role in achieving strategic advantage. *California Management Review, 57*(1), 44–66. doi:10.1525/cmr.2014.57.1.44

Beugelsdijk, S., & Noorderhaven, N. (2005). Personality characteristics of self-employed; an empirical study. *Small Business Economics, 24*(2), 159–167. doi:10.100711187-003-3806-3

Beyer, J. M., & Browning, L. D. (1999). Transforming an industry in crisis: Charisma, routinization, and supportive cultural leadership. *The Leadership Quarterly, 10*(3), 483–520. doi:10.1016/S1048-9843(99)00026-0

Bhutiani, D., Flicker, K., Nair, P., & Groen, A. (2012). Is Social Entrepreneurship transformational Leadership in action? In J. Kickul & S. Bacq (Eds.), *Patterns in Social Entrepreneurship Research*. Edward Elgar Publishing.

Bilgin, P., & Elis, B. (2008). Hard power, soft power: Toward a more realistic power analysis. *Insight Turkey, 10*(2), 5–20.

Bjartmarz, T., & Pedersen, E. R. G. (2015). CSR Standards and Social Accounting. In C. S. Responsibility (pp. 103–123). Croydon: SAGE.

Bledow, R., Frese, M., Anderson, N., Erez, M., & Farr, J. (2009). A Dialectic Perspective on Innovation: Conflicting Demands, Multiple Pathways, and Ambidexterity. *Industrial and Organizational Psychology: Perspectives on Science and Practice*, *2*(3), 305–337. doi:10.1111/j.1754-9434.2009.01154.x

Blenkhorn, D., & MacKenzie, H. (2017). Categorizing corporate social responsibility (CSR) initiatives in B2B markets: The why, when and how. *Journal of Business and Industrial Marketing*, *32*(8), 1172–1181. doi:10.1108/JBIM-02-2016-0036

Block, P. (1993). *Stewardship: Choosing service over self-interest*. San Francisco, CA: Berrett-Koehler.

Bloomberg, L. D., & Volpe, M. (2016). *Completing your qualitative dissertation: A road map from beginning to end*. Sage Publications.

Bocock, G. B. J. (1986). *Hegemony*. London: Tavistock.

Boddy, C. (2011). The corporate psychopaths theory of the global financial crisis. *Journal of Business Ethics*, *102*(2), 255–259. doi:10.100710551-011-0810-4

Boekhorst, J. (2015). The role of authentic leadership in fostering workplace inclusion: A social information processing perspective. *Human Resource Management*, *54*(2), 241–264. doi:10.1002/hrm.21669

Bogers, M., Hadar, R., & Bilberg, A. (2016). Additive manufacturing for consumer-centric business models: Implications for supply chains in consumer goods manufacturing. *Technological Forecasting and Social Change*, *102*, 225–239. doi:10.1016/j.techfore.2015.07.024

Bolin, A., & Heatherly, L. (2001). Predictors of employee deviance: The relationship between bad attitudes and bad behaviour. *Journal of Business and Psychology*, *15*(3), 405–418. doi:10.1023/A:1007818616389

Bonn, I., & Fisher, J. (2011). Sustainability: The missing ingredient in strategy. *The Journal of Business Strategy*, *32*(1), 5–14. doi:10.1108/02756661111100274

Bono, J. E., & Ilies, R. (2006). Charisma, positive emotions and mood contagion. *The Leadership Quarterly*, *17*(4), 317–334. doi:10.1016/j.leaqua.2006.04.008

Boschee, J. (1998). *Merging mission and money: A board member's guide to social entrepreneurship*. The National Center for Nonprofit Boards.

Bowen, H. R. (1953). *Social responsibilities of the businessman*. New York: Harper & Row.

Bowie, N. E. (2000). Business ethics, philosophy and the next 25 years. *Business Ethics Quarterly*, *10*(1), 7–20. doi:10.2307/3857690

Box, R. C. (2007). Introduction: The Importance of Democracy for Public Administration. In Democracy and Public Administration. M.E. Sharpe (Routledge).

Boyack, K. W., & Klavans, R. (2010). Co-citation analysis, bibliographic coupling, and direct citation: Which citation approach represents the research front most accurately? *Journal of the Association for Information Science and Technology*, *61*(12), 2389–2404.

Boyatzis, R., & McKee, A. (2006). Inspiring others through resonant leadership. *Business Strategy Review*, *17*(2), 15–19. doi:10.1111/j.0955-6419.2006.00394.x

Bozkurt, A. (2017). Türkiye'de e-devlet Uygulamalari. In Bilgi Toplumu ve E-Devlet. Eskişehir: Anadolu Üniversitesi Yayini No: 3559.

Bredmar, K. (2017). Digitalisation of Enterprises Brings New Opportunities to Traditional Management Control. *Business Systems Research*, *8*(2), 115–125. doi:10.1515/bsrj-2017-0020

Breslin, D., & Jones, C. (2012). The evolution of entrepreneurial learning. *The International Journal of Organizational Analysis*, *20*(3), 294–308. doi:10.1108/19348831211243811

Brissette, I., Scheier, M. F., & Carver, C. S. (2002). The role of optimism in social network development, coping, and psychological adjustment during a life transition. *Journal of Personality and Social Psychology*, *82*(1), 102–111. doi:10.1037/0022-3514.82.1.102 PMID:11811628

Brown, M. E., & Treviño, L. K. (2006). Ethical leadership: A review and future directions. *The Leadership Quarterly*, *17*(6), 595–616. doi:10.1016/j.leaqua.2006.10.004

Brown, M. E., Trevino, L. K., & Harrison, D. (2005). Ethical leadership: A social learning perspective for construct development and testing. *Organizational Behavior and Human Decision Processes*, *97*(2), 117–134. doi:10.1016/j.obhdp.2005.03.002

Brown, T. (2009). *Change by design: How design thinking transforms organisations and inspires innovation*. New York: HarperCollins.

Brown, T., & Wyatt, J. (2010). Design thinking for social innovation. *Development Outreach*, *12*(1), 29–43. doi:10.1596/1020-797X_12_1_29

Brumback, G. B. (1999). The power of servant leadership: A critical but appreciative review. *The Leadership Quarterly*, *13*(4), 397–419.

Bryson, J. M., Crosby, B. C., & Stone, M. M. (2006). The design and implementation of Cross-Sector collaborations: Propositions from the literature. *Public Administration Review*, *66*(s1), 44–55. doi:10.1111/j.1540-6210.2006.00665.x

Buchen, I. H. (1998). Servant leadership: A model for future faculty and future institutions. *The Journal of Leadership Studies*, *5*(1), 125–128. doi:10.1177/107179199800500111

Buddelmeyer, H., Jensen, P. H., & Webster, E. (2009). Innovation and the determinants of company survival. *Oxford Economic Papers*, *62*(2), 261–285. doi:10.1093/oep/gpp012

Bull, M. (2008). Challenging tensions: Critical, theoretical and empirical perspectives on social enterprise. *International Journal of Entrepreneurial Behaviour & Research*, *14*(5), 268–275. doi:10.1108/13552550810897641

Burnes, B. (2005). Complexity theories and organizational change. *International Journal of Management Reviews*, *7*(2), 73–90. doi:10.1111/j.1468-2370.2005.00107.x

Burns, J. (1996). Defining leadership: Can we see the forest for the trees? *The Journal of Leadership Studies*, *3*(2), 148–157. doi:10.1177/107179199600300212

Burns, J. M. (1978). *Leadership*. New York: Harper & Row.

Burt, R. S. (2004). Structural holes and good ideas. *American Journal of Sociology*, *110*(2), 349–399. doi:10.1086/421787

Bush, J., LePine, J., & Newton, D. (2018). Teams in transition: An integrative review and synthesis of research on team task transitions and propositions for future research. *Human Resource Management Review*, *28*(4), 423–433. doi:10.1016/j.hrmr.2017.06.005

Butkevičienė, E. (2009). Social innovations in rural communities: Methodological framework and empirical evidence. *Social Sciences*, *63*(1), 80–87.

Bycio, P., Hackett, R. D., & Allen, J. S. (1995). Further assessments of bass's (1985) conceptualization of transactional and transformational leadership. *The Journal of Applied Psychology*, *80*(4), 468–478. doi:10.1037/0021-9010.80.4.468

Byrne, D. (2005). Complexity, configurations and cases. *Theory, Culture & Society*, *22*(5), 95–111. doi:10.1177/0263276405057194

Cajaiba-Santana, G. (2014). Social innovation: Moving the field forward. A conceptual framework. *Technological Forecasting and Social Change*, *82*, 42–51. doi:10.1016/j.techfore.2013.05.008

Cameron, K. S., Dutton, J. E., & Quinn, R. E. (2003). *Positive organizational scholarship*. San Francisco, CA: Berrett- Koehler Publishers.

Cameron, K. S., & Quinn, R. E. (1999). *Diagnosing and changing organizational culture: Based on the competing values framework*. Reading, MA: Addison-Wesley.

Camilleri, M. A. (2017). *Corporate Sustainability, Social Responsibility and Environmental Management: An Introduction to Theory and Practice with Case Studies*. Heidelberg, Germany: Springer International. doi:10.1007/978-3-319-46849-5

Campbell, J. (2018). 2017 Decade Award Invited Article Reflections on the 2017 Decade Award: Corporate Social Responsibility and the Financial Crisis. *Academy of Management Review*, *43*(4), 546–556. doi:10.5465/amr.2018.0057

Campbell, J. L. (2007). Why would corporations behave in socially responsible ways? An institutional theory of corporate social responsibility. *Academy of Management Review*, *32*(3), 946–967. doi:10.5465/amr.2007.25275684

Campbell, S. (1998). Social entrepreneurship: How to develop new social-purpose business ventures. *Health Care Strategic Management*, *16*(5), 17–18. PMID:10179045

Canadian Centre for Social Entrepreneurship (CCSE). (2001). *Social entrepreneurship discussion paper*, (1). Author.

Cao, Q., Gedajlovic, E., & Zhang, H. (2009). Unpacking Organizational Ambidexterity: Dimensions, Contingencies and Synergistic Effects. *Organization Science*, *20*(4), 781–796. doi:10.1287/orsc.1090.0426

Cao, Q., & Şimşek, Z. ve Zhang, H. (2010). Modelling the joint impact of the CEO and the TMT on organizational ambidexterity. *Journal of Management Studies*, *47*, 1272–1296.

Carland, J. W., Hoy, F., Boulton, W. R., & Carland, J. A. C. (1984). Differentiating entrepreneurs from small business owners: A conceptualization. *Academy of Management Review*, *9*(2), 354–359. doi:10.5465/amr.1984.4277721

Carlborg, P., Kindström, D., & Kowalkowski, C. (2014). The evolution of service innovation research: A critical review and synthesis. *Service Industries Journal*, *34*(5), 373–398. doi:10.1080/02642069.2013.780044

Carless, D. (2012). Innovation in language teaching and learning. In C. Chapelle (Ed.), *The encyclopedia of applied linguistics* (pp. 1–12). Oxford, UK: Wiley-Blackwell. doi:10.1002/9781405198431.wbeal0540

Carmeli, A., & Halevi, M. Y. (2009). How top management team behavioral integration and behavioral complexity enable organizational ambidexterity: The moderating role of contextual ambidexterity. *The Leadership Quarterly*, *20*(2), 207–218. doi:10.1016/j.leaqua.2009.01.011

Carroll, A. (1979). A three-dimensional conceptual model of corporate performance. *Academy of Management Review*, *4*(4), 497–505. doi:10.5465/amr.1979.4498296

Carroll, A. (2008). A history of corporate social responsibility; Concepts and practices. In A. Crane, A. McWilliams, D. Matten, J. Moon, & D. Siegel (Eds.), *The Oxford handbook of corporate social responsibility* (pp. 19–46). Oxford, UK: Oxford University Press.

Carver, C. S., Scheier, M. F., & Segerstrom, S. C. (2010). Optimism. *Clinical Psychology Review*, *30*(7), 879–889. doi:10.1016/j.cpr.2010.01.006 PMID:20170998

Casson, M., & Giusta, M. D. (2007). Entrepreneurship and social capital: Analysing the impact of social networks on entrepreneurial activity from a rational action perspective. *International Small Business Journal*, *25*(3), 220–244. doi:10.1177/0266242607076524

Castellö, I., Morsing, M., & Schultz, F. (2013). Communicative Dynamics and the Polyphony of Corporate Social Responsibility in the Network Society. *Journal of Business Ethics*, *118*(4), 683–694. doi:10.100710551-013-1954-1

Caulier-Grice, J., Davies, A., Patrick, R., & Norman, W. (2012). Defining Social Innovation (D. Research, Trans.). In E. C.-t. F. Programme (Ed.), The theoretical, empirical and policy foundations for building social innovation in Europe (TEPSIE) (pp. 1-43). Brussels: European Commission.

Cefis, E., & Marsili, O. (2006). Survivor: The role of innovation in firms' survival. *Research Policy*, *35*(5), 626–641. doi:10.1016/j.respol.2006.02.006

Çekmecelioğlu, H. G. (2014). Göreve ve İnsana Yönelik Liderlik Tarzlarının Örgütsel Bağlılık, İş Performansı ve İşten Ayrılma Niyeti Üzerindeki Etkileri. *Kocaeli Üniversitesi Sosyal Bilimler Dergisi*, *28*, 21–34.

Çekmecelioğlu, H. G., & Ülker, F. (2014). Lider - Üye Etkileşimi ve Çalışan Tutumları Üzerindeki Etkisi: Eğitim Sektöründe Bir Araştırma. *Kocaeli Üniversitesi Sosyal Bilimler Dergisi*, *28*, 35–58.

Çevik, H. H. (2012). *Kamu Yönetimi Kavramlar – Sorunlar – Tartişmalar, 2. Baski*. Ankara: Seçkin Yayincilik.

Chadwick, A. (n.d.a). *E-democracy*. Retrieved from https://www.britannica.com/topic/e-democracy

Chadwick, A. (n.d.b). *E-government*. Retrieved from https://www.britannica.com/topic/e-government

Chalmers, D. (2012). Social innovation: An exploration of the barriers faced by innovating organizations in the social economy. *Local Economy*, *28*(1), 17–34. doi:10.1177/0269094212463677

Chang, Y. (2015). Strategic Human Resource Management, Transformational Leadership Organizational Ambidexterity: Evidence From Taiwan. *Asia Pacific Business Review*, *21*(4), 517–533. doi:10.1080/13602381.2015.1029298

Chapman, B., Edward, C., Page, E. C., & Mosher, F. C. (n.d.). *Public Administration*. Retrieved from https://www.britannica.com/topic/public-administration

Chasmer, J. (2016, Nov. 15). PacketSled CEO Matt Harrigan resigns over Trump assassination threat. *The Washington Times*.

Chebbi, H., Yahiaoul, D., Vrontis, D., & Thrassou, A. (2017). The Impact of Ambidextrous Leadership on the Internationalization of Emerging-Market Firms: The Case of India. *Thunderbird International Business Review*, *59*(3), 421–436. doi:10.1002/tie.21882

Chell, E., Nicolopoulou, K., & Karataş-Özkan, M. (2010). Social entrepreneurship and enterprise: International and innovation perspectives. *Entrepreneurship and Regional Development*, *22*(6), 485–493. doi:10.1080/08985626.2010.488396

Chemers, M. M., Watson, C. B., & May, S. T. (2000). Dispositional affect and leadership effectiveness: A comparison of self-esteem, optimism, and efficacy. *Personality and Social Psychology Bulletin*, *26*(3), 267–277. doi:10.1177/0146167200265001

Cheng, B. S., Chou, L. F., Wu, T. Y., Huang, M. P., & Farh, J. L. (2004). Paternalistic leadership and subordinates responses: Establishing a leadership model in Chinese organizations. *Asian Journal of Social Psychology*, *7*(1), 89–117. doi:10.1111/j.1467-839X.2004.00137.x

Cheng, C., & Yang, M. (2019). Creative process engagement and new product performance: The role of new product development speed and leadership encouragement of creativity. *Journal of Business Research, 99*, 215–225. doi:10.1016/j.jbusres.2019.02.067

Chen, Y., Tang, G., Cooke, F. L., & Jın, J. (2016). How Does Executive Strategic Human Resource Management Link To Organizational Ambidexterity? An Empirical Examination Of Manufacturing Firms In China. *Human Resource Management, 55*(5), 919–943. doi:10.1002/hrm.21797

Chesbrough, H. (2006). Open innovation: A new paradigm for understanding industrial innovation. In H. Chesbrough, W. Vanhaverbeke & J. West (Eds.), Open Innovation: Researching a New Paradigm (pp. 1-14). Oxford, UK: Oxford University Press.

Chetkovich, C., & Frumkin, P. (2003). Balancing margin and mission: Nonprofit competition in charitable versus fee-based programs. *Administration & Society, 35*(5), 564–596. doi:10.1177/0095399703256162

Chhokar, J. S., Brodbeck, F. C., & House, R. J. (Eds.). (2013). *Culture and leadership across the world: The GLOBE book of in-depth studies of 25 societies*. Routledge.

Chin, D. T., & Smith, W. A. (2006). *An inductive model of servant leadership: The considered difference to transformational and charismatic leadership*. Melbourne, Australia: Monash University.

Christensen, C. M., Suárez, F. F., & Utterback, J. M. (1998). Strategies for survival in fast-changing industries. *Management Science, 44*(12-part-2), S207-S220.

Christensen, C. M., Marx, M., & Stevenson, H. H. (2006). The tools of cooperation and change. *Harvard Business Review, 84*(10), 72–80. PMID:17040041

Christman, R. (2007). Servant-leadership and power in positional-led organizations. In *Servant Leadership Research Roundtable presentations*. School of Global Leadership and Entrepreneurship, Regent University. Retrieved from http://www.regant.edu/acad/global/publications/sl-proceedings/2007/christman.pdf

Chua, C. K., Wong, C. H., & Yeong, W. Y. (2017). *Chapter Nine-Quality Management Framework in Additive Manufacturing*. Academic Press.

Chua, J. (2012). The e-Transformation Journey of Singapore. In National Strategies to Harness Information Technology, Seeking Transformation in Singapore, Finland, The Philippines, and South Africa. New York: Springer Science Business Media, LLC. doi:10.1007/978-1-4614-2086-6_2

Çiftçi, G. E. (2017). Örgütsel Ustalık Çalışmalarının Personel Güçlendirme Üzerinde Etkisi. *International Journal of Academic Value Studies, 3*(10), 188–202.

Ciulla, J. B. (2010). Being there: Why leaders should not "fiddle" while Rome burns. *Presidential Studies Quarterly, 40*(1), 38–56. doi:10.1111/j.1741-5705.2009.03753.x

Civitillo, S., Juang, L. P., & Schachner, M. (2018). Challenging beliefs about cultural diversity in education: A synthesis and critical review of trainings with pre-service teachers. *Educational Research Review*, *24*(1), 67–83. doi:10.1016/j.edurev.2018.01.003

Clegg, S. (1989). *Frameworks of power*. Newbury Park, CA: Sage. doi:10.4135/9781446279267

Clements, C., & Washbush, J. B. (1999). The two faces of leadership: Considering the dark side of leader-follower dynamics. *Journal of Workplace Learning*, *11*(5), 170–176. doi:10.1108/13665629910279509

Coetzer, M. F., Bussin, M., & Geldenhuys, M. (2017). The functions of a servant leader [review]. *Administrative Sciences*, *7*(5), 1–32.

Cogliser, C. C., & Brigham, K. H. (2004). The intersection of leadership and entrepreneurship: Mutual lessons to be learned. *The Leadership Quarterly*, *15*(6), 771–799. doi:10.1016/j.leaqua.2004.09.004

Cohen, W. M., & Levinthal, D. A. (1990). Absorptive capacity: A new perspective on learning and innovation. *Administrative Science Quarterly*, *35*(1), 128–152. doi:10.2307/2393553

Coker, K. K., Flight, R. L., & Valle, K. N. (2017). Social entrepreneurship: The role of national leadership culture. *Journal of Research in Marketing and Entrepreneurship*, *19*(2), 125–139. doi:10.1108/JRME-10-2016-0040

Conger, J. A. (1990). The dark side of leadership. *Organizational Dynamics*, *19*(2), 44–55. doi:10.1016/0090-2616(90)90070-6

Conger, J. A., & Kanungo, R. N. (1987). Toward a Behavioral Theory of Charismatic Leadership in Organizational Settings. *Academy of Management Review*, *12*(4), 637–647. doi:10.5465/amr.1987.4306715

Connellan, J. (2014). Big disruptive and here to stay: The impact of the National Disability Insurance Scheme on not for profit housing and homelessness agencies. *Parity*, *27*(5), 23–24.

Corrigan, P. W., & Garman, A. W. (1999). Transformational and transactional leadership skills for mental health teams. *Community Mental Health Journal*, *35*(4), 301–312. doi:10.1023/A:1018757706316 PMID:10452698

Coutu, D. L. (2002). How resilience works. *Harvard Business Review*, *80*(5), 46. PMID:12024758

Covey, S. R. (2004). *The 8th Habit: From effectiveness to greatness*. New York: Free Press.

Crane, D. (1972). *Invisible Colleges: Diffusion of Knowledge in Scientific Communities*. Chicago, IL: University of Chicago Press.

Crawford, J. (2015). *Authentic leadership in student leaders: An empirical study in an Australian university*. First Class Honours Thesis. Launceston, Australia: University of Tasmania.

Crawford, J., Dawkins, S., Martin, A., & Lewis, G. (2018). Conceptualising authentic followers and developing a future research agenda. In Authentic Leadership and Followership (pp. 271-293). Palgrave Macmillan. doi:10.1007/978-3-319-65307-5_11

Crawford, J., Dawkins, S., Martin, A., & Lewis, G. (2017). Understanding the organizational climate of unethical leadership in the Australian Football League. *The Journal of Leadership Studies*, *11*(2), 52–54. doi:10.1002/jls.21525

Crawford, J., Dawkins, S., Martin, A., & Lewis, G. (2019). Putting the leader back into authentic leadership: Reconceptualizing and rethinking leaders. *Australian Journal of Management*, *10*. doi:10.1177/0312896219836460

Crawford, J., & Kelder, J.-A. (2019). Do we measure leadership effectively? Articulating and evaluating scale development psychometrics for best practice. *The Leadership Quarterly*, *30*(1), 133–144. doi:10.1016/j.leaqua.2018.07.001

Crawford, J., & Newstead, T. (2018). The cascading effect of unethical leadership. In S. McShane & M. Von Glinow (Eds.), *Organizational behavior* (p. 384). Sydney, Australia: McGraw-Hill.

Crippen, C. (2005). Inclusive education: A servant leadership perspective. *Education Canada*, *45*(4), 19–22.

Crook, T. R., Todd, S. Y., Combs, J. G., Woehr, D. J., & Ketchen, D. J. (2011). Does human capital matter? a meta-analysis of the relationship between human capital and firm performance. *The Journal of Applied Psychology*, *96*(3), 443–456. doi:10.1037/a0022147 PMID:21244126

Crossan, M., Vera, D., & Nanjad, L. (2008). Transcendent leadership: Strategic leadership in dynamic environments. *The Leadership Quarterly*, *19*(5), 569–581. doi:10.1016/j.leaqua.2008.07.008

Csikszentmihalyi, M., & Seligman, M. (2000). Positive psychology: An introduction. *The American Psychologist*, *55*(1), 5–14. doi:10.1037/0003-066X.55.1.5 PMID:11392865

Culler, J. (1983). *On deconstruction: Theory and criticism after structuralism*. London: Routledge.

Culture Map Tool. (2018). *Country comparison: Mexico, France and Japan*. Available on line: www.erinmeyer.com/tools/

Cunningham, J. B., & Lischeron, J. (1993). Entrepreneurship: Some popular impressions and new information. *Journal of Small Business and Entrepreneurship*, *10*(2), 46–50. doi:10.1080/21692610.1993.11683545

D'Cruz, P., & Noronha, P. (2013). Breathers, releases, outlets, and pauses: Employee resistance in the context of depersonalized bullying. *Qualitative Report*, *18*(72), 1–24.

Dacin, P. A., Dacin, M. T., & Matear, M. (2010). Social entrepreneurship: Why we don't need a new theory and how we move forward from here. *The Academy of Management Perspectives*, *24*(3), 37–57.

Daghfous, N., Petrof, J. V., & Pons, F. (1999). Values and adoption of innovations: A cross-cultural study. *Journal of Consumer Marketing, 16*(4), 314–331. doi:10.1108/07363769910277102

Dahl, R. A. (1961). *Who governs? Democracy and power in an American city*. London: Yale University Press.

Dahlsrud, A. (2008). How corporate social responsibility is defined: An analysis of 37 definitions. *Corporate Social Responsibility and Environmental Management, 15*(1), 1–13. doi:10.1002/csr.132

Dai, Y., Du, K., Byun, G., & Zhu, X. (2017). Ambidexterity In New Ventures:The Impact Of New Product Development Alliances And Transactive Memory Systems. *Journal of Business Research, 75*, 77–85. doi:10.1016/j.jbusres.2017.02.009

Dansereau, F. Jr, Graen, G. B., & Haga, W. J. (1975). A Vertical Dyad Linkage Approach to Leadership Within Formal Organizations. *Organizational Behavior and Human Performance, 13*(1), 46–78. doi:10.1016/0030-5073(75)90005-7

Dart, R. (2004). The legitimacy of social enterprise. *Nonprofit Management & Leadership, 14*(4), 411–424. doi:10.1002/nml.43

Davies, A., Mulgan, G., Norman, W., Pulford, L., Patrick, R., & Simon, J. (2012). *Systemic Innovation*. Social Innovation Europe.

Davies, B. J., & Davies, B. (2004). Strategic leadership. *School Leadership & Management, 24*(1), 29–38. doi:10.1080/1363243042000172804

Davis, J. H., Schoorman, F. D., & Donaldson, L. (1997). Towards a stewardship theory of management. *Academy of Management Review, 22*(1), 20–47. doi:10.5465/amr.1997.9707180258

Davis, K. (1973). The case for and against business assumption of social responsibilities. *Academy of Management Journal, 16*(2), 312–322.

Dawson, P., & Daniel, L. (2010). Understanding social innovation: A provisional framework. *International Journal of Technology Management, 51*(1), 9–21. doi:10.1504/IJTM.2010.033125

Day, R. C., & Hamblin, R. L. (1964). Some effects of close and punitive styles of supervision. *American Journal of Sociology, 69*(5), 499–510. doi:10.1086/223653

De Cremer, D. (2006). Affective and motivational consequences of leader self-sacrifice: The moderating effect of autocratic leadership. *The Leadership Quarterly, 17*(1), 79–93. doi:10.1016/j.leaqua.2005.10.005

De Cremer, D., & Knippenberg, D. V. (2004). Leader self-sacrifice and leadership effectiveness: The moderating role of leader self-confidence. *Organizational Behavior and Human Decision Processes, 95*(2), 140–155. doi:10.1016/j.obhdp.2004.04.002

De Hoogh, A. H. B., Greer, L. L., & Den Hartog, D. N. (2015). Diabolical dictators or capable commanders? An investigation of the differential effects of autocratic leadership on team performance. *The Leadership Quarterly, 26*(5), 687–701. doi:10.1016/j.leaqua.2015.01.001

De Pree, M. (1992). *Leadership Jazz*. New York: Dell Publishing.

De Vries, M. (1989). Can you survive an entrepreneur? In J. Kao (Ed.), *Entrepreneurship, creativity and organization* (pp. 157–165). Englewood Cliffs, NJ: Prentice-Hall.

de Zilwa, D. (2016). The strengths and capacities of Authentic Followership. *Leadership and Organization Development Journal*, *37*(3), 310–324. doi:10.1108/LODJ-01-2014-0010

Dedahanov, A. T., Lee, D. H., Rhee, J., & Yoon, J. (2016). Entrepreneur's paternalistic leadership style and creativity: The mediating role of employee voice. *Management Decision*, *54*(9), 2310–2324. doi:10.1108/MD-11-2015-0537

Dees, J. (2007). Taking social entrepreneurship seriously. *Society*, *44*(3), 24–31. doi:10.1007/BF02819936

Dees, J. (2017). 1 The Meaning of Social Entrepreneurship. In J. Hamschmidt & M. Pirson (Eds.), *Case studies in social entrepreneurship and sustainability* (pp. 34–42). Abingdon, UK: Routledge.

Dees, J. G. (1998). *The meaning of social entrepreneurship*. Durham, NC: Center for the Advancement of Social Entrepreneurship, Fuqua School of Business, Duke University. Available at https://centers.fuqua.duke.edu/case/wp-content/uploads/sites/7/2015/03/Article_Dees_MeaningsofSocialEntrepreneurship_2001.pdf

Defourny, J., & Nyssens, M. (2010). Conceptions of social enterprise and social entrepreneurship in Europe and the United States: Convergences and divergences. *Journal of Social Entrepreneurship*, *1*(1), 32–53. doi:10.1080/19420670903442053

Demirhan, Y., & Türkoğlu, İ. (2014). Türkiye'de E-Devlet Uygulamalarinin Bazi Yönetim Süreçlerine Etkisinin Örnek Projeler Bağlaminda Değerlendirilmesi. *Uluslararasi Yönetim İktisat ve İşletme Dergisi*, *10*(22), 235–256.

Den Hartog, D. N., Van Muijen, J. J., & Koopman, P. K. (1997). Transactional versus transformational leadership: An analysis of the MLQ. *Journal of Occupational and Organizational Psychology*, *70*(1), 19–34. doi:10.1111/j.2044-8325.1997.tb00628.x

Dennis, R. S., & Bocarnea, M. (2005). Development of a servant leadership assessment instrument. *Leadership and Organization Development Journal*, *26*(8), 600–615. doi:10.1108/01437730510633692

Denson, T. F., Pedersen, W. C., Friese, M., Hahm, A., & Roberts, L. (2011). Understanding impulsive aggression: Angry rumination and reduced self-control capacity are mechanisms underlying the provocation – aggression relationship. *Personality and Social Psychology Bulletin*, *37*(6), 850–862. doi:10.1177/0146167211401420 PMID:21421767

DePree, M. (1992). *Leadership jazz*. New York, NY: Dell.

Dervitsiotis, K. N. (2010). Developing full-spectrum innovation capability for survival and success in the global economy. *Total Quality Management*, *21*(2), 159–170. doi:10.1080/14783360903549865

Dery, K., Sebastian, I., & van der Meulen, N. (2017). The Digital Workplace is Key to Digital Innovation. *MIS Quarterly Executive*, *16*(2), 135–152.

Dew, A., & Boydell, K. M. (2017). Knowledge translation: Bridging the disability researchto-practice gap. *Research and Practice in Intellectual and Developmental Disabilities*, *4*(2), 142–157.

Dewar, R. D., & Dutton, J. E. (1986). The adoption of radical and incremental innovations: An empirical analysis. *Management Science*, *32*(11), 1422–1433. doi:10.1287/mnsc.32.11.1422

Dey, P., & Steyaert, C. (2012). Social entrepreneurship: Critique and the radical enactment of the social. *Social Enterprise Journal*, *8*(2), 90–107. doi:10.1108/17508611211252828

Di Stefano, J. J. (1995). Tracing the vision and impact of Robert A. Greenleaf. In L. C. Spears (Ed.), *Reflections on Leadership: How Robert. K Greenleaf's theory of servant leadership influenced today's top management thinkers* (pp. 61–78). New York: John Wiley.

Diaz-Fernandez, M., Pasamar-Reyes, S., & Valle-Cabrera, R. (2017). Human Capital and Human Resource Management to Achieve Ambidextrous Learning: A Structural Perspective. *BRQ Business Research Quarterly*, *20*(1), 63–77. doi:10.1016/j.brq.2016.03.002

Dickson, M. W., Castaño, N., Magomaeva, A., & Den Hartog, D. N. (2012). Conceptualizing leadership across cultures. *Journal of World Business*, *47*(4), 483–492. doi:10.1016/j.jwb.2012.01.002

Dilberoglu, U. M., Gharehpapagh, B., Yaman, U., & Dolen, M. (2017). The Role of Additive Manufacturing in the Era of Industry 4.0. *Procedia Manufacturing*, *11*, 545–554. doi:10.1016/j.promfg.2017.07.148

DiMaggio, P., & Powell, W. (1983). The iron cage revisited: Institutional isomorphism and collective rationality in organizational fields. *American Sociological Review*, *48*(2), 147–160. doi:10.2307/2095101

Doğan, A., Schroevers, S., & Venter, I. (2016). *The Global Leadership-Lab: A comparative analysis of leadership styles across nations*. CCBS Press.

Dorado, S., & Ventresca, M. (2013). Crescive entrepreneurship in complex social problems: Institutional conditions for entrepreneurial engagement. *Journal of Business Venturing*, *28*(1), 69–82. doi:10.1016/j.jbusvent.2012.02.002

Dorfman, P. W., & Howell, J. P. (1997). Managerial leadership in the United States and Mexico: Distant neighbors or close cousins? In C. S. Granrose & S. Oskamp (Eds.), Cross-cultural workgroups (pp. 234-264). Thousand Oaks, CA: Sage Publications.

Dorst, K. (2015). *Frame innovation: Create new thinking by design*. MIT Press. doi:10.7551/mitpress/10096.001.0001

Douglas, A., Doris, J., & Johnson, B. (2004). Corporate social reporting in Irish financial institutions. *The TQM Magazine*, *16*(6), 387–395. doi:10.1108/09544780410563301

Driver, M. (2012). An interview with Michael Porter: Social entrepreneurship and the transformation of capitalism. *Academy of Management Learning & Education, 11*(3), 421–431. doi:10.5465/amle.2011.0002a

Duncan, R. B. (1976). The Ambidextrous Organization: Designing Dual Structures for Innovation. In R. H. Kilmann, L. R. Pondy, & D. Slevin (Eds.), *The Management of Organization*. New York: North-Holland.

Dunlop, P. D., & Lee, K. (2004). Work place deviance, organization citizenship behaviour and business unit performance: The bad apples do spoil the whole barrel. *Journal of Organizational Behavior, 25*(1), 67–80. doi:10.1002/job.243

Dym, B., & Hutson, H. (2005). *Leadership in nonprofit organizations: Lessons from the third sector*. Thousand Oaks, CA: Sage Publications.

Edmondson, A. (1999). Psychological safety and learning behaviour in work teams. *Administrative Science Quarterly, 44*(2), 350–383. doi:10.2307/2666999

Einarsen, S., Aasland, M. S., & Skogstad, A. (2007). Destructive leadership behavior: A definition and conceptual model. *The Leadership Quarterly, 18*(3), 207–216. doi:10.1016/j.leaqua.2007.03.002

Elkington, J., & Hartigan, P. (2008). *Power of Unreasonable People: How Social Entrepreneurs Create Markets that Change the World*. Cambridge, MA: Harvard Business School Press Books.

Engle, E. M., & Lord, R. G. (1997). Implicit theories, self-schemas, and leader-member exchange. *Academy of Management Journal, 40*(4), 988–1010.

Eppel, E., Matheson, A., & Walton, M. (2011). Applying complexity theory to New Zealand public policy: Principles for practice. *Police Quarterly, 7*(1), 48–55.

Erdoğan Tosun, G. (2004). *Elektronik Demokrasi. In Çağdaş Kamu Yönetimi II (Konular-Kuramlar-Kavramlar)* (pp. 413–440). Ankara: Nobel Yayinevi.

Eren, B. A. (2018). Şikayete yol açan hataların ve müşterilerin bankadan ayrılmasını engelleyen faktörlerin belirlenmesi: Kamu ve özel mevduat banka müşterilerine yönelik bir araştırma. *Uluslararası Sosyal Araştırmalar Dergisi, 11*(57), 586–600.

Erickson, R. J. (1995). The importance of authenticity for self and society. *Symbolic Interaction, 18*(2), 121–144. doi:10.1525i.1995.18.2.121

Eriksen, M. (2009). Authentic leadership: Practical reflexivity, self awareness, and self-authorship. *Journal of Management Education, 33*(6), 747–771. doi:10.1177/1052562909339307

European Commission (EC). (2019). *Commission Staff Working Document Corporate Social Responsibility, Responsible Business Conduct, and Business & Human Rights: Overview of Progress*. Retrieved from: https://ec.europa.eu/docsroom/documents/34482

European Union. (2019). *European Innovation Scoreboard*. Retrieved from https://www.interactivetool.eu

European, C. (1999). *Public Sector Information: A Key Resource For Europe*, Green Paper On Public Sector Information. In *The Information Society, COM(1998) 585*. Belgium: Office for Official Publications of the European Communities.

Eva, N., Mulyadi, R., Sandjaya, S., Direndonck, D., & Liden, R. (2018). Servant leadership: A systematic review and call for future research. *The Leadership Quarterly*. doi:10.1016/j.leaqua.2018.07.004

Evers, A., & Laville, J. (2004). Defining the third sector in Europe. In A. Evers & J. Laville (Eds.), *The third sector in Europe: Globalization and welfare* (pp. 11–42). Cheltenham, UK: Edward Elgar.

Fagence, M. (1977). *Citizen Participation in Planning*. Pergamon Press.

FAHCSIA (Department of Families, Housing, Community Services and Indigenous Affairs). (2011). *2010-2020 National Disability Strategy*. Canberra: Commonwealth of Australia.

Fairweather, G. W. (1967). *Methods for experimental social innovation*. Oxford, UK: Wiley and Sons.

Farh, J. L., & Cheng, B. S. (2000). A cultural analysis of paternalistic leadership in Chinese organizations. In *Management and organizations in the Chinese context* (pp. 84–127). London: Palgrave Macmillan. doi:10.1057/9780230511590_5

Farling, M. L., Stone, A. G., & Winston, B. E. (1999). Servant leadership: Setting the stage for empirical research. *The Journal of Leadership Studies*, *6*(1-2), 49–72. doi:10.1177/107179199900600104

Fayol, H. (1917). *Administration Industrielle et Générale*. Paris: H. Dunod Et E.Pinat Editeurs.

Ferch, S.R. (2003). *Servant leadership, forgiveness and social justice*. Retrieved from Ferch Essay 5/03 guweb 2.gonzaga.edu/orgl/orgl532/essays/ferch.pdf

Fernandez, C., & Awamleh, R. (2004). The impact of transformational and transactional leadership styles on employee's satisfaction and performance: An empirical test in a multicultural environment. *International Business and Economics Research*, *3*(8), 65–76.

Firat, M. (2017). e-Devlet Altyapi ve Teknolojileri. In Bilgi Toplumu ve E-Devlet. Eskişehir: Anadolu Üniversitesi Yayini No: 3559.

Fiss, P. (2007). A set-theoretic approach to organizational configurations. *Academy of Management Review*, *32*(4), 1180–1198. doi:10.5465/amr.2007.26586092

Fontaine, M. (2013). Corporate Social Responsibility and Sustainability: The New Bottom Line? *International Journal of Business and Social Science*, *4*(4), 110–119.

Ford, S. L. (2014). Additive manufacturing technology: potential implications for US manufacturing competitiveness. *Journal of International Commerce and Economics*. Retrieved from http://www.usitc.gov/journals/Vol_VI_Article4_Additive_Manufacturing_Technology.pdf

Foss, N. J. (1999). Networks, capabilities, and competitive advantage. *Scandinavian Journal of Management*, *15*(1), 1–15. doi:10.1016/S0956-5221(98)00030-X

Foucault, M. (1977). *Discipline and punish*. London: Penguin.

Fowler, A. (2000). NGDOs as a moment in history: Beyond aid to social entrepreneurship or civic innovation? *Third World Quarterly*, *21*(4), 637–654. doi:10.1080/713701063

Fox, N. (2003). Practice-based evidence: Towards collaborative and transgressive research. *Sociology*, *37*(1), 81–102. doi:10.1177/0038038503037001388

Frantz, E., Kendall-Taylor, A., & Ezrow, N. (2014). Autocratic fate: How leaders' post-tenure expectations influence the behavior of dictatorships. *Seton Hall Journal of Diplomacy and International Relations*, *15*(1), 39.

Fraser, C. (2009). E-Government: The Canadian Experience. *Dalhousie Journal of Interdisciplinary Management*, *5*, 1–14.

Frederiksborg, A., & Fort, T. (2014). The paradox of pharmaceutical CSR: The sincerity nexus. *Business Horizons*, *57*(2), 151–160. doi:10.1016/j.bushor.2013.10.006

Fredrickson, B. L. (2001). The role of positive emotions in positive psychology: The broaden-and-build theory of positive emotions. *The American Psychologist*, *56*(3), 218–226. doi:10.1037/0003-066X.56.3.218 PMID:11315248

Friedman, M. (1970). The social responsibility of business is to increase its profits. *The New York Times Magazine*, *32–33*(122), 126-132.

Friedman, M. (2005). Making Philanthropy Out of Obscenity. *Reason*. Retrieved from http://www2.pitt.edu/~woon/courses/reason.pdf

Fung, B. (2014). The Demand and Need for Transparency and Disclosure in Corporate Governance. *Universal Journal of Management*, *2*(2), 72–80.

Gable, S. L., & Haidt, J. (2005). What (and why) is positive psychology? *Review of General Psychology*, *9*(2), 103–110. doi:10.1037/1089-2680.9.2.103

Gachter, S., & Renner, E. (2007). *The role of leadership and beliefs in the voluntary provision of public goods*. University of Nottingham.

Gaddis, B., Connelly, S., & Mumford, M. D. (2004). Failure feedback as an affective event: Influences of leader affect on subordinate attitudes and performance. *The Leadership Quarterly*, *15*(5), 663–686. doi:10.1016/j.leaqua.2004.05.011

Galbreath, J. (2009). Building corporate social responsibility into strategy. *European Business Review*, *21*(2), 109–127. doi:10.1108/09555340910940123

Galpin, T., & Whittington, L. J. (2012). Sustainability leadership: From strategy to results. *The Journal of Business Strategy*, *33*(4), 40–48. doi:10.1108/02756661211242690

Gangloff, T. (2018). *Corporate Social Responsibility and Digitization*. Retrieved from: https://novunex.com/corporate-social-responsibility-and-digitization/

Garcia, R., & Calantone, R. (2002). A critical look at technological innovation typology and innovativeness terminology: A literature review. *Journal of Product Innovation Management*, *19*(2), 110–132. doi:10.1111/1540-5885.1920110

Gardner, W. L., Avolio, B. J., Luthans, F., May, D. R., & Walumbwa, F. (2005). Can you see the real me? A self based model of authentic leader and follower development. *The Leadership Quarterly*, *16*(3), 343–372. doi:10.1016/j.leaqua.2005.03.003

Gardner, W. L., Cogliser, C. C., Davis, K. M., & Dickens, M. P. (2011). Authentic leadership: A review of the literature and research agenda. *The Leadership Quarterly*, *22*(6), 1120–1145. doi:10.1016/j.leaqua.2011.09.007

Garg, G., & Krishnan, V. R. (2003). Transformational leadership and organizational structure: The role of value-based leadership. *Transformational leadership: Value-based management for Indian organizations*, 82-100.

Garson, G. D. (2004). The Promise of Digital Government. In A. Pavlichev & G. David Garson (Eds.), *Digital Government: Principles and Best Practices* (pp. 1–15). Idea Group Publishing. doi:10.4018/978-1-59140-122-3.ch001

Gaventa, J. (1980). *Power and powerlessness: Quiescence and rebellion in an Appalachian Valley*. University of Illinois Press.

Gaventa, J. (2006). Finding the spaces for change: A power analysis. *IDS Bulletin*, *37*(6), 23–29. doi:10.1111/j.1759-5436.2006.tb00320.x

Geels, F. W. (2005). *Technological Transitions and System Innovations: a Co-evolutionary and Socio-technical Analysis*. Edward Elgar Publishing. doi:10.4337/9781845424596

Geels, F. W., & Schot, J. (2007). Typology of sociotechnical transition pathways. *Research Policy*, *36*(3), 399–417. doi:10.1016/j.respol.2007.01.003

George, G. (2014). From the editors – Rethinking management scholarship. *Academy of Management Journal*, *57*(1), 1–6. doi:10.5465/amj.2014.4001

George, J. M. (2000). Emotions and leadership: The role of emotional intelligence. *Human Relations*, *53*(8), 1027–1055. doi:10.1177/0018726700538001

Gertler, M. (2003). Tacit knowledge and the economic geography of context, or The undefinable tacitness of being (there). *Journal of Economic Geography*, *3*(1), 75–99. doi:10.1093/jeg/3.1.75

Ghaffar, S. H., Corker, J., & Fan, M. (2018). Additive manufacturing technology and its implementation in construction as an eco-innovative solution. *Automation in Construction*, *93*, 1–11. doi:10.1016/j.autcon.2018.05.005

Ghobadian, A., Talavera, I., Bhattacharya, A., Kumar, V., Garza-Reyes, J. A., & O'Regan, N. (2018). Examining legitimatisation of additive manufacturing in the interplay between innovation, lean manufacturing and sustainability. *International Journal of Production Economics*. doi:10.1016/j.ijpe.2018.06.001

Giacomin, O., Janssen, F., Pruett, M., Shinnar, R. S., Llopis, F., & Toney, B. (2011). Entrepreneurial intentions, motivations and barriers: Differences among American, Asian and European students. *The International Entrepreneurship and Management Journal*, *7*(2), 219–238. doi:10.100711365-010-0155-y

Giampetro-Meyer, A., Timothy-Brown, S., Browne, M.N., & Kubasek, N. (1998). Do we really want more leaders in business? *Journal of Business Ethics*, *17*(15), 1728–1738.

Gibb, A. A. (2006). *Entrepreneurship. Unique Solutions for Unique Environments. Can this be achieved with the existing paradigm?* Paper as background to Plenary presentation ICSB World Conference, Melbourne, Australia.

Gibson, C. B., & Birkinshaw, J. (2004). The Antecedents, consequences and mediating role of organizational ambidexterity. Academy of Management Journal, 47, 209-226.

Gibson, C. B., & Birkinshaw, J. (2004). The Antecedents, Consequences, And Mediating Role Of Organizational Ambidexterity. *Academy of Management Journal*, *47*(2), 209–226.

Gilson, L. L., & Shalley, C. E. (2004). A little creativity goes a long way: An examination of teams' engagement in creative processes. *Journal of Management*, *30*(4), 453–470. doi:10.1016/j.jm.2003.07.001

Gkorezis, P., Petridou, E., & Krouklidou, T. (2015). The detrimental effect of machiavellian leadership on employees' emotional exhaustion: Organizational cynicism as a mediator. *Europe's Journal of Psychology*, *11*(4), 619–631. doi:10.5964/ejop.v11i4.988 PMID:27247681

Glaister, K. W., Ahammad, M. F., & Junni, P. (2015). Organizational Ambidexterity and Human Resource Practices. *International Journal of Human Resource Management*, *26*(13), 1785–1789. doi:10.1080/09585192.2015.1028757

Glaser, B., & Strauss, A. (1967). *The discovery of grounded theory.* London: Weidenfield & Nicolson.

GLOBE. (2014). *Globe CEO study 2014*. Available online: http://globeproject.com/study_2014

Goffman, E. (1959). *The presentation of self in everyday life*. New York, NY: Doubleday.

Goggin, G., & Wadiwel, D. (2014, Sept.). Australian disability reform and political participation. *Australian Review of Public Affairs*, 1–8.

Goldstein, J., Hazy, J. K., & Silberstang, J. (2010). A complexity science model of social innovation in social enterprise. *Journal of Social Entrepreneurship*, *1*(1), 101–125. doi:10.1080/19420671003629763

Goodwin, N. (2006). *Leadership in Health Care: A European Perspective*. Oxfordshire, UK: Routledge.

Googins, B. K., & Rochlin, S. A. (2000). Creating the partnership society: Understanding the rhetoric and reality of cross-sectoral partnerships. *Business and Society Review*, *105*(1), 127–144. doi:10.1111/0045-3609.00068

Govern, J., & Marsch, L. (2001). Development and validation of the situational self-awareness scale. *Consciousness and Cognition*, *10*(3), 366–378. doi:10.1006/ccog.2001.0506 PMID:11697870

Graber, D. R., & Kilpatrick, A. O. (2008). Establishing values-based leadership and value systems in healthcare organizations. *Journal of Health and Human Services Administration*, 179–197. PMID:18998522

Graham, J. W. (1991). Servant-leadership in organizations: Inspirational and moral. *The Leadership Quarterly*, *5*(1), 105–119. doi:10.1016/1048-9843(91)90025-W

Gramsci, A. (1971). *Selections from the prison notebooks of Antonio Gramsci*. London: Lawrence and Wishant.

Granovetter, M. (1985). Economic action and social structure: The problem of embeddedness. *American Journal of Sociology*, *91*(3), 481–510. doi:10.1086/228311

Granovetter, M. S. (1973). The Strength of Weak Ties. *American Journal of Sociology*, *78*(6), 1360–1380. doi:10.1086/225469

Gray, B. (1989). *Collaborating: Finding common ground for multiparty problems*. Academic Press.

Green, J., & Mears, J. (2014). The implementation of the NDIS: Who wins, who loses? *Cosmopolitan Civil Societies Journal*, *6*(2), 25–39.

Greenleaf Center for Servant Leadership. (2010). *What Is Servant Leadership?* Retrieved from http://www.greenleaf.org/whatissl

Greenleaf, R. (1977). *Servant Leadership: A Journey into the Nature of Legitimate Power and Greatness*. Mahwah, NJ: Paulist Press.

Greenleaf, R. (1977). *Servant leadership: A journey into the nature of legitimate power and greatness*. New York: Paulist Press.

Greenleaf, R. K. (1977). *Servant leadership: A journey into the natüre of legitimate power and greatness*. New York: Paulist Press.

Greenleaf, R., & Spears, L. (2002). *Servant leadership: A journey into the nature of legitimate power and greatness, 25th Anniversary Edition*. New York, NY: Paulist Press.

Green, S. G., Gavin, M. B., & Aiman-Smith, L. (1995). Assessing a multidimensional measure of radical technological innovation. *IEEE Transactions on Engineering Management, 42*(3), 203–214. doi:10.1109/17.403738

Grenny, J., Switzler, A., & Mc Millan, R. (2002). *Crucial conversations: Tools for talking when stakes are high.* US Mc Graw Hill Publications.

GRI. (2019). *GRI Database.* Retrieved from: http://database.globalreporting.org

Grojean, M. W., Christian, J. R., Marcus, W. D., & Smith, D. B. (2004). Leaders, values, and organizational climate: Examining leadership strategies for establishing and organizational climate regarding ethics. *Journal of Business Ethics, 55*(3), 223–241. doi:10.100710551-004-1275-5

Grote, D. (2001). Discipline without punishment. *Across the Board, 38*(5), 52–58.

Groves, K. S., & LaRocca, M. A. (2011). Responsible leadership outcomes via stakeholder CSR values: Testing a values-centered model of transformational leadership. *Journal of Business Ethics, 98*(S1), 37–55. doi:10.100710551-011-1019-2

Guerrero, S., Lapalme, M. È., & Séguin, M. (2015). Board Chair Authentic Leadership and Nonexecutives' Motivation and Commitment. *Journal of Leadership & Organizational Studies, 22*(1), 88–101. doi:10.1177/1548051814531825

Gültekin, Y., Özutku, H., & Cevrioğlu, E. (2008). Lider-Üye Etkileşimine Çok Boyutlu Yaklaşım: Liden ve Maslyn'in Dört Boyutlu Lider-Üye Etkileşim Ölçeğinin Psikometrik Özelliklerine Yönelik Görgül Bir Araştırma. *Akademik İncelemeler, 3*(1), 95–123.

Gumusluoglu, L., & Ilsev, A. (2009). Transformational leadership, creativity, and organizational innovation. *Journal of Business Research, 62*(4), 461–473. doi:10.1016/j.jbusres.2007.07.032

Gupta, A., Smith, K., & Shalley, C. (2006). The interplay between exploration and exploitation. *Academy of Management Journal, 49*(4), 693–706. doi:10.5465/amj.2006.22083026

Gupta, V., MacMillan, I. C., & Surie, G. (2004). Entrepreneurial leadership: Developing and measuring a cross-cultural construct. *Journal of Business Venturing, 19*(2), 241–260. doi:10.1016/S0883-9026(03)00040-5

Gurtvis, N., Nikitina-Kalame, M., & Sidorova, I. (2015). Social enterprise in Estonia: Present situation and the perspectives of future development, survey of Estonian opinion. *Procedia: Social and Behavioral Sciences, 213*, 497–502. doi:10.1016/j.sbspro.2015.11.440

Guth, W., Levati, M. V., Sutter, M., & Van der Heijden, E. (2007). Leading by example with and without exclusion power in voluntary contribution experiments. *Journal of Public Economics, 91*(5), 1023–1042. doi:10.1016/j.jpubeco.2006.10.007

Güttel, W. H., & Konlechner, S. W. (2009). Continuously hanging by a thread: Managing contextually ambidextrous organizations. *Schmalenbach Business Review, 61*(2), 149–171. doi:10.1007/BF03396782

Hague, B. N., & Loader, B. D. (1999). Digital Democracy: An Introduction. In B. N. Hague & B. D. Loader (Eds.), *Digital Democracy: Discourse and Decision Making in the Information Age* (pp. 3–23). New York: Routledge.

Haig, B. D. (2005). An abductive theory of scientific method. *Psychological Methods, 10*(4), 371–388. doi:10.1037/1082-989X.10.4.371 PMID:16392993

Hallahan, L. (2013). In all its unfitness: The public's framing of the NDIS. *International Journal of Innovation. Creativity and Change, 1*(1), 1–13.

Hambrick, D. C., & Mason, P. A. (1984). Upper echelons: The organization as a reflection of its top managers. *Academy of Management Review, 9*(2), 193–206. doi:10.5465/amr.1984.4277628

Hamilton, M. R. (2007). Democracy and Public Service. In Democracy and Public Administration. M.E. Sharpe (Routledge).

Hamilton, F., & Bean, C. J. (2005). The importance of context, beliefs and values in leadership development. *Business Ethics (Oxford, England), 14*(4), 336–347. doi:10.1111/j.1467-8608.2005.00415.x

Hammack, D. C. (1995). Accountability and nonprofit organizations: A historical perspective. *Nonprofit Management & Leadership, 6*(2), 127–139. doi:10.1002/nml.4130060203

Harding, S. (1987). *Feminism and methodology*. Bloomington, IN: Indiana University Press.

Hargreaves, T., Hielscher, S., Seyfang, G., & Smith, A. (2013). Grassroots innovations in community energy: The role of intermediaries in niche development. *Global Environmental Change, 23*(5), 868–880. doi:10.1016/j.gloenvcha.2013.02.008

Harris, J. D., Sapienza, H. J., & Bowie, N. E. (2009). Ethics and entrepreneurship. *Journal of Business Venturing, 24*(5), 407–418. doi:10.1016/j.jbusvent.2009.06.001

Harrisson, D., Klein, J. L., & Browne, P. L. (2011). Social innovation, social enterprise and services. In F. Gallouj & F. Djellal (Eds.), *The handbook of innovation and services: A multi-disciplinary perspective* (pp. 197–218). Edward Elgar Publishing.

Harter, S. (2002). Authenticity. In C. S. Snyder & S. J. Lopez (Eds.), *Handbook of positive psychology* (pp. 382–394). Oxford, UK: Oxford University Press.

Hazy, J. K., & Uhl-Bien, M. (2015). Towards operationalizating complexity leadership: How generative, administrative and community-building leadership practices enact organizational outcomes. *Leadership, 11*(1), 79–104. doi:10.1177/1742715013511483

Heeks, R. (2006). *Implementing and Managing eGovernment: An International Text.* SAGE Publications.

Heikkurinen, P. (2010). Image differentiation with corporate environmental responsibility. *Corporate Social Responsibility and Environmental Management, 17*(3), 142–152. doi:10.1002/csr.225

Henderson, J. E., & Brookhart, S. M. (1996). *Organizational Leader Authenticity Scale*. OLAS.

Henderson, R., & Robertson, M. (2000). Who wants to be an entrepreneur? Young adult attitudes to entrepreneurship as a career. *Career Development International*, *5*(6), 279–287.

Hennessey, R. (2015, Nov. 18). It's time for Sean Rad to leave Tinder. For good. *Entrepreneur*.

Herrera, M. E. B. (2015). Creating competitive advantage by institutionalizing corporate social innovation. *Journal of Business Research*, *68*(7), 1468–1474. doi:10.1016/j.jbusres.2015.01.036

Herrington, V., & Schafer, J. (2019). Preparing individuals for leadership in Australasia, the United States, and the UK. *Police Practice and Research*, *20*(3), 240–258. doi:10.1080/15614263.2019.1598069

Hersey, P., Blanchard, K. H., & Natemeyer, W. E. (1979). Situational leadership, perception, and the impact of power. *Group & Organization Studies*, *4*(4), 418–428. doi:10.1177/105960117900400404

He, Z., & Wong, P. (2004). Exploration vs. Exploitation: An Empirical Test of the Ambidexterity Hypothesis. *Organization Science*, *15*(4), 481–494. doi:10.1287/orsc.1040.0078

Hill, C. W. (1995). National institutional structures, transaction cost economizing and competitive advantage: The case of Japan. *Organization Science*, *6*(1), 119–131. doi:10.1287/orsc.6.1.119

Hobday, M. (1998). Product complexity, innovation and industrial organisation. *Research Policy*, *26*(6), 689–710. doi:10.1016/S0048-7333(97)00044-9

Hoel, H., & Salin, D. (2002). Organisational antecedents of workplace bullying. In Bullying and emotional abuse in the workplace (pp. 221-236). CRC Press.

Hoel, H., Glaso, L., Hetland, J., Cooper, C. L., & Einarsen, S. (2009). 2010). Leadership styles as predictors of self-reported and observed workplace bullying. *British Journal of Management*, *21*(2), 453.

Hofstede Insights. (2018). *Country comparison tool*. Available on line: www.hofstede-insights.com/product/compare-countries/

Hoftede, G., Hofstede, G. J., & Minkov, M. (2010). *Cultures and organizations: software of the mind: intercultural cooperation and its importance for survival*. New York: McGraw-Hill.

Holmes, T., & Schmitz, J. Jr. (2001). A gain from trade: From unproductive to productive entrepreneurship. *Journal of Monetary Economics*, *47*(2), 417–446. doi:10.1016/S0304-3932(01)00044-7

Holt, S., & Marques, J. (2012). Empathy in leadership: Appropriate or misplaced? An empirical study on a topic that is asking for attention. *Journal of Business Ethics*, *105*(1), 105. doi:10.100710551-011-0951-5

Holzer, M., & Schwester, R. W. (2011). *Public Administration: An Introduction*. Routledge.

Hooijberg, R., & Lane, N. (2005). *Leader effectiveness and integrity: Wishful thinking?* Retrieved from https://www.imd.org/research/publications/upload/Hooijberg_lane_wp_2005_1_level_

House, R., Javidan, M., Hanges, P., & Dorfman, P. (2002). Understanding cultures and implicit leadership theories across the globe: An introduction to project GLOBE. *Journal of World Business, 37*(1), 3–10. doi:10.1016/S1090-9516(01)00069-4

Howaldt, J., & Schwarz, M. (2010). *Social innovation: Concepts, research fields and international trends. International Monitoring*. IMA/ ZLW.

Howaldt, J., & Schwarz, M. (2010). *Social Innovation: Concepts, research fields, and international trends*. Sozialforschungsstelle Dortmund.

Howell, J. M., & Hall-Merenda, K. E. (1999). The ties that bind: The impact of leader-member exchange, transformational and transactional leadership, and distance on predicting follower performance. *The Journal of Applied Psychology, 84*(5), 680–694. doi:10.1037/0021-9010.84.5.680

Hsiung, H. (2012). Authentic leadership and employee voice behavior: A multi-level psychological process. *Journal of Business Ethics, 107*(3), 349–361. doi:10.100710551-011-1043-2

Huang, Y., Leu, M. C., Mazumder, J., & Donmez, A. (2015). Additive manufacturing: Current state, future potential, gaps and needs, and recommendations. *Journal of Manufacturing Science and Engineering, 137*(1), 014001. doi:10.1115/1.4028725

Hughes, D., Lee, A., Tian, A., Newman, A., & Legood, A. (2018). Leadership, creativity, and innovation: A critical review and practical recommendations. *The Leadership Quarterly, 29*(5), 549–569. doi:10.1016/j.leaqua.2018.03.001

Hussain, S. T., Abbas, J., Lei, S., Jamal Haider, M., & Akram, T. (2017). Transactional leadership and organizational creativity: Examining the mediating role of knowledge sharing behavior. *Cogent Business & Management, 4*(1), 1–11. doi:10.1080/23311975.2017.1361663

Husted, B. W. (2015). Corporate social responsibility practice from 1800–1914: Past initiatives and current debates. *Business Ethics Quarterly, 25*(1), 125–141. doi:10.1017/beq.2014.1

Husted, B. W., & Allen, D. B. (2011). *Corporate Social Strategy, Stakeholder Engagement and Competitive Advantage*. Cambridge, UK: Cambridge University Press.

Hutchinson, D. S. (1995). Ethics. In J. Barnes (Ed.), *The Cambridge companion to Aristotle* (pp. 195–232). Cambridge, UK: Cambridge University Press.

Idowu, A. (2014). Corporate Social Responsibility in Nigerian Banking Industry: When Will the Lip-Service Games End? *Journal of Economics and Sustainable Development, 5*(22), 21–26.

Ilies, R., Morgeson, F. P., & Nahrgang, J. D. (2005). Authentic leadership and eudaemonic well-being: Understanding leader- follower outcomes. *The Leadership Quarterly, 76*(3), 373–394. doi:10.1016/j.leaqua.2005.03.002

Imoukhuede, O. (2019). *The impact of entrepreneurial leadership on authentic followership in Nigeria and the United States* (Doctoral dissertation). Regent University.

İnce, M. N. (2001). *Elektronik Devlet Kamu Hizmetlerinin Sunulmasinda Yeni İmkanlar*. Ankara: TC Başbakanlik Devlet Planlama Teşkilati.

Ishikawa, J. (2012). Transformational leadership and gatekeeping leadership: The rules of norm for maintaining consensus and shared leadership in team performance. *Asia Pacific Journal of Management, 29*(2), 265–283. doi:10.100710490-012-9282-z

Isik, Z., Arditi, D., Dikmen, I., & Birgonul, M. T. (2009). Impact of corporate strengths/weaknesses on project management competencies. *International Journal of Project Management, 27*(6), 629–637. doi:10.1016/j.ijproman.2008.10.002

Iyengar, V. (2014). For goodness, we will change-An ultimate goal of the entrepreneurs for effecting social enterprise. *Procedia Economics and Finance, 11*, 767–774. doi:10.1016/S2212-5671(14)00240-8

Jaeger, P. T. (2003). The Endless Wire: E-Government As Global Phenomenon. *Government Information Quarterly, 20*(4), 323–331. doi:10.1016/j.giq.2003.08.003

Jain, V., & Kesar, S. (2007). E-Government Implementation Challenges in UK: A Case Study of Trading Standards Department. *Americas' Conference on Information Systems (AMCIS) 2007 Proceedings*, 135. Retrieved from http://aisel.aisnet.org/amcis2007/135.

Jansen, J. J. P., Vera, D., & Crossan, M. (2009). Strategic leadership for exploration and exploitation: The moderating role of environmental dynamism. *The Leadership Quarterly, 20*(1), 5–18. doi:10.1016/j.leaqua.2008.11.008

Janssen, C., Sen, S., & Bhattacharya, C. (2015). Corporate crises in the age of corporate social responsibility. *Business Horizons, 58*(2), 183–192. doi:10.1016/j.bushor.2014.11.002

Janssen, O., & Van Yperen, N. W. (2004). Employees' Goal Orientations, The Quality of Leader Member Exchange, and the Outcomes of Job Performance and Job Satisfaction. *Academy of Management Journal, 47*(3), 368–384.

Jensen, S. M., & Luthans, F. (2006). Entrepreneurs as authentic leaders: Impact on employees' attitudes. *Leadership and Organization Development Journal, 27*(8), 646–666. doi:10.1108/01437730610709273

Jensen, S. M., & Luthans, F. (2006). Relationship between entrepreneurs' psychological capital and their authentic leadership. *Journal of Managerial Issues, 18*(2), 254–273.

Jensen, S., & Luthans, F. (2006). Relationship between Entrepreneurs ' Psychological Capital and Their Authentic Leadership. *Journal of Managerial Issues, 18*(2), 254–273.

Jit, R., Sharma, C. S., & Kawatra, M. (2016). Servant Leadership and Conflict Resolution: A qualitative study. *International Journal of Conflict Management, 27*(4), 591–612. doi:10.1108/IJCMA-12-2015-0086

Jit, R., Sharma, C. S., & Kawatra, M. (2017). Healing a broken spirit: Role of servant leadership. *Vikalpa, 42*(2), 80–94. doi:10.1177/0256090917703754

Johne, A. (1999). Successful market innovation. *European Journal of Innovation Management, 2*(1), 6–11. doi:10.1108/14601069910248838

Johnson, B. T. (1998). *Implementing progressive discipline policies to minimize liability and improve employee performance*. Retrieved from www.fwlaw.com/news/146-implementing-progressive-discipline-policies-minimize-liability-improve-employee-performance

Johnson, C. E. (2001). *Meeting the ethical challenges of leadership* (3rd ed.). Thousand Oaks, CA: Sage.

Johnson, N. J., & Klee, T. (2007). Passive-aggressive behaviour and leadership styles in organizations. *Journal of Leadership & Organizational Studies, 14*(2), 130–142. doi:10.1177/1071791907308044

Johnson, S. (2000). Literature review on social entrepreneurship. *Canadian Centre for Social Entrepreneurship, 16*(23), 96–106.

Johnson, S. (2000). *Literature review on social entrepreneurship*. Canadian Centre for Social Entrepreneurship.

Joseph, E. E., & Winston, B. E. (2005). A correlation of servant leadership, leader trust, and organizational trust. *Leadership and Organization Development Journal, 26*(1), 6–22. doi:10.1108/01437730510575552

Joyner, B., Payne, D., & Raiborn, C. (2002). Building values, business ethics and corporate social responsibility into the developing organization. *Journal of Developmental Entrepreneurship, 7*(1), 113–131.

Judge, T. A., & Piccolo, R. F. (2004). Transformational and transactional leadership: A meta analytic test of relative validity. *The Journal of Applied Psychology, 89*(5), 755–768. doi:10.1037/0021-9010.89.5.755 PMID:15506858

Jung, D. I. (2001). Transformational and Transactional Leadership and Their Effects on Creativity in Groups. *Creativity Research Journal*, 13.

Kabeer, N. (2005). *Inclusive Citizenship*. London: Sage.

Kamu Harcama ve Muhasebe Bilişim Sistemi (KBS) Nedir? (n.d.). Retrieved from https://www.muhasebat.gov.tr/kbs/kbs-nedir

Kania, J., & Kramer, M. (2011). *Collective impact*. Academic Press.

Kanungo, R. N. (2001). Ethical values of transactional and transformational leaders. *Canadian Journal of Administrative Sciences, 18*(4), 257–265. doi:10.1111/j.1936-4490.2001.tb00261.x

Kanungo, R. N., & Conger, J. A. (1993). Promoting altruism as a corporate goal. *The Academy of Management Executive, 7*(3), 37–48. doi:10.5465/ame.1993.9411302345

Karip, E. (1998). Dönüşümcü Liderlik. *Kuram ve Uygulamada Eğitim Yönetimi, 16*, 1–23.

Katz, D., & Kahn, R. L. (1978). *The Social Psychology of Organizations* (2nd ed.). New York: Wiley.

Kauppila, O. P., & Tempelaar, M. P. (2016). The social-cognitive underpinnings of employees' ambidextrous behavior and the supportive role of group managers' leadership. *Journal of Management Studies, 53*(6), 1019–1044. doi:10.1111/joms.12192

Kaya Bensghir, T. (2013). E-devlet ve Türkiye'de Uygulamalar. In Kamu Yönetiminde Çağdaş Yaklaşimlar. Eskişehir: Anadolu Üniversitesi Yayini No: 2997.

Kearney, E., Shemla, M., van Knippenberg, D., & Scholz, F. A. (2019). A paradox perspective on the interactive effects of visionary and empowering leadership. *Organizational Behavior and Human Decision Processes*. doi:10.1016/j.obhdp.2019.01.001

Kellerman, B. (1984). *Leadership: Multidisciplinary perspectives*. Englewood Cliffs, NJ: Prentice-Hall.

Kelley, T., & Littman, J. (2001). *The art of innovation: Lessons in creativity from IDEO America's leading design firm*. London, UK: Profile Books.

Kernis, M. H., & Goldman, B. M. (2006). A multicomponent conceptualization of authenticity: Theory and research. In M. P. Zanna (Ed.), Advances in experimental social psychology (Vol. 38, pp. 283–357). San Diego, CA: Academic Press. doi:10.1016/S0065-2601(06)38006-9

Kessler, A. (2008). Success factors for new businesses in Austria and the Czech Republic. *Entrepreneurship and Regional Development, 19*(5), 381–403. doi:10.1080/08985620701439959

Kessler, M. M. (1963). Bibliographic coupling between scientific papers. *Journal of the Association for Information Science and Technology, 14*(1), 10–25.

Khan, H. A. (2008). *An Introduction to Public Administration*. University Press of America.

Khan, H. U. Z., Halabi, A. K., & Samy, M. (2009). Corporate social responsibility (CSR) reporting: A study of selected banking companies in Bangladesh. *Social Responsibility Journal, 5*(3), 344–357. doi:10.1108/17471110910977276

Kim, N. (2016). *Servant leadership style: What is it and what are the benefits and contrasts to other leadership styles?* (Unpublished Masters Thesis). University of North Carolina at Chapel Hill.

King, A. (2007). Cooperation between corporations and environmental groups: A transaction cost perspective. *Academy of Management Review, 32*(3), 889–900. doi:10.5465/amr.2007.25275680

Kirçova, İ. (2003). *E-Devlet Uygulamalari ve Ekonomiye Etkileri, Yayin No: 2003-38*. İstanbul: İstanbul Ticaret Odasi Yayinlari.

Kirkpatrick, S. A., & Locke, E. A. (1996). Direct and indirect effects of three core charismatic leadership components on performance and attitudes. *The Journal of Applied Psychology, 81*(1), 36–51. doi:10.1037/0021-9010.81.1.36

KK, V. D., & Kumar, M. P. (2004). Augmenting subordinates' competencies: Leaders matter. *Journal of Management Research*, *4*(3), 164.

Klepper, S., & Simons, K. L. (2000). The making of an oligopoly: Firm survival and technological change in the evolution of the US tire industry. *Journal of Political Economy*, *108*(4), 728–760. doi:10.1086/316100

Klotz, A. C., & Neubaum, D. O. (2016). Research on the dark side of personality traits in entrepreneurship: Observations from an organizational behavior perspective. *Entrepreneurship Theory and Practice*, *40*(1), 7–17. doi:10.1111/etap.12214

Kluckhohn, F. R., & Strodtbeck, F. L. (1961). *Variations in value orientations*. Evanston, IL: Row, Peterson.

Knaut, A. (2017). How CSR Should Understand Digitalization. In T. Osburg & C. Lohrmann (Eds.), *Sustainability in a Digital World. CSR, Sustainability, Ethics & Governance* (pp. 249–256). Cham: Springer. doi:10.1007/978-3-319-54603-2_21

Korosec, K. (2017, Oct. 5). Jessica Alba's Honest Co. cuts its valuation in new funding round. *Fortune*.

Kostopoulos, K. C., Bozionelos, N., & Syrigos, E. (2015). Ambidexterity and Unit Performance: Intellectual Capital Antecedents and Crosslevel Moderating Effects of Human Resource Practices. *Human Resource Management*, *54*(S1), S111–S132. doi:10.1002/hrm.21705

Kouzes, J. M., & Posner, B. Z. (1993). *Credibility: How leaders gain and lose it, why people demand it*. San Francisco: Jossey-Bass.

Kouzes, J. M., & Posner, B. Z. (1995). *The leadership challenge: How to keep getting extraordinary things done in the organizations*. San Francisco, CA: Jossey-Bass Publishers.

Kouzes, J. M., & Posner, B. Z. (2003). *Encouraging the heart: A leader's guide to rewarding and recognizing others*. San Francisco, CA: Jossey-Bass Publishers.

Kovacs, A., Van Looy, B., & Cassiman, B. (2015). Exploring the scope of open innovation: A bibliometric review of a decade of research. *Scientometrics*, *104*(3), 951–983. doi:10.100711192-015-1628-0

Kovács, G., & Spens, K. (2005). Abductive reasoning in logistics research. *International Journal of Physical Distribution & Logistics Management*, *35*(2), 132–144. doi:10.1108/09600030510590318

KPMG. (2017). *The road ahead, The KPMG Survey of Corporate Responsibility Reporting, 2017*. Retrieved from: http://www.kpmg.com/crreporting

Krackhardt, D. (1995). Entrepreneurial opportunities in an entrepreneurial firm: A structural approach. *Entrepreneurship Theory and Practice*, *19*(3), 53–69. doi:10.1177/104225879501900305

Kuhnert, K. W., & Lewis, P. (1987). Transactional and Transformational Leadership: A Constructive/Developmental Analysis. *Academy of Management Review, 12*(4), 648–657. doi:10.5465/amr.1987.4306717

Kuran, N. H. (2005). *Türkiye İçin E-Devlet Modeli, Devlet Baba'dan E-Devlet'e, Analiz ve Model Önerisi*. İstanbul: İstanbul Bilgi Üniversitesi Yayinlari.

Laplume, A. O., Petersen, B., & Pearce, J. M. (2016). Global value chains from a 3D printing perspective. *Journal of International Business Studies, 47*(5), 595–609. doi:10.1057/jibs.2015.47

Larson, M., & Luthans, F. (2006). Potential Added Value of Psychological Capital in Predicting Work Attitudes. *Journal of Leadership & Organizational Studies, 13*(2), 75–92. doi:10.1177/107179190701300206 01

Laub, J. A. (1999). Assessing the Servant Organization: Development of the Servant Organizational Leadership Assessment (SOLA) Instrument. *Dissertation Abstracts International, 60*(2), 921–922.

Laursen, K., & Salter, A. (2006). Open for innovation: The role of openness in explaining innovation performance among UK manufacturing firms. *Strategic Management Journal, 27*(2), 131–150. doi:10.1002mj.507

Lavoie, M., & Addis, J. L. (2018). Harnessing the potential of additive manufacturing technologies: Challenges and opportunities for entrepreneurial strategies. *International Journal of Innovation Studies, 2*(4), 123–136. doi:10.1016/j.ijis.2018.11.001

Le Ber, M. J., & Branzei, O. (2010). (Re) forming strategic cross-sector partnerships relational processes of social innovation. *Business & Society, 49*(1), 140–172. doi:10.1177/0007650309345457

Lee, B., & Kelly, L. (2019). Cultural leadership ideals and social entrepreneurship: An international study. *Journal of Social Entrepreneurship, 10*(1), 108–128. doi:10.1080/19420676.2018.1541005

Lee, H., Chiang, H., & Kuo, H. (2019). Relationship between authentic leadership and nurses' intent to leave: The mediating role of work environment and burnout. *Journal of Nursing Management, 27*(1), 52–65. doi:10.1111/jonm.12648 PMID:30238670

Lee, K. H. (2009). Why and how to adopt green management into business organizations? The case study of Korean SMEs in manufacturing industry. *Management Decision, 47*(7), 1101–1121. doi:10.1108/00251740910978322

Lee, M.-D. P. (2011). Configuration of external influences: The combined effects of institutions and stakeholders on corporate social responsibility strategies. *Journal of Business Ethics, 102*(2), 281–298. doi:10.100710551-011-0814-0

Leitch, C., McMullan, C., & Harrison, R. (2013). The development of entrepreneurial leadership: The role of human, social and institutional capital. *British Journal of Management, 24*(3), 347–366. doi:10.1111/j.1467-8551.2011.00808.x

Lemoine, G., Hartnell, C., & Leroy, H. (2019). Taking Stock of Moral Approaches to Leadership: An Integrative Review of Ethical, Authentic, and Servant Leadership. *The Academy of Management Annals*, *13*(1), 148–187. doi:10.5465/annals.2016.0121

Leon, R.-D. (2013). From the sustainable organization to sustainable knowledge-based organization. *Economic Insight – Trends and Challenges*, *2*(2), 63-73.

Leroy, H., Anseel, F., Gardner, W., & Sels, L. (2015). Authentic leadership, authentic followership, basic need satisfaction, and work role performance. *Journal of Management*, *41*(6), 1677–1697. doi:10.1177/0149206312457822

Leroy, H., Palanski, M. E., & Simons, T. (2012). Authentic Leadership and Behavioral Integrity as Drivers of Follower Commitment and Performance. *Journal of Business Ethics*, *107*(3), 255–264. doi:10.100710551-011-1036-1

Levering, R., & Moskowitz, M. (2000). The 100 best companies to work for in America. *Fortune*, *141*(1), 82–110. PMID:10176767

Lewis, R. D. (2006). *When cultures collide*. Boston: Nicholas Brealey.

Liberman, J. K. (1981). *The litigious society*. New York: Basic Books.

Li, C.-R., Lin, C.-J., & Tien, Y.-H. (2015). CEO transformational leadership and top manager ambidexterity: An empirical study in Taiwan SMEs. *Leadership and Organization Development Journal*, *36*(8), 927–954. doi:10.1108/LODJ-03-2014-0054

Liden, R. C., Wayne, S. J., Zhao, H., & Henderson, D. (2008). Servant leadership: Development of a multidimensional measure and multilevel assessment. *The Leadership Quarterly*, *19*(2), 161–177. doi:10.1016/j.leaqua.2008.01.006

Liñán, F., & Chen, Y. W. (2009). Development and cross–cultural application of a specific instrument to measure entrepreneurial intentions. *Entrepreneurship Theory and Practice*, *33*(3), 593–617. doi:10.1111/j.1540-6520.2009.00318.x

Linley, A. P., Joseph, S., Harrington, S., & Wood, A. M. (2006). Positive psychology: Past, present, and (possible) future. *The Journal of Positive Psychology*, *1*(1), 3–16. doi:10.1080/17439760500372796

Litzky, B. E., Godshalk, V. M., & Walton-Bongers, C. (2010). Social entrepreneurship and community leadership: A service-learning model for management education. *Journal of Management Education*, *34*(1), 142–162. doi:10.1177/1052562909338038

Loi, M., Castriotta, M., & Di Guardo, M. C. (2016). The theoretical foundations of entrepreneurship education: How co-citations are shaping the field. *International Small Business Journal*, *34*(7), 948–971. doi:10.1177/0266242615602322

Lopez, S. J., & Snyder, C. (2009). *Oxford handbook of positive psychology*. New York: Oxford University Press. doi:10.1093/oxfordhb/9780195187243.001.0001

Lowe, L., Mills, A., & Mullen, J. (2002). Gendering the silences: Psychoanalysis, gender and organization studies. *Journal of Managerial Psychology, 17*(5), 422–432. doi:10.1108/02683940210432655

Loya, S. (2016). *Liderazgo en el comportamiento organizacional [Leadership and Organizational Behavior]*. Trillas.

Lubatkin, M. H., Şimşek, Z., Ling, Y., & Veiga, J. F. (2006). Ambidexterity and performance in small- to medium-sized firms: The pivotal role of top management team behavioral integration. *Journal of Management, 32*(5), 646–672. doi:10.1177/0149206306290712

Lubit, R. (2001). The keys to sustainable competitive advantage. *Organizational Dynamics, 29*(3), 164–178. doi:10.1016/S0090-2616(01)00026-2

Ludescher, J., & Mahsud, R. (2010). Opening Pandora's box: Corporate social responsibility exposed. *Independent Review, 15*(1), 123–131.

Lukes, S. (1974). *Power: A radical view*. London: Macmillan. doi:10.1007/978-1-349-02248-9

Luthans, F., & Avolio, B. J. (2003). Authentic leadership development. *Positive organizational scholarship,* 241-258.

Luthans, F., Youssef, C. M., & Avolio, B. J. (2007). Psychological Capital: Developing the Human Competitive Edge. In Psychological Capital: Developing the Human Competitive Edge. Academic Press.

Luthans, F. (2002). The need for and meaning of positive organizational behavior. *Journal of Organizational Behavior, 23*(6), 695–706. doi:10.1002/job.165

Luthans, F., Avey, J. B., Avolio, B. J., & Peterson, S. J. (2010). The development and resulting performance impact of positive psychological capital. *Human Resource Development Quarterly, 21*(1), 41–67. doi:10.1002/hrdq.20034

Luthans, F., & Avolio, B. J. (2003). Authentic leadership development. In K. S. Cameron, J. E. Dutton, & R. E. Quinn (Eds.), *Positive organizational scholarship: Foundations of a new discipline* (pp. 241–261). San Francisco: Barrett-Koehler.

Luthans, F., & Avolio, B. J. (2009). The "point" of positive organizational behavior. *Journal of Organizational Behavior, 30*(2), 291–307. doi:10.1002/job.589

Luthans, F., Avolio, B., Walumbwa, F., & Li, W. (2005). The psychological capital of Chinese workers: Exploring the relationship with performance. *Management and Organization Review, 1*(2), 249–271. doi:10.1111/j.1740-8784.2005.00011.x

Luthans, F., Luthans, K. W., & Luthans, B. C. (2004). Positive psychological capital: Beyond human and social capital. *Business Horizons, 47*(1), 45–50. doi:10.1016/j.bushor.2003.11.007

Luthans, F., Norman, S. M., Avolio, B. J., & Avey, J. B. (2008). The mediating role of psychological capital in the supportive organizational climate—Employee performance relationship. *Journal of Organizational Behavior, 29*(2), 219–238. doi:10.1002/job.507

Luthans, F., & Youssef, C. M. (2007). Emerging positive organizational behavior. *Journal of Management, 33*(3), 321–349. doi:10.1177/0149206307300814

Luttikhuis, M. O. (2014). *Influence of leadership behaviors and attitudes on innovativeness' of employees.* 1st IBA Bachelor Thesis Conference, Enschede, The Netherlands.

Luu, T. (2019). *Linking authentic leadership to salespeople's service performance: The roles of job crafting and human resource flexibility. Industrial Marketing Management.* Advanced Online Publication.

Lyons, M., & Passey, A. (2006). Need public policy ignore the third sector? Government policy in Australia and the United Kingdom. *Australian Journal of Public Administration, 65*(3), 90–102. doi:10.1111/j.1467-8500.2006.00496a.x

Maak, T., & Stoetter, N. (2012). Social entrepreneurs as responsible leaders:'Fundación Paraguaya'and the case of Martin Burt. *Journal of Business Ethics, 111*(3), 413–430. doi:10.100710551-012-1417-0

Magoni, F. (2003). Pyramids and paradigms. *Pastoral Forum, 20*(1), 14.

Mair, J., & Noboa, E. (2003). *Social entrepreneurship: How intentions to create a social enterprise get formed.* IESE Working Paper No. D/521. Available at SSRN: https://ssrn.com/abstract=462283

Mair, J., & Marti, I. (2006). Social entrepreneurship research: A source of explanation, prediction and delight. *Journal of World Business, 41*(1), 36–44. doi:10.1016/j.jwb.2005.09.002

Mair, J., Robinson, J., & Hockerts, K. (2006). Introduction. In J. Mair, J. Robinson, & K. Hockerts (Eds.), *Social entrepreneurship.* New York: Palgrave Macmillan. doi:10.1057/9780230625655_1

Manz, C. C., & Sims, H. P. (1987). Leading workers to lead themselves: The external leadership of self-managing work teams. *Administrative Science Quarterly, 32*(1), 106–129. doi:10.2307/2392745

March, J. G. (1991). Exploration And Exploitation in Organizational Learning. *Organization Science, 2*(1), 71–87. doi:10.1287/orsc.2.1.71

Marcy, R. T., & Mumford, M. D. (2007). Social innovation: Enhancing creative performance through causal analysis. *Creativity Research Journal, 19*(2-3), 123–140. doi:10.1080/10400410701397180

Martin, A., Sanderson, K., Warnecke, E., Dawkins, S., Bartlett, L., Memish, K., … Peebles, D. (2017). *An integrated approach to workplace mental health.* Hobart, Australia: University of Tasmania.

Martinez, P. G. (2005). Paternalism as a positive form of leadership in the Latin American context: Leader benevolence, decision-making control and human resource management practices. In M. Elvira & A. Davila (Eds.), Managing human resources in Latin America: An agenda for international leaders (pp. 75-93). Oxford, UK: Routledge.

Martino, J. (2019). Coaching Greatness: An Application of Authentic Leadership Development Theory to Wooden and Lombardi. *The Journal of Values Based Leadership*, *12*(1), 13. doi:10.22543/0733.121.1258

Martin, R. (2009). *The design of business: Why design thinking is the next competitive advantage?* Harvard Business School Press.

Martin, R., & Osberg, S. (2007, Spring). Social entrepreneurship: The case for definition. *Stanford Social Innovation Review*, 28–39.

Martin, S., Liao, H., & Campbell-Bush, E. (2012). Directive versus empowering leadership: A field experiment comparing the impact on task proficiency and pro-activity. *Academy of Management Journal*. doi:10.5465/amj.2011.0113

Marx, K. (1976). Capital: Vol. 1. *Introduced by E. Mandel* (B. Fowkes, Trans.). Harmondsworth, UK: Penguin, New York.

Mason, C. (2013). It's not you, it's me! Breaking up social entrepreneurship identity. In S. Denny & F. Seddon (Eds.), *Social enterprise* (pp. 66–80). Abingdon, UK: Routledge.

Matheson, A. (2016). Reducing social inequalities in obesity: Complexity and power relationships. *Journal of Public Health*, *38*(4), 826–829. PMID:28158692

Matheson, A., Dew, K., & Cumming, J. (2009). Complexity, evaluation and the effectiveness of community-based interventions to reduce health inequalities. *Health Promotion Journal of Australia*, *20*(3), 221–226. doi:10.1071/HE09221 PMID:19951243

Mathieu, C., Neumann, C. S., Hare, R. D., & Babiak, P. (2014). A dark side of leadership: Corporate psychopathy and its influence on employee well-being and job satisfaction. *Personality and Individual Differences*, *59*, 83–88. doi:10.1016/j.paid.2013.11.010

Mayall, W. H. (1979). *Principles in design*. Design Council.

Mayer, D. M., Bardes, M., & Piccolo, R. F. (2008). Do servant leaders help satisfy follower needs? An organizational justice perspective. *European Journal of Work and Organizational Psychology*, *17*(2), 180–197. doi:10.1080/13594320701743558

Mazdar, S. (2005). Subordinates information inquiry in undertain times: A cross cultural consideration of leadership style effect. *International Journal of Cross Cultural Management*, *5*(3), 255–275. doi:10.1177/1470595805058412

Mc Cleskey, J. A. (2014). Situation, transformational and transactional leadership and leadership development. *Journal of Business Studies Quarterly*, *5*(4), 117–130.

McCarthy, M. (2019). Cross-cultural differences in creativity: A process-based view through a prism of cognition, motivation and attribution. *Thinking Skills and Creativity*, *32*, 82–91. doi:10.1016/j.tsc.2019.04.002

McClelland, D. C. (1961). *Achieving society*. New York: The Free Press. doi:10.1037/14359-000

McCormack, J., Dorin, A., & Innocent, T. (2004). *Generative design: a paradigm for design research. Proceedings of Futureground*. Melbourne: Design Research Society.

McGee-Cooper, A., & Looper, G. (2001). *The essentials of servant leadership: Principles in practice*. Waltham, MA: Pegasus Communications.

McKenna, S. (1996). The darker side of the entrepreneur. *Leadership and Organization Development Journal*, *17*(6), 41–45. doi:10.1108/01437739610130519

Melcher, D. E., & Bosco, S. M. (2010). Achieving high organisation performance through servant leadership. *Journal of Business Enquiry*, *9*(2), 74–88.

Meloen, J. D. (1999). Authoritarianism in the Netherlands: Mission completed? Downward trends in authoritarianism in the Netherlands 1970-1992 with an international comparison of world data. *Social Thought & Research*, *22*(1/2), 45–93.

Merriman, S. B. (1998). *Qualitative research and case study applications in education*. San Francisco, CA: Jossey-Bass.

Meyer, E. (2014). *The Culture Map*. Available online: https://www.erinmeyer.com/tools/culture-map-premium/

Meyer, J. P., & Allen, N. J. (1997). *Commitment in the Workplace: Theory, Research and Application*. SAGE Publication Ltd.

Meyer, J. P., Allen, N. J., & Smith, C. A. (1993). Commitment to organizations and occupations: Extension and test of a three-component conceptualization. *The Journal of Applied Psychology*, *78*(4), 538–551. doi:10.1037/0021-9010.78.4.538

Michel, J. W., Lyons, B. D., & Cho, J. (2011). Is the Full-Range Model of Leadership Really a Full-Range Model of Effective Leader Behavior? *Journal of Leadership & Organizational Studies*, *18*(4), 493–507. doi:10.1177/1548051810377764

Mikolajuk, Z., & Yeh Gar-On, A. (2002). Sustainable Development and Decision Support Systems. In G. E. Kersten, Z. Mikolajuk, & A. Gar-On Yeh (Eds.), *Decision Support Systems for Sustainable Development* (pp. 13–27). Boston: Springer. doi:10.1007/0-306-47542-1_2

Milgron, P., & Holmstrom, B. (1991). *Incentive contracts, asset ownership and job design*. London: Prentice Hall.

Miller, D. (2015). A downside to the entrepreneurial personality? *Entrepreneurship Theory and Practice*, *39*(1), 1–8. doi:10.1111/etap.12130

Miller, T. L., Grimes, M. G., McMullen, J., & Vogus, T. (2012). Venturing for others with heart and head: How compassion encourages social entrepreneurship. *Academy of Management Review*, *37*(4), 616–640. doi:10.5465/amr.2010.0456

Mills, J. D. (1869). *The subjection of women*. London: Longmans, Green, Reader & Dyer. doi:10.1037/12288-000

Miloloza, I. (2018). Impact of leadership style to financial performance of enterprises. *Business Systems Research*, *9*(1), 93–106. doi:10.2478/bsrj-2018-0008

Minnis, S., & Callahan, J. (2010). *Servant Leadership in question: A critical review of power within servant leadership*. In The 17th International UFHRD Conference, Manchester, UK.

Mittal, R., & Dorfman, P. W. (2012). Servant leadership across cultures. *Journal of World Business*, *47*(4), 555–570. doi:10.1016/j.jwb.2012.01.009

Mohiuddin, Z. A. (2017). Influence of leadership style on employee's performance: Evidence from literatures. *Journal of Marketing Management*, *8*(1), 18.

Moneva, J. M., & Ortas, E. (2010). Corporate environmental and financial performance: A multivariate approach. *Industrial Management & Data Systems*, *110*(2), 193–210. doi:10.1108/02635571011020304

Morrison, K. (2005). Structuration theory, habitus and complexity theory: Elective affinities or old wine in new bottles? *British Journal of Sociology of Education*, *26*(3), 311–326. doi:10.1080/01425690500128809

Mort, S. G., Weerawardena, J., & Carnegie, K. (2003). Social entrepreneurship: Towards conceptualisation. *International Journal of Nonprofit and Voluntary Sector Marketing*, *8*(1), 76–88. doi:10.1002/nvsm.202

Moulaert, F., Martinelli, F., González, S., & Swyngedouw, E. (2007). *Introduction: social innovation and governance in European cities: urban development between path dependency and radical innovation*. Academic Press.

Moulaert, F. (2009). Social innovation: Institutionally embedded, territorially (re)produced. In D. MacCallum, F. Moulaert, J. Hillier, & S. V. Haddock (Eds.), *Social Innovation and Territorial Development*. London: Ashgate.

Moulaert, F. (Ed.). (2013). *The international handbook on social innovation: collective action, social learning and transdisciplinary research*. Edward Elgar Publishing. doi:10.4337/9781849809993

Moulaert, F., MacCallum, D., & Hillier, J. (2013). Social innovation: Intuition, precept, concept, theory and practice. In F. Moulaert, D. MacCallum, A. Mehmood, & A. Hamdouch (Eds.), *The International Handbook on Social Innovation* (pp. 13–24). Cheltenham, UK: Edward Elgar. doi:10.4337/9781849809993.00011

Moulaert, F., Martinelli, F., Swyngedouw, E., & Gonzalez, S. (2005). Towards alternative model(s) of local innovation. *Urban Studies (Edinburgh, Scotland)*, *42*(11), 1969–1990. doi:10.1080/00420980500279893

Moulaert, F., & Nussbaumer, J. (2005). The social region: Beyond the territorial dynamics of the learning economy. *European Urban and Regional Studies*, *12*(1), 45–64. doi:10.1177/0969776405048500

Muirhead, S. (1999). *Corporate contributions: The view from 50 Years*. New York: The Conference Board.

Mulgan, G., Tucker, S., Ali, R., & Sanders, B. (2007). *Social innovation: what it is, why it matters and how it can be accelerated*. Academic Press.

Mulgan, G., Tucker, S., Ali, R., & Sanders, B. (2007). *Social innovation: what it is, why it matters and how it can be accelerated: Working paper*. Oxford, UK: Oxford Said Business School.

Mulgan, G. (2006). The process of social innovation. *Innovations: Technology, Governance, Globalization*, *1*(2), 145–162. doi:10.1162/itgg.2006.1.2.145

Mumford, M. D. (2002). Social innovation: Ten cases from Benjamin Franklin. *Creativity Research Journal*, *14*(2), 253–266. doi:10.1207/S15326934CRJ1402_11

Mumford, M. D., & Moertl, P. (2003). Cases of social innovation: Lessons from two innovations in the 20th century. *Creativity Research Journal*, *15*(2-3), 261–266. doi:10.1080/10400419.2003.9651418

Muralidharan, E., & Pathak, S. (2018). Sustainability, transformational leadership, and social entrepreneurship. *Sustainability*, *10*(2), 567. doi:10.3390u10020567

Murphy, S. E., & Ensher, E. A. (1999). The effects of leader and subordinate characteristics in the development of leader–member exchange quality. *Journal of Applied Social Psychology*, *29*(7), 1371–1394. doi:10.1111/j.1559-1816.1999.tb00144.x

Murray, R., Caulier-Grice, J., & Mulgan, G. (2010). The open book of social innovation. London: National Endowment for Science, Technology and the Art.

Muse, L., Harris, S. G., Giles, W. F., & Feild, H. S. (2008). Work-life benefits and positive organizational behavior: Is there a connection? *Journal of Organizational Behavior*, *29*(2), 171–192. doi:10.1002/job.506

Naldi, L., Nordqvist, M., Sjöberg, K., & Wiklund, J. (2007). Entrepreneurial orientation, risk taking, and performance in family firms. *Family Business Review*, *20*(1), 33–47. doi:10.1111/j.1741-6248.2007.00082.x

Nazari, K., Saud, N., Emami, M., Regagholiyan, Z., & Asi Soltain, H. (2012). Analysis of leadership styles in different countries. *Journal of Basic and Applied Scientific Research*, *2*(8), 7921–7129.

Nemme, A., & Walden, R. (in press). *Integrating generative design and topology optimisation with product design values*. Paper presented at the Technology Education Conference (TENZ), Auckland, New Zealand.

Neumeier, S. (2012). Why do social innovations in rural development matter and should they be considered more seriously in rural development research? - Proposal for a stronger focus on social innovations in rural development research. *Sociologia Ruralis*, *52*(1), 48–69. doi:10.1111/j.1467-9523.2011.00553.x

Neuschel, R.P., (1992). *The servant leader: Unleashing the power of your people.* East Lansing, MI: Vision Sports Management.

Newman, A., Neesham, C., Manville, G., & Tse, H. H. (2018). Examining the influence of servant and entrepreneurial leadership on the work outcomes of employees in social enterprises. *International Journal of Human Resource Management*, *29*(20), 2905–2926. doi:10.1080/09585192.2017.1359792

Newman, L. (2005). Uncertainty, innovation, and dynamic sustainable development. *Sustainability: Science. Practice and Policy*, *1*(2), 25–31.

Newstead, T., Dawkins, S., Macklin, R., & Martin, A. (2019). The virtues project: An approach to developing good leaders. *Journal of Business Ethics*, *15*, 1–18.

Ng, S., & Terlep, S. (2016, Sept. 15). Unilever is in talks to acquire Jessica Alba's Honest Co. *Wall Street Journal.*

Nicholls, A. (2008). *Social entrepreneurship: New models of sustainable social change*. Oxford, UK: Oxford University Press.

Nicholls, A., & Murdock, A. (2012). The nature of social innovation. In A. Nicholls & A. Murdock (Eds.), *Social innovation: Blurring boundaries to reconfigure markets* (pp. 1–32). Hampshire: Palgrave MacMillan. doi:10.1057/9780230367098_1

Nicholls, A., & Murdock, A. (Eds.). (2011). *Social innovation: Blurring boundaries to reconfigure markets*. Springer. doi:10.1057/9780230367098.0006

Nobel Prize. (2006). *Grameen Bank: Facts.* The Nobel Prize. Retrieved from https://www.nobelprize.org/prizes/peace/2006/grameen/facts/

Noussair, C., Offerman, T., Suetens, S., Van de ven, J., Van Leeuwen, B., & Van Veelen, M. (2014). Predictably angry: Facial cues provide a credible signal of destructive behaviour. *IAST Working Papers from Institute of Advanced Study in Toulouse (IAST).*

NRHA. (2013). Discussion paper: Issues relating to the NDIS in rural and remote areas of Australia. *Australian Journal of Rural Health.*

Nussbaum, A. (2009). Ethical corporate social responsibility (CSR) and the pharmaceutical industry: A happy couple? *Journal of Medical Marketing*, *9*(1), 67–76. doi:10.1057/jmm.2008.33

Nwogu, O. (2004). The role of follower self-esteem, emotional intelligence, and attributions on organisation effectiveness. The 2004 proceedings of the servant leadership round table at Regent University.

Nye, J. S. (1990). Soft power. *Foreign Policy*, *80*(80), 153–171. doi:10.2307/1148580

Nystedt, L. (1997). Who should rule? Does personality matter? *European Journal of Personality*, *11*(1), 1–14. doi:10.1002/(SICI)1099-0984(199703)11:1<1::AID-PER275>3.0.CO;2-H

O'Reilly, C. A., & Tushman, M. L. (2004). The Ambidextrous Organization. *Harvard Business Review*, *82*(4), 74–83. PMID:15077368

Odabaş, H. (2009). *E-Devlet Sürecinde Elektronik Belge Yönetimi, Yayin No: 35*. İstanbul: Türk Kütüphaneciler Derneği İstanbul Şubesi Yayinlari.

Oeij, P., van der Torre, W., Vaas, F., & Dhondt, S. (2019). Understanding social innovation as an innovation process: Applying the innovation journey model. *Journal of Business Research*, *101*(1), 243–254. doi:10.1016/j.jbusres.2019.04.028

Oh, Y., Zhou, C., & Behdad, S. (2018). Part decomposition and assembly-based (Re) design for additive manufacturing: A review. *Additive Manufacturing*, *22*, 230–242. doi:10.1016/j.addma.2018.04.018

Okoye, A. (2009). Theorising Corporate Social Responsibility as an Essentially Contested Concept: Is a Definition Necessary? *Journal of Business Ethics*, *89*(4), 613–627. doi:10.100710551-008-0021-9

Oliver, C. (1997). Sustainable competitive advantage: Combining institutional and resource-based views. *Strategic Management Journal*, *18*(9), 697–713. doi:10.1002/(SICI)1097-0266(199710)18:9<697::AID-SMJ909>3.0.CO;2-C

Onesimo, A. (2011). What motivates entrepreneurs? A Study of the value systems of Filipino entrepreneur. *International Journal of Entrepreneurship*, *15*(1), 73–97.

Ouellette, P. M., Lazaer, K., & Chamber, K. (1999). Action leadership: The development of an approach to leadership enhancement for grassroots community leaders in children's mental health. *The Journal of Behavioral Health Services & Research*, *26*(2), 171–184. doi:10.1007/BF02287489 PMID:10230145

Özdevecioğlu, M., & Kanıgür, S. (2009). Çalışanların İlişki ve Görev Yönelimli Liderlik Algılamalarının Performansları Üzerindeki Etkileri. *KMU İİBF Dergisi*, *11*(16), 53–82.

Öztekin, A. (2012). *Yönetim Bilimi, Gözde Geçirilmiş ve Genişletilmiş 5. Baski*. Ankara: Siyasal Kitabevi.

Page, D., & Wong, P. (2002). A conceptual framework for measuring servant leadership. In S. Adjibolosoo (Ed.), *The human factor in shaping the course of history and development*. Boston: University Press of America.

Page, D., & Wong, T. P. (2000). A conceptual framework for measuring servant leadership. In S. Adjibolosoo (Ed.), *The Human Factor in Shaping the Course of History and Development*. Lanham, MD: University Press of America.

Paris, M. (2005). Local E-Government and Devolution: Electronic Service Delivery in Northern Ireland. *Local Government Studies*, *31*(3), 307–319. doi:10.1080/03003930500095137

Parke, H., & Wormell, D. (1956). *The Delphic Oracle*. Oxford, UK: Blackwell.

Parlak, B. (2011). *Yönetim Bilimi ve Çağdaş Yönetim Teknikleri*. İstanbul: Beta Yayinevi.

Paşaoğlu, D. (2017). Dünyada e-Devlet Uygulamalari. In Bilgi Toplumu ve E-Devlet. Eskişehir: Anadolu Üniversitesi Yayini No: 3559.

Patokorpi, E., & Ahvenainen, M. (2009). Developing an abduction-based method for futures research. *Futures*, *41*(3), 126–139. doi:10.1016/j.futures.2008.09.019

Patterson, K. A. (2003). *Servant leadership: a theoretical model* (Doctoral dissertation). Regent University.

Patterson, K. A. (2003). *Servant leadership: A theoretical model* (Doctoral Dissertation). Regent University.

Pava, M. L. (2003). *Leading with meaning: Using covenantal leadership to build a better organization*. New York: Palgrave, MacMillan.

Peck, M. S. (2004). *The road less travelled, 25th anniversary edition: A New Psychology of Love, Traditional Values and Spiritual Growth*. New York: Touchstone.

Peck, S. (1997). *The different drum: Community making and peace*. New York: Simon and Shuster.

Pedersen, E. R. G. (2015). The Anatomy of CSR. In C. S. Responsibility. Croydon: SAGE.

Peirce, C. S. (1955). *Philosophical writings of Peirce*. New York: Dover.

Peirce, C. S. (1997). *Pragmatism as a principle and method of right thinking: The 1903 Harvard lectures on pragmatism*. New York: State University of New York Press.

Pejić-Bach, M., Omazić, M. A., Aleksić, A., & Zoroja, J. (2018). Massive Open Online Courses (MOOCs) as an education tool for corporate social responsibility. In A. Stachowicz-Stanusch & W. Amann (Eds.), *Management Education for Corporate Social Performance* (pp. 157–179). Charlotte, NC: IAP.

Pekerti, A. A., & Sendjaya, S. (2010). Exploring servant leadership across cultures: A comparative study in Australia and Indonesia. *International Journal of Human Resource Management*, *21*(5), 754–780. doi:10.1080/09585191003658920

Pellegrini, E. K., & Scandura, T. A. (2005). Construct equivalence across groups: An unexplored issue in mentoring research. *Educational and Psychological Measurement*, *65*(2), 323–335. doi:10.1177/0013164404268665

Pennings, J. M., & Buitendam, A. (1987). *New technology and organizational innovation: The development and diffusion of micro-electronics*. Cambridge: Ballinger Publishing.

Peredo, A. M., & McLean, M. (2006). Social entrepreneurship: A critical review of the concept. *Journal of World Business*, *41*(10), 56–65. doi:10.1016/j.jwb.2005.10.007

Perrini, F., & Vurro, C. (2006). Social entrepreneurship: Innovation and social change across theory and practice. In J. Mair, J. Robinson, & K. Hockerts (Eds.), *Social entrepreneurship* (pp. 57–85). New York: Palgrave Macmillan. doi:10.1057/9780230625655_5

Peteraf, M. A. (1993). The cornerstones of competitive advantage: A resource-based view. *Strategic Management Journal*, *14*(3), 179–191. doi:10.1002mj.4250140303

Peters, G., Marland, G., Le Quéré, C., Boden, T., Canadell, J., & Raupach, M. (2012). Rapid growth in CO 2 emissions after the 2008–2009 global financial crisis. *Nature Climate Change*, *2*(1), 2–4. doi:10.1038/nclimate1332

Peterson, S. J., Galvin, B. M., & Lange, D. (2012). CEO servant Leadership: Exploring executive characteristics and firm performance. *Personnel Psychology*, *65*(3), 656–596. doi:10.1111/j.1744-6570.2012.01253.x

Peterson, S. J., & Luthans, F. (2002). Does the manager's level of hope matter? Preliminary research evidence of a positive impact. *Proceedings of the Midwest Academy of Management*.

Petrovskaya, I., & Mirakyan, A. (2018). A mission of service: Social entrepreneur as a servant leader. *International Journal of Entrepreneurial Behaviour & Research*, *24*(1), 755–767. doi:10.1108/IJEBR-02-2016-0057

Peus, C., Wesche, J. S., Streicher, B., Braun, S., & Frey, D. (2012). Authentic Leadership: An Empirical Test of Its Antecedents, Consequences, and Mediating Mechanisms. *Journal of Business Ethics*, *107*(3), 331–348. doi:10.100710551-011-1042-3

Pfeffer, J. (1998). *The human equation: Building power by putting people first*. Cambridge, MA: Harvard Business School Press.

Pfeffer, J., & Salancik, G. (2003). *The External Control of Organizations: A Resource Dependence Perspective*. Stanford, CA: Stanford University Press.

Phillips, W., Lee, H., Ghobadian, A., O'Regan, N., & James, P. (2015). Social innovation and social entrepreneurship: A systematic review. *Group & Organization Management*, *40*(3), 428–461. doi:10.1177/1059601114560063

Phills, J. A., Deiglmeier, K., & Miller, D. T. (2008). Rediscovering social innovation. *Stanford Social Innovation Review*, *6*(4), 34–43.

Pirola-Merlo, A., Härtel, C., Mann, L., & Hirst, G. (2002). How leaders influence the impact of affective events on team climate and performance in R&D teams. *The Leadership Quarterly*, *13*(5), 561–581. doi:10.1016/S1048-9843(02)00144-3

Pless, N. M. (2007). Understanding responsible leadership: Role identity and motivational drivers. *Journal of Business Ethics*, *74*(4), 437–456. doi:10.100710551-007-9518-x

Plowman, D. A., Baker, L. T., Beck, T. E., Kulkarni, M., Solansky, S. T., & Travis, D. V. (2007). Radical change accidentally: The emergence and amplification of small change. *Academy of Management Journal*, *50*(3), 515–543. doi:10.5465/amj.2007.25525647

Podsakoff, P., Mac Kenzie, S., & Boomer, W. (1996). Transformational leader behaviours substitute for leadership as determinants of employee satisfaction, commitment, trust and organizational citizenship behaviour. *Journal of Management*, *22*(2), 259–298.

Podsakoff, P., Niehoff, B., Mac Kenzie, S., & Williams, M. (1993). Do substitutes for leadership really substitute for leadership? An empirical examination of Kerr and Jermier's situational leadership model. *Organizational Behavior and Human Decision Processes*, *54*(1), 1–44. doi:10.1006/obhd.1993.1001

Pol, E., & Ville, S. (2009). Social innovation: Buzz word or enduring term? *Journal of Socio-Economics*, *38*(6), 878–885. doi:10.1016/j.socec.2009.02.011

Pomerantz, M. (2003). The business of social entrepreneurship in a down economy. In Business, 25(2), 25-28.

Porter, M. E. (1985). Technology and competitive advantage. *The Journal of Business Strategy*, *5*(3), 60–78. doi:10.1108/eb039075

Porter, M. E. (1996). What is Strategy? *Harvard Business Review*, *74*(6), 61–78. PMID:10158474

Porter, M. E. (2011). *Competitive advantage of nations: creating and sustaining superior performance*. New York: Simon and Schuster.

Porter, M. E., & Kramer, M. R. (2006). Strategy and society: The link between competitive advantage and corporate social responsibility. *Harvard Business Review*, *84*(12), 78–92. PMID:17183795

Porter, M. E., & Kramer, M. R. (2006). Strategy and society: The link between corporate social responsibility and competitive advantage. *Harvard Business Review*, *84*(12), 78–92. PMID:17183795

Porter, M. E., & Kramer, M. R. (2011). The Big Idea: Creating Shared Value. How to reinvent capitalism—and unleash a wave of innovation and growth. *Harvard Business Review*, *89*(1-2).

Porter, M., & Kramer, M. (2006). The link between competitive advantage and corporate social responsibility. *Harvard Business Review*, *84*(12), 78–92. PMID:17183795

Prabhu, G. N. (1999). Social entrepreneurial leadership. *Career Development International*, *4*(3), 140–145. doi:10.1108/13620439910262796

Pricto-Pastor, I., & Martin-Perez, V. (2015). Does HRM Generate Ambidextrous Employees for Ambidextrous Learning? The Moderating Role of Management Support. *XXIII Congreso Nacional de Acede*.

Pride, W., & Ferrell, O. (2006). *Marketing: Concepts and strategies*. Boston, MA: Houghton, Mifflin.

Prieto, L., Phipps, S., & Friedrich, T. (2012). Social entrepreneur development: An integration of critical pedagogy, the theory of planned behavior and the ACS model. *Academy of Entrepreneurship Journal, 18*(2), 1–82.

Prilleltensky, I. (2000). Value-based leadership in organizations: Balancing values, interests, and power among citizens, workers, and leaders. *Ethics & Behavior, 10*(2), 139–158. doi:10.1207/S15327019EB1002_03

Productivity Commission. (2010). *Contribution of the Not-for-Profit Sector* (Vol. Research Report). Canberra: Commonwealth of Australia.

Productivity Commission. (2011). *Productivity Commission Inquiry Report: Disability Care and Support* (Vol. 1). Canberra: Commonwealth of Australia.

Pruett, M., Shinnar, R., Toney, B., Llopis, F., & Fox, J. (2009). Explaining entrepreneurial intentions of university students: A cross-cultural study. *International Journal of Entrepreneurial Behaviour & Research, 15*(6), 571–594. doi:10.1108/13552550910995443

Pruzan, P., & Miller, W. C. (2006). Spirituality as the basis of responsible leaders and responsible companies. In *Responsible leadership* (pp. 88–112). Routledge.

Qiu, S., Alizadeh, A., Dooley, L., & Zhang, R. (2019). The effects of authentic leadership on trust in leaders, organizational citizenship behavior, and service quality in the Chinese hospitality industry. *Journal of Hospitality and Tourism Management, 40*(1), 77–87. doi:10.1016/j.jhtm.2019.06.004

Quinn, L., & Dalton, M. (2009). Leading for sustainability: implementing the tasks of leadership. Corporate Governance. *The International Journal of Business in Society, 9*(1), 21-38.

Ragin, C. C. (1987). *The Comparative Method*. Berkeley, CA: University of California Press.

Ragin, C. C. (2000). *Fuzzy-set Social Science*. Chicago: University of Chicago Press.

Ragin, C. C. (2008). *Redesigning Social Inquiry: Fuzzy Sets and Beyond*. Chicago: University of Chicago Press. doi:10.7208/chicago/9780226702797.001.0001

Raisch, S., & Birkinshaw, J. (2008). Organizational Ambidexterity: Antecedents Outcomes, and Moderators. *Journal of Management, 34*(3), 375–409. doi:10.1177/0149206308316058

Rast, D. E., Hogg, M. A., & Geisser, S. R. (2013). Self-uncertainty and support for autocratic leadership. *Self and Identity, 12*(6), 635–649. doi:10.1080/15298868.2012.718864

Reddy, A. V., & Kamesh, A. V. S. (2016). Integrating servant leadership and ethical leadership. In M. Chatterji & L. Zsolnai (Eds.), *Ethical leadership*. London: Palgrave Macmillan. doi:10.1057/978-1-137-60194-0_7

Reed, L. L., Cohen, D. V., & Colwell, S. R. (2011). A new scale to measure executive servant leadership: Development, analysis, and implications for research. *Journal of Business Ethics*, *101*(3), 41–434.

Rego, A., Sousa, F., Marques, C., & Cunha, M. P. e. (2012). Authentic leadership promoting employees' psychological capital and creativity. *Journal of Business Research*, *65*(3), 429–437. doi:10.1016/j.jbusres.2011.10.003

Reid, S. W., Anglin, A. H., Baur, J. E., Short, J. C., & Buckley, M. R. (2018). Blazing new trails or opportunity lost? Evaluating research at the intersection of leadership and entrepreneurship. *The Leadership Quarterly*, *29*(1), 150–164. doi:10.1016/j.leaqua.2017.11.005

Reinke, S. J. (2003). Does the form really matter? Leadership, trust and acceptance of the performance appraisal process. *Review of Public Personnel Administration*, *23*(1), 23–37. doi:10.1177/0734371X02250109

Reischauer, E. O. (1989). *Japan: The story of a nation* (4th ed.). New York: Knopf.

Reissman, C. K. (1993). *Narrative Analysis*. London: Sage Publications.

Reynolds, M., & Yuthas, K. (2008). Moral discourse and corporate social responsibility reporting. *Journal of Business Ethics*, *78*(1-2), 47–64. doi:10.100710551-006-9316-x

Rhodes M., & Donnelly-Cox G. (2008, Sept. 30). Social entrepreneurship as a performance landscape: The case of 'Front Line'. *Emergence: Complexity and Organization*, 1–17.

Ribeiro, N., Duarte, A., Filipe, R., & Torres de Oliveira, R. (2019). How authentic leadership promotes individual creativity: The mediating role of affective commitment. *Journal of Leadership & Organizational Studies*; Advanced Online Publication. doi:10.1177/1548051819842796

Ringham, K., & Miles, S. (2018). The boundary of corporate social responsibility reporting: The case of the airline industry. *Journal of Sustainable Tourism*, *26*(7), 1043–1062. doi:10.1080/09669582.2017.1423317

Ring, P. S., & Van de Ven, A. H. (1994). Developmental processes of cooperative inter-organizational relationships. *Academy of Management Review*, *19*(1), 90–118. doi:10.5465/amr.1994.9410122009

Rittel, H., & Webber, M. (1973). Dilemmas in a general theory of planning. *Policy Sciences*, *4*(2), 155–169. doi:10.1007/BF01405730

Rivera, R. G., Santos, D., Martín-Fernández, M., Requero, B., & Cancela, A. (2018). Predicting attitudes and behavioural intentions towards social entrepreneurship: The role of servant leadership in young people/Predicción de las actitudes y las intenciones conductuales hacia el emprendimiento social: el papel del liderazgo de servicio en los jóvenes. *Revista de Psicología Social*, *33*(3), 650–681. doi:10.1080/02134748.2018.1482057

Roberts, D., & Woods, C. (2005). Changing the world on a shoestring: The concept of social entrepreneurship. *University of Auckland Business Review*, *7*(1), 45–51.

Rochat, P. (2003). Five levels of self-awareness as they unfold early in life. *Consciousness and Cognition, 12*(4), 717–731. doi:10.1016/S1053-8100(03)00081-3 PMID:14656513

Rodríguez-Rubio, A., & Kiser, A. I. T. (2013). An examination of servant leadership in the United States and Mexico. Do age and gender make a difference? *The Global Studies Journal, 5*(2), 127–149. doi:10.18848/1835-4432/CGP/v05i02/40848

Roe, R. & Dalton, P. (n.d.). *Giving Hope: The Journey of the For-Purpose Organisation and Its Quest for Success*. Gateway East, Singapore: Springer Nature.

Roehrich, J. K., Davies, A., Frederiksen, L., & Sergeeva, N. (2018). Management innovation in complex products and systems: The case of integrated project teams. *Industrial Marketing Management.*

Rondinelli, D. A., & London, T. (2003). How corporations and environmental groups cooperate: Assessing cross-sector alliances and collaborations. *The Academy of Management Executive, 17*(1), 61–76. doi:10.5465/ame.2003.9474812

Roper, J., & Cheney, G. (2005). Leadership, learning and human resources management: The meanings of social entrepreneurship today. *Corporate Governance*, *5*(3), 95–104. doi:10.1108/14720700510604733

Roscoe, A. (2015). *The Digital Future of Corporate Responsibility*. Retrieved from: https://ssir.org/articles/entry/the_digital_future_of_corporate_responsibility#

Rosing, K., Frese, M., & Bausch, A. (2011). Explaining the heterogeneity of the leadership-innovation relationship: Ambidextrous leadership. *The Leadership Quarterly*, *22*(5), 956–974. doi:10.1016/j.leaqua.2011.07.014

Rosing, K., Rosenbusch, N., & Frese, M. (2010). Ambidextrous Leadership in the Innovation Process. In *Innovation and International Corporate Growth* (pp. 191–204). Springer Berlin Heidelberg. doi:10.1007/978-3-642-10823-5_12

Rowley, C., Ishikawa, J., & Oh, I. (2019). Changing facets of leadership in East Asia: Globalization, innovation and performance in Japan, South Korea and China. *Asia Pacific Business Review*, *25*(2), 161–172. doi:10.1080/13602381.2018.1557425

Rowley, C., Oh, I., & Jang, W. (2019). New perspectives on East Asian leadership in the age of globalization: Local grounding and historical comparisons in the Asia Pacific region. *Asia Pacific Business Review*, *25*(2), 307–315. doi:10.1080/13602381.2018.1557424

Rubin, R. S., Munz, D. C., & Bommer, W. H. (2005). Leading from within: The effects of emotion recognition and personality on transformational leadership behavior. *Academy of Management Journal*, *48*(5), 845–858. doi:10.5465/amj.2005.18803926

Rubio-Sánchez, A., Bosco, S. M., & Melchar, D. E. (2013). Servant leadership and world values. *The Global Studies Journal*, *5*(3), 19–30. doi:10.18848/1835-4432/CGP/v05i03/40803

Rumelt, R. P. (1984). Competitive strategic management. In R. Lamb (Ed.), Toward a Strategic Theory of the Firm, (pp. 556-570). Prentice Hall.

Rundle-Thiele, S. (2009). Social gain: Is corporate social responsibility enough? *Australasian Marketing Journal, 17*(4), 204–210. doi:10.1016/j.ausmj.2009.06.006

Ruschman, N. L. (2002). *Servant leadership and the best companies to work for in America. In Focus on Leadership: Servant leadership for the 21st century* (pp. 123–140). New York: Wiley.

Russell, R. F. (2000). *Exploring the values and attributes of servant leaders*. Doctoral dissertation, 5359435.

Russell, R. F. (2001). The role of values in servant leadership. *Leadership and Organization Development Journal, 22*(2), 76–84. doi:10.1108/01437730110382631

Russell, R. F., & Stone, A. G. (2002). A review of servant leadership attributes: Developing a practical model. *Leadership and Organization Development Journal, 23*(3), 145–157. doi:10.1108/01437730210424

Ryan, R. M., & Deci, E. L. (2000). Self-determination theory and the facilitation of intrinsic motivation, social development and well-being. *The American Psychologist, 55*(1), 68–78. doi:10.1037/0003-066X.55.1.68 PMID:11392867

Sabella, A. R., & Eid, N. L. (2016). A strategic perspective of social enterprise sustainability. *Journal of General Management, 41*(4), 71–89. doi:10.1177/030630701604100405

Saldinger, A. (2015). Can redefining humanity be Muhammad Yunus legacy? an interview with Muhammad Yunus. *Yunus Centre*. Available at http://muhammadyunus.org/index.php/news-media/in-the-media/1451-can-redefining-humanity-be-muhammad-yunus-legacy

Santora, J. C., Seaton, W., & Sarros, J. C. (1999). Changing times: Entrepreneurial leadership in a community-based nonprofit organization. *The Journal of Leadership Studies, 6*(3-4), 101–109. doi:10.1177/107179199900600308

Santos, R. A., & Guevera, A. J. (2012). Compliance and leadership: The susceptibility of leader to the risk of corruption in the organizations. *Einstein (Sao Paulo, Brazil), 10*(1), 1–10. doi:10.1590/S1679-45082012000100003 PMID:23045818

Santos, S. C., Caetano, A., Baron, R., & Curral, L. (2015). Prototype models of opportunity recognition and the decision to launch a new venture: Identifying the basic dimensions. *International Journal of Entrepreneurial Behaviour & Research, 21*(4), 510–538. doi:10.1108/IJEBR-04-2014-0058

Sanzo, M. J., Álvarez, L. I., Rey, M., & García, N. (2015). Business-nonprofit partnerships: A new form of collaboration in a corporate responsibility and social innovation context. *Service Business, 9*(4), 611–636. doi:10.100711628-014-0242-1

Schein, E. H. (2010). *Organizational culture and leadership* (Vol. 2). New York, NY: Wiley.

Schilling, J. (2009). From ineffectiveness to destruction: A qualitative study on the meaning of negative leadership. *Leadership*, *5*(1), 102–128. doi:10.1177/1742715008098312

Schindler, A. (2015). *Ambidextrous Leadership* (Master Thesis). Enschede: University of Twente.

Schneider, S. (2001). In search of realistic optimism: Meaning, knowledge, and warm fuzziness. *The American Psychologist*, *56*(3), 250–263. doi:10.1037/0003-066X.56.3.250 PMID:11315251

Scholl, J. (2013). Inclusive Business Models as a Key Driver for Social Innovation. In T. Osburg & R. Schmidpeter (Eds.), *Social Innovation. CSR, Sustainability, Ethics & Governance*. Berlin: Springer. doi:10.1007/978-3-642-36540-9_9

Schreck, P. (2013). Disclosure (CSR Reporting). In *Encyclopedia of Corporate Social Responsibility* (pp. 801–810). Berlin: Springer-Verlag. doi:10.1007/978-3-642-28036-8_145

Schrempf-Stirling, J., Palazzo, G., & Phillips, R. (2016). Historic corporate social responsibility. *Academy of Management Review*, *41*(4), 700–719. doi:10.5465/amr.2014.0137

Schulman, P. (1999). Applying learned optimism to increase sales productivity. *Journal of Personal Selling & Sales Management*, *19*(1), 31–37.

Schumacher, P., & Krish, S. (2010). *Teaching Generative Design Strategies for Industrial Design*. Academic Press.

Schumpeter, J. A. (1934). *The theory of economic develeopment*. Cambridge, MA: Harvard University Press.

Schumpeter, J. A. (1942). *Capitalism, socialism and democracy*. New York: Harper and Row.

Schussler, D., Stooksberry, L., & Bercaw, L. (2010). Understanding teacher candidate dispositions: Reflecting to build self-awareness. *Journal of Teacher Education*, *61*(4), 350–363. doi:10.1177/0022487110371377

Schwantes, M. (2015). Servant leadership. The case for the best leadership philosophy. *Leadership Excellence Essentials*, *32*(5), 30–31.

Schwartz, B. (2012). *Rippling: How Social Entrepreneurs Spread Innovation throughout the World*. San Francisco: Jossey-Bass.

Seelos, C., & Mair, J. (2005). Social entrepreneurship: Creating new business models to serve the poor. *Business Horizons*, *48*(3), 241–246. doi:10.1016/j.bushor.2004.11.006

Seferoğlu, S. S., Çelen, F. K., & Çelik, A. (2011). Türkiye'de E-Devlet Uygulamalari: Sorunlar ve Çözüm Önerileri Üstüne Bir Analiz. In Türkiye'de E-Öğrenme: Gelişmeler ve Uygulamalar II. Eskişehir: Anadolu Üniversitesi İletişim Bilimleri Fakültesi Yayinlari, Yayin No:77.

Seifert, J. W., & Chung, J. (2009). Using E-Government to Reinforce Government–Citizen Relationships Comparing Government Reform in the United States and China. *Social Science Computer Review*, *27*(1), 3–23. doi:10.1177/0894439308316404

Seitanidi, M. M., & Crane, A. (2009). Implementing CSR through partnerships: Understanding the selection, design, and institutionalization of nonprofit-business partnerships. *Journal of Business Ethics*, *85*(2), 413–429. doi:10.100710551-008-9743-y

Seligman, M. (2004). *Authentic Happiness: Using the new positive psychology to realize your potential for lasting fulfillment*. Simon and Schuster.

Selsky, J. W., & Parker, B. (2005). Cross-sector partnerships to address social issues: Challenges to theory and practice. *Journal of Management*, *31*(6), 849–873. doi:10.1177/0149206305279601

Semedo, A., Coelho, A., & Ribeiro, N. (2019). *Authentic leadership, happiness at work and affective commitment: An empirical study in Cape Verde. European Business Review*. Advanced Online Publication. doi:10.1108/EBR-01-2018-0034

Sendjaya, S., & Pekerti, A. (2010). Servant leadership as antecedent of trust in organizations. *Leadership and Organization Development Journal*, *31*(7), 643–663. doi:10.1108/01437731011079673

Sendjaya, S., & Sarros, J. (2002). Servant leadership: Its origin, development and application in organizations. *Journal of Leadership & Organizational Studies*, *9*(2), 57–64. doi:10.1177/107179190200900205

Sendjaya, S., Sarros, J. C., & Santora, J. C. (2008). Defining and measuring servant leadership behaviour in organizations. *Journal of Management Studies*, *45*(2), 402–424. doi:10.1111/j.1467-6486.2007.00761.x

Senge, P. (1990). The leader's new work: Building learning organizations. *Sloan Management Review*, *32*(1), 7–24.

Sethi, S. P., Martell, T. F., & Demir, M. (2017). An evaluation of the quality of corporate social responsibility reports by some of the world's largest financial institutions. *Journal of Business Ethics*, *140*(4), 787–805. doi:10.100710551-015-2878-8

Seyfang, G., & Longhurst, N. (2013). Desperately seeking niches: Grassroots innovations and niche development in the community currency field. *Global Environmental Change*, *23*(5), 881–891. doi:10.1016/j.gloenvcha.2013.02.007

Seyfang, G., & Smith, A. (2007). Grassroots innovations for sustainable development: Towards a new research and policy agenda. *Environmental Politics*, *16*(4), 584–603. doi:10.1080/09644010701419121

Shahnawaz, M., & Hassan, M. (2009). Psychological Capital as Predictors of Organizational Commitment and Organizational Citizenship Behaviour. *Journal of the Indian Academy of Applied Psychology*, *35*(October), 78–84.

Shamir, B. (2007). From passive recipients to active co-producers: The roles of followers in the leadership process. In B. Shamir, R. Pillai, M. Bligh, & M. Uhl-Bien (Eds.), *Follower-centered perspectives on leadership: A tribute to the memory of James R. Meindl*. Stanford, CT: Information Age.

Shamir, B., & Eilam, G. (2005). What's your story? A life-stories approach to authentic leadership development. *The Leadership Quarterly*, *16*(3), 395–417. doi:10.1016/j.leaqua.2005.03.005

Shamir, B., House, R. J., & Arthur, M. B. (1993). The Motivational Effects of Charismatic Leadership: A Self-Concept Theory. *Organization Science*, *4*(4), 1–17. doi:10.1287/orsc.4.4.577

Shane, S., & Venkataraman, S. (2000). The promise of entrepreneurship as a field of research. *Academy of Management Review*, *25*(1), 217–226.

Sharma, G. & Pokharel, M. (2016). The Strategic Role of Information Technology In E-Government: A Case Of Nepal. *International Journal Of eBusiness And eGovernment Studies*, *8*(1), 19-39.

Shaw, E., & Carter, S. (2007). Social entrepreneurship: Theoretical antecedents and empirical analysis of entrepreneurial processes and outcomes. *Journal of Small Business and Enterprise Development*, *14*(3), 418–434. doi:10.1108/14626000710773529

Shea, K., Aish, R., & Gourtovaia, M. (2005). Towards integrated performance-driven generative design tools. *Automation in Construction*, *14*(2), 253–264. doi:10.1016/j.autcon.2004.07.002

Sheehan, S. (2018). *Calling and Authentic Followership* (Doctoral dissertation). Regent University.

Sheldon, K. M., & King, L. (2001). Why positive psychology is necessary. *The American Psychologist*, *56*(3), 216–217. doi:10.1037/0003-066X.56.3.216 PMID:11315247

Shim, K., & Yang, S. (2016). The effect of bad reputation: The occurrence of crisis, corporate social responsibility, and perceptions of hypocrisy and attitudes toward a company. *Public Relations Review*, *42*(1), 68–78. doi:10.1016/j.pubrev.2015.11.009

Shirey, M., White-Williams, C., & Hites, L. (2019). Integration of Authentic Leadership Lens for Building High Performing Interprofessional Collaborative Practice Teams. *Nursing Administration Quarterly*, *43*(2), 101–112. doi:10.1097/NAQ.0000000000000339 PMID:30839447

Short, J. C., Moss, T. W., & Lumpkin, G. T. (2009). Research in social entrepreneurship: Past contributions and future opportunities. *Strategic Entrepreneurship Journal*, *3*(2), 161–194. doi:10.1002ej.69

Skelton, C. (2016, Sept.). NDIS: Is your approach to governance ready for the new environment? *Governance Directions*, 458–460.

Skogstad, A., & Einarsen, S. (1999). The İmportance of A Change-Centred Leadership Style in Four Organizational Cultures. *Scandinavian Journal of Management*, *15*(3), 289–306. doi:10.1016/S0956-5221(98)00028-1

Small, H. (1973). Co-citation in the scientific literature: A new measure of the relationship between two documents. *Journal of the Association for Information Science and Technology*, *24*(4), 265–269.

Smith, A., & Raven, R. (2012). What is protective space? Reconsidering niches in transitions to sustainability. *Research Policy*, *41*(6), 1025–1036. doi:10.1016/j.respol.2011.12.012

Smith-Merry, J., Hancock, N., Bresnan, A., Yen, I., Gilroy, J., & Llewellyn, G. (2018). *Mind the Gap: The National Disability Insurance Scheme and psychosocial disability*. Final Report: Stakeholder identified gaps and solutions. University of Sydney: Lidcombe.

Smith, P. A. C., & Sharicz, C. (2011). The shift needed for Sustainability. *The Learning Organization, 18*(1), 73–86. doi:10.1108/09696471111096019

Smith, W. K., Besharov, M. L., Wessels, A. K., & Chertok, M. (2012). A paradoxical leadership model for social entrepreneurs: Challenges, leadership skills, and pedagogical tools for managing social and commercial demands. *Academy of Management Learning & Education, 11*(3), 463–478. doi:10.5465/amle.2011.0021

Snyder, C. R., Feldman, D. B., Taylor, J. D., Schroeder, L. L., & Adams, V. H. III. (2000). The roles of hopeful thinking in preventing problems and enhancing strengths. *Applied & Preventive Psychology, 9*(4), 249–269. doi:10.1016/S0962-1849(00)80003-7

Snyder, C. R., Harris, C., Anderson, J. R., Holleran, S., Irving, L., Sigmon, S., & Harney, P. (1991). The Will and the Ways: Development and Validation of an Individual-Differences Measure of Hope. *Journal of Personality and Social Psychology, 60*(4), 570–585. doi:10.1037/0022-3514.60.4.570 PMID:2037968

Sosik, J., Jung, D., & Dinger, S. (2009). Values in authentic action: Examining the roots and rewards of altruistic leadership. *Group & Organization Management, 34*(4), 395–431. doi:10.1177/1059601108329212

Sousa, M., & van Dierendonck, D. (2017). Servant Leadership and the Effect of the Interaction Between Humility, Action, and Hierarchical Power on Follower Engagement. *Journal of Business Ethics, 141*(1), 13–25. doi:10.100710551-015-2725-y

Spanos, L. (2018). Complementarity and Interconnection Between CSR and Crowdfunding: A Case Study in Greece. In G. Grigore, A. Stancu, & D. McQueen (Eds.), *Corporate Responsibility and Digital Communities. Palgrave Studies in Governance, Leadership and Responsibility* (pp. 29–49). Cham: Palgrave Macmillan. doi:10.1007/978-3-319-63480-7_3

Spears, L. (1995). *Reflections on leadership: How Robert K. Greenleaf's theory of servant leadership influenced today's top management thinkers*. New York: John Wiley.

Spears, L. C. (1995). *Reflections on leadership: How Robert K. Greenleaf's theory of servant-leadership influenced today's top management thinkers*. John Wiley & Sons.

Spears, L. C. (1996). Reflections on Robert K Greenleaf and servant leadership. *Leadership and Organization Development Journal, 17*(7), 33–35. doi:10.1108/01437739610148367

Spears, L. C. (1998). *Insights on leadership: Service, stewardship, spirit and servant-leadership*. New York: John Wiley and Sons.

Spears, L. C. (2004). Practicing servant-leadership. *Leader to Leader, 34*(34), 7–11. doi:10.1002/ltl.94

Spicker, P. (2009). The Nature of a Public Service. *International Journal of Public Administration, 32*(11), 970–991. doi:10.1080/01900690903050927

Srinivas, M. N. (1962). *Caste in modern India and other essays.* London: Asia Publishing House.

Stacey, R. D. (2003). *Strategic Management and Organisational Dynamics: The Challenge of Complexity.* Harlow: FT/Prentice Hall.

Stainton, T. (2000). What is self-determination? *Proceedings of the First International Conference on Self Determination and Individualized Funding.*

Stainton, T. (2002). Taking rights structurally: Disability, rights and social worker responses to direct payments. *British Journal of Social Work, 32*(6), 751–763. doi:10.1093/bjsw/32.6.751

Stajkovic, A. D., & Luthans, F. (1998). Self-Efficacy and Work-Related Performance: A Meta-Analysis. *Psychological Bulletin, 124*(2), 240–261. doi:10.1037/0033-2909.124.2.240

Stake, R. E. (1995). *The art of case study research.* Thousand Oaks, CA: Sage.

Starbuck, W. H. (2004). Why I stopped trying to understand the real world. *Organization Studies, 25*(7), 1233–1254. doi:10.1177/0170840604046361

Steers, R. M. (1991). *Introduction to Organizational Behavior* (4th ed.). New York, NY: Harper Collins Publishers.

Steers, R. M., Sánchez-Runde, C., & Nardon, L. (2012). Leadership in a global context: New directions in research and theory development. *Journal of World Business, 47*(4), 479–489. doi:10.1016/j.jwb.2012.01.001

Steinbauer, R., Renn, R. W., Taylor, R. R., & Njoroge, P. K. (2014). Ethical leadership and followers' moral judgement: The role of followers perceived accountability and self-leadership. *Journal of Business Ethics, 120*(3), 381–392. doi:10.100710551-013-1662-x

Stephan, U., & Drencheva, A. (2017). The person in social entrepreneurship: A systematic review of research on the social entrepreneurial personality. In G. Ahmetoglu, T. Chamorro-Premuzic, B. Klinger, & T. Karcisky (Eds.), *The Wiley Handbook of Entrepreneurship* (pp. 205–229). Hoboken, NJ: John Wiley & Sons. doi:10.1002/9781118970812.ch10

Stone, A. G., Russell, R. F., & Patterson, K. (2004). Transformational versus servant leadership: A difference in leader focus. *Leadership and Organization Development Journal, 25*(4), 349–361. doi:10.1108/01437730410538671

Strielkowski, W., & Chigisheva, O. (2018). Social, Economic, and Academic Leadership for Sustainable Development of Business and Education in the Future: An Introduction. In W. Strielkowski & O. Chigisheva (Eds.), *Leadership for the Future Sustainable Development of Business and Education* (pp. 3–8). Cham: Springer. doi:10.1007/978-3-319-74216-8_1

Suchman, M. C. (1995). Managing legitimacy: Strategic and institutional approaches. *Academy of Management Review, 20*(3), 571–610. doi:10.5465/amr.1995.9508080331

Svennevig, J. (2001). Abduction as a methodological approach to the study of spoken interaction. *Norskrift, 103*, 1–22.

Sweetman, D., Luthans, F., Avey, J. B., & Luthans, B. C. (2011). Relationship between positive psychological capital and creative performance. *Canadian Journal of Administrative Sciences, 28*(1), 4–13. doi:10.1002/cjas.175

Swyngedouw, E. (2005). Governance innovation and the citizen: The Janus face of governance-beyond-the-state. *Urban Studies (Edinburgh, Scotland), 42*(11), 1991–2006. doi:10.1080/00420980500279869

Sy, T., Côté, S., & Saavedra, R. (2005). The contagious leader: Impact of the leader's mood on the mood of group members, group affective tone, and group processes. *The Journal of Applied Psychology, 90*(2), 295–305. doi:10.1037/0021-9010.90.2.295 PMID:15769239

Székely, F., & Knirsch, M. (2005). Responsible leadership and corporate social responsibility: Metrics for sustainable performance. *European Management Journal, 23*(6), 628–647. doi:10.1016/j.emj.2005.10.009

Sztop-Rutkowska, K., Zagulski, S., Maszkowska, A., Ignatowska, B., Holat, R., Oraloglu, Z., . . . Ballesta, A. G. (2019). E-Government Closer To The People, E-Citizen- Effective Inclusion Through E-Governance Services. Bialystok.

Takahashi, K., Ishikawa, J., & Kanai, T. (2012). Qualitative and quantitative leadership studies in multinational settings: Meta analytics and cross cultural reviews. *Journal of World Business, 47*(4), 530–538. doi:10.1016/j.jwb.2012.01.006

TAKBİS Nedir? (n.d.). Retrieved from https://www.tkgm.gov.tr/tr/sayfa/tapu-ve-kadastro-bilgi-sistemi-takbis

Taleghani, G., Salmani, D., & Taatian, A. (2010). Survey of leadership styles in different countries. *Iranian Journal of Management Studies, 3*(3), 91–111.

Tammy, E. (2012). The biggest mistake you (probably) make with teams. *Harvard Business Review*, (April): 5.

Tan, W. L., Williams, J., & Tan, T. M. (2005). Defining the 'social' in 'social entrepreneurship': Altruism and Entrepreneurship. *The International Entrepreneurship and Management Journal, 1*(3), 353–365. doi:10.100711365-005-2600-x

Tapsell, P., & Woods, C. (2010). Social entrepreneurship and innovation: Self-organization in an indigenous context. *Entrepreneurship and Regional Development, 22*(6), 535–556. doi:10.1080/08985626.2010.488403

Taylor, A., & Helfat, C. E. (2009). Organizational Linkages for Surviving Technological Change: Complementary Assets, Middle Management, and Ambidexterity. *Organization Science, 20*(4), 718–739. doi:10.1287/orsc.1090.0429

Taylor, J. B. (1970). Introducing social innovation. *The Journal of Applied Behavioral Science*, *6*(1), 69–77. doi:10.1177/002188637000600104

Taylor, R., Torugsa, N., & Arundel, A. (2018a). Leaping into real-world relevance: An 'abduction' process for nonprofit research. *Nonprofit and Voluntary Sector Quarterly*, *47*(1), 206–227. doi:10.1177/0899764017718635

Taylor, R., Torugsa, N., & Arundel, A. (2018b). Thriving within the chaos: A complexity theorizing approach to social innovation by nonprofit organizations. In C. Dogru (Ed.), *Handbook of Research on Contemporary Approaches in Management and Organizational Strategy* (pp. 36–61). IGI Global.

Taylor, R., Torugsa, N., & Arundel, A. (2019a). Organizational pathways for social innovation and societal impacts in disability nonprofits. *Voluntas*. doi:10.100711266-019-00113-8

Taylor, R., Torugsa, N., & Arundel, A. (2019b). Social innovation in disability nonprofits: An abductive study of capabilities for social change. *Nonprofit and Voluntary Sector Quarterly*. doi:10.1177/0899764019873965

Teece, D. J. (1986). Profiting from technological innovation: Implications for integration, collaboration, licensing and public policy. *Research Policy*, *15*(6), 285–305. doi:10.1016/0048-7333(86)90027-2

Thakur, R., Hsu, S., & Fontenot, G. (2012). Innovation in healthcare: Issues and future trends. *Journal of Business Research*, *65*(4), 562–569. doi:10.1016/j.jbusres.2011.02.022

The Economist Intelligence Unit. (2016). *Old problems, new solutions: Measuring the capacity for social innovation across the world. (Social innovation index 2016)*. London, UK: The Economist Intelligence Unit.

The Greenleaf Center for Servant Leadership. (1997, Fall). *The Servant Leader*.

The Holy Bible (New International Version). (1984). Grand Rapid, MI: Zondervan Publishing House.

Thompson, J. L. (2008). Social enterprise and social entrepreneurship: Where have we reached? A summary of issues and discussion points. *Social Enterprise Journal*, *4*(2), 149–161. doi:10.1108/17508610810902039

Thompson, J., Alvy, G., & Lees, A. (2000). Social entrepreneurship–a new look at the people and the potential. *Management Decision*, *38*(5), 328–338. doi:10.1108/00251740010340517

Thornton, D., & Campbell, L. (2017). Improving The Management of Digital Government. Institute for Government 2017.

Thoroughgood, C. N., Padilla, A., Hunter, S. T., & Tate, B. W. (2012). The susceptible circle: A taxonomy of followers associated with destructive leadership. *The Leadership Quarterly*, *23*(5), 897–917. doi:10.1016/j.leaqua.2012.05.007

Thorpe, A., & Gamman, L. (2011). Design with society: Why socially responsive design is good enough. *CoDesign*, *7*(3-4), 217–230. doi:10.1080/15710882.2011.630477

Tilly, C. (1991). Domination, resistance, compliance...discourse. *Sociological Forum*, *6*(3), 593–602. doi:10.1007/BF01114480

Timmons, J. A. (1978). Characteristics and role demands of entrepreneurship. *American Journal of Small Business*, *3*(1), 5–17. doi:10.1177/104225877800300102

Tobak, S. (2017, Jan. 4). The 10 worst entrepreneurs of 2016. *Entrepreneur.*

Toepler, S. (2003). Grassroots associations versus larger nonprofits: New evidence from a community case study in arts and culture. *Nonprofit and Voluntary Sector Quarterly*, *32*(2), 236–251. doi:10.1177/0899764003032002004

Toft, B. J. (2018). *Competitive advantage in a global market: The Case of the Norwegian Salmon Farming Industry A comparative study of the Chilean salmon farming industry* (Master's thesis). University of Agder.

Tonge, A., Greer, L., & Lawton, A. (2003). The Enron story: You can fool some of the people some of the time.... *Business Ethics (Oxford, England)*, *12*(1), 4–22. doi:10.1111/1467-8608.00301

Topaloğlu, C., & Dalgın, T. (2013). Algılanan Liderlik Tarzı ve Örgütsel Bağlılık İlişkisi, Marmaris'te Yer Alan Beş Yıldızlı Otel İşletmeleri Üzerinde Bir Uygulama. Balikesir University The Journal of Social Sciences Institute, 16(29), 277-301.

Tortop, N., İsbir, E. G., Aykaç, B., Yayman, H., & Özer, M. A. (2017). Yönetim Bilimi, 11. Ankara: Nobel Yayinevi.

Tracey, P., & Stott, N. (2017). Social innovation: A window on alternative ways of organizing and innovating. *Innovation*, *19*(1), 51–60. doi:10.1080/14479338.2016.1268924

Trevino, L. K., Hartman, L. P., & Brown, M. (2000). Moral person and moral manager: How executives develop a reputation for ethical leadership. *California Management Review*, *42*(2), 128–142. doi:10.2307/41166057

Trice, H. M., & Beyer, J. M. (1984). Studying organizational cultures through rites and ceremonials. *Academy of Management Review*, *9*(4), 653–669. doi:10.5465/amr.1984.4277391

Trilling, L. (1972). *Sincerity and authenticity*. Cambridge, MA: Harvard University Press.

Trompenaars, A., & Voerman, E. (2010). *Servant-leadership across cultures: Harnessing the strength of the world's most powerful management philosophy*. Oxford, UK: Infinite Limited Ideas.

Tuan Luu, T. (2017). Ambidextrous leadership, entrepreneurial orientation, and operational performance: Organizational social capital as a moderator. *Leadership and Organization Development Journal*, *38*(2), 229–253. doi:10.1108/LODJ-09-2015-0191

Tuan, L. T. (2017). Reform in Public Organizations: The Roles of Ambidextrous Leadership and Moderating Mechanisms. *Public Management Review*, *19*(4), 518–541. doi:10.1080/14719037.2016.1195438

Tung, F. C. (2016). Does Transformational, Ambidextrous, Transactional Leadership Promote Employee Creativity? Mediating Effects of Empowerment and Promotion Focus. *International Journal of Manpower*, *37*(8), 1250–1263. doi:10.1108/IJM-09-2014-0177

Turnip, K.; Lubis, A. H., & Lubis, M. S. (2018). A Review of ICT in Government Bureaucracy: Psychological and Technology Skill Perspectives. *International Journal of Civil Engineering and Technology, 9*(9), 1309–1319.

Turulja, L., & Bajgorić, N. (2018). Knowing Means Existing: Organizational Learning Dimensions and Knowledge Management Capability. *Business Systems Research Journal*, *9*(1), 1–18. doi:10.2478/bsrj-2018-0001

Tushman, M. L., Smith, W. K. & Binns, A. (2011, June). The Ambidextrous CEO. *Harvard Business Review*, 74-80.

Tushman, M. L., & O'Reilly, C. A. III. (1996). Ambidextrous Organizations: Managing Evolutionary and Revolutionary Change. *California Management Review*, *38*(4), 1–23. doi:10.2307/41165852

Uça Güneş, E. P. (2017). e-Devlet Tanimi ve Özellikleri. In Bilgi Toplumu ve E-Devlet. Eskişehir: Anadolu Üniversitesi Yayini No: 3559.

Uğurluoğlu, Ö., & Çelik, Y. (2009). Örgütlerde Stratejik Liderlik ve Özellikleri. *Hacettepe Sağlık İdaresi Dergisi*, *12*(2), 121–156.

Uhl-Bien, M., Marion, R., & McKelvey, B. (2007). Complexity leadership theory: Shifting leadership from the industrial age to the knowledge era. *The Leadership Quarterly*, *18*(4), 298–318. doi:10.1016/j.leaqua.2007.04.002

Ullrich, D., Koleňák, J., Ambrozová, E., Pokorný, V., & Milichovský, F. (2019). Global X-tream Index and its Partial Parameters for Identifying the Level of Potential Individual Characteristics in the Challenging Conditions of a Modern Corporate and Security Environment. *Sustainability*, *11*(12), 3325. doi:10.3390u11123325

United Nations Development Program. (2017). *Country profiles: Mexico, Japan, and France*. Available online: http://hdr.undp.org/en/countries

United Nations. (2015). *Resolution adopted by the General Assembly on 25 September 2015*. United Nations. Retrieved from https://www.un.org/ga/search/view_doc.asp?symbol=A/RES/70/1&Lang=E

Urhal, P., Weightman, A., Diver, C., & Bartolo, P. (2019). Robot assisted additive manufacturing: A review. *Robotics and Computer-integrated Manufacturing*, *59*, 335–345. doi:10.1016/j.rcim.2019.05.005

Van de Ven, A. H. (2015). Welcome to the Academy of Management Discoveries (AMD). *Academy of Management Discoveries, 1*(1), 1–4. doi:10.5465/amd.2014.0143

Van de Vliert, E. (2006). Autocratic leadership around the globe - do climate and wealth drive leadership culture? *Journal of Cross-Cultural Psychology, 37*(1), 42–59. doi:10.1177/0022022105282294

Van der Have, R. P., & Rubalcaba, L. (2016). Social innovation research: An emerging area of innovation studies? *Research Policy, 45*(9), 1923–1935. doi:10.1016/j.respol.2016.06.010

Van Dierendonck, D. (2011). Servant leadership: A review and synthesis. *Journal of Management, 37*(4), 1228–1261. doi:10.1177/0149206310380462

Van Dierendonck, D., & Nuijten, I. (2011). The servant leadership survey: Development and validation of multidimensional measure. *Journal of Business and Psychology, 26*(3), 249–267. doi:10.100710869-010-9194-1 PMID:21949466

Van Dierendonck, D., Nuijten, I., & Heeren, I. (2009). Servant leadership, key to follower well-being. In D. Tjosvold & B. Wisse (Eds.), *Power and interdependence in organizations*. Cambridge, UK: Cambridge University Press. doi:10.1017/CBO9780511626562.020

Van Dierendonck, D., Sousa, M., Gunnarsdóttir, S., Bobbio, A., Hakanen, J., Armin, P. V., & Rodriguez-Carvajal, R. (2017). The cross-cultural invariance of the servant leadership survey: A comparative study across eight countries. *Administrative Sciences, 7*(2), 8. doi:10.3390/admsci7020008

van Eck, N. J., Waltman, L., Dekker, R., & van den Berg, J. (2010). A comparison of two techniques for bibliometric mapping: Multidimensional scaling and VOS. *Journal of the American Society for Information Science and Technology, 61*(12), 2405–2416. doi:10.1002/asi.21421

Van Knippenberg, B., & Van Knippenberg, D. (2005). Leader self-sacrifice and leadership effectiveness: The moderating role of leader prototypicality. *The Journal of Applied Psychology, 90*(1), 25–37. doi:10.1037/0021-9010.90.1.25 PMID:15641888

Van Manen, M. (1998). *Researching lived experience: Human science for an action sensitive pedagogy*. Albany, NY: SUNY Press.

Van Marrewijk, M. (2003). Concepts and definitions of CSR and corporate sustainability: Between agency and communion. *Journal of Business Ethics, 44*(2-3), 95–105. doi:10.1023/A:1023331212247

Vanstone, C., & Graham, B. (2016). NDIS readiness and beyond. Adelaide: TACSI (The Australian Centre for Social Innovation).

VanWhy, L. (2015). *Development of the Authentic Followership Profile (AFP) Test Instrument* (Doctoral dissertation). Regent University.

Vartia, M. (1996). The sources of bullying: Psychological work environment and organizational climate. *European Journal of Work and Organizational Psychology*, *5*(2), 203–214. doi:10.1080/13594329608414855

Verba, S., Nie, N. H., & Kim, J. (1987). *Participation and Political Equality: A Seven-Nation Comparison*. Chicago: The University of Chicago Press.

Vintar, M., Dečman, M., & Kunstelj, M. (2001). The Use Of Ict In The Public Sector And Its Influence On Communication With Citizens In Slovenia. *Global Co-Operation in the New Millennium, The 9th European Conference on Information Systems*, 923-934.

Vito, F., Higgins, G. E., & Denney, A. (2014). Transactional and transformational leadership: An examination of the leadership challenge model. *Policing: An International Journal of Police Strategies & Management*, *37*(4), 809–822. doi:10.1108/PIJPSM-01-2014-0008

Voegtlin, C., Patzer, M., & Scherer, A. G. (2012). Responsible leadership in global business: A new approach to leadership and its multi-level outcomes. *Journal of Business Ethics*, *105*(1), 1–16. doi:10.100710551-011-0952-4

von Hippel, E. (1986). Lead users: A source of novel product concepts. *Management Science*, *32*(7), 791–805. doi:10.1287/mnsc.32.7.791

von Hippel, E. (2005). *Democratizing Innovation*. Cambridge, MA: MIT Press. doi:10.7551/mitpress/2333.001.0001

Vroom, V. H., & Yetton, P. W. (1973). *Leadership and decision making*. Pittsburgh, PA: University of Pittsburgh Press. doi:10.2307/j.ctt6wrc8r

Waddock, S. A. (1989). Understanding social partnerships: An evolutionary model of partnership organizations. *Administration & Society*, *21*(1), 78–100. doi:10.1177/009539978902100105

Waddock, S. A., & Post, J. E. (1991). Social entrepreneurs and catalytic change. *Public Administration Review*, *51*(5), 393–401. doi:10.2307/976408

Wagner, T., Lutz, R., & Weitz, B. (2009). Corporate hypocrisy: Overcoming the threat of inconsistent corporate social responsibility perceptions. *Journal of Marketing*, *73*(6), 77–91. doi:10.1509/jmkg.73.6.77

Walden, R., Pandolfo, B., Lie, S., & Nemme, A. (2018b). *Adaptable practices for next-generation design in manufacturing SMEs informed by university-industry collaboration*. Paper presented at the Design Thinking Research Symposium DTRS, UNIST.

Walden, R., & Lie, S. (2019). *Design Knowledge Development and Additive Manufacturing Systems: How Does Design Knowledge Change With Design for AM?* Academic Press. doi:10.4018/978-1-5225-7152-0.ch011

Walden, R., Lie, S., Pandolfo, B., Lee, T., & Lockhart, C. (2018a). Design Research Units and Small to Medium Enterprises (SMEs): An Approach for Advancing Technology and Competitive Strength in Australia. *The Design Journal*, *21*(2), 247–265. doi:10.1080/14606925.2018.1427913

Waldman, D. A., Bass, B. M., & Einstein, W. O. (1987). Leadership and outcomes of performance appraisal processes. *Journal of Occupational Psychology*, *60*(3), 177–186. doi:10.1111/j.2044-8325.1987.tb00251.x

Waldman, D. A., Siegel, D. S., & Javidan, M. (2006). Components of CEO transformational leadership and corporate social responsibility. *Journal of Management Studies*, *43*(8), 1703–1725. doi:10.1111/j.1467-6486.2006.00642.x

Waldrop, M. (2016). The chips are down for Moore's law. *NATNews*, *530*(7589), 144. PMID:26863965

Walumbwa, F. O., Avolio, B. J., Gardner, W. L., Wernsing, T. S., & Peterson, S. J. (2008). Authentic leadership: Development and validation of a theory-based measure. *Journal of Management*, *34*(1), 89–126. doi:10.1177/0149206307308913

Walumbwa, F. O., Christensen, A. L., & Hailey, F. (2011). Authentic leadership and the knowledge economy. *Organizational Dynamics*, *40*(2), 110–118. doi:10.1016/j.orgdyn.2011.01.005

Walumbwa, F. O., Luthans, F., Avey, J. B., & Oke, A. (2011). Retracted : Authentically leading groups: The mediating role of collective psychological capital and trust. *Journal of Organizational Behavior*, *32*(1), 4–24. doi:10.1002/job.653

Walumbwa, F. O., Wang, P., Wang, H., Schaubroeck, J., & Avolio, B. J. (2010). Psychological processes linking authentic leadership to follower behaviors. *The Leadership Quarterly*, *21*(5), 901–914. doi:10.1016/j.leaqua.2010.07.015

Wang, B. (2018). The Future of Manufacturing: A New Perspective. *Engineering*, *4*(5), 722–728. doi:10.1016/j.eng.2018.07.020

Wang, H., Tong, L., Takeuchi, R., & George, G. (2016). Corporate social responsibility: An overview and new research directions. *Academy of Management Journal*, *59*(2), 534–544. doi:10.5465/amj.2016.5001

Wang, J., Zhuang, X., Yang, J., & Sheng, J. (2014). The effects of optimism bias in teams. *Applied Economics*, *46*(32), 3980–3994. doi:10.1080/00036846.2014.948678

Ward, T. B. (2004). Cognition, creativity, and entrepreneurship. *Journal of Business Venturing*, *19*(2), 173–188. doi:10.1016/S0883-9026(03)00005-3

Warrick, D. (1981). Leadership styles and their consequences. *Journal of Experiential Learning and Simulation*, *3-4*, 155–172.

Waterman, H. (2011). Principles of 'servant leadership' and how they can enhance practice. *Nursing Management*, *17*(9), 24–26. doi:10.7748/nm2011.02.17.9.24.c8299 PMID:21473215

Web of Science. (2018). *Clarivate analytics, sear on servant, transactional and autocratic leadership.* Available online: clarivate.com/products/web-of-science/

Weber, E., & Khademian, A. (2008). Wicked problems, knowledge challenges, and collaborative capacity builders in network settings. *Public Administration Review*, *68*(2), 334–349. doi:10.1111/j.1540-6210.2007.00866.x

Weber, J. M. (2012). Social innovation and social enterprise in the classroom: Frances Westley on bringing clarity and rigor to program design. *Academy of Management Learning & Education*, *11*(3), 409–418. doi:10.5465/amle.2011.0403

Weerawardena, J., & Mort, G. S. (2001). Learning, innovation and competitive advantage in not-for-profit aged care marketing: A conceptual model and research propositions. *Journal of Nonprofit & Public Sector Marketing*, *9*(3), 53–73. doi:10.1300/J054v09n03_04

Weerawardena, J., & Mort, G. S. (2006). Investigating social entrepreneurship: A multidimensional model. *Journal of World Business*, *41*(1), 21–35. doi:10.1016/j.jwb.2005.09.001

Weerawardena, J., & Mort, G. S. (2012). Competitive strategy in socially entrepreneurial nonprofit organizations: Innovation and differentiation. *Journal of Public Policy & Marketing*, *31*(1), 91–101. doi:10.1509/jppm.11.034

Weick, K. E. (1989). Theory Construction as Disciplined Imagination. *Academy of Management Review*, *14*(4), 516–531. doi:10.5465/amr.1989.4308376

Weiss, M., Razinskas, S., Backmann, J., & Hoegl, M. (2018). Authentic leadership and leaders' mental well-being: An experience sampling study. *The Leadership Quarterly*, *29*(2), 309–321. doi:10.1016/j.leaqua.2017.05.007

Wensveen, S., & Matthews, B. (2015). *Prototypes and prototyping in design research. In The Routledge Companion to Design Research*. Taylor & Francis.

Wernerfelt, B. (1984). A resource-based view of the firm. *Strategic Management Journal*, *5*(2), 171–180. doi:10.1002mj.4250050207

West, D. M. (2007). *Digital Government: Technology and Public Sector Performance*. Princeton University Press.

West, D. M. (2008). *State and Federal Electronic Government in the United States, 2008*. Governance Studies at Brookings.

Westley, F., Antadze, N., Riddell, D. J., Robinson, K., & Geobey, S. (2014). Five configurations for scaling up social innovation: Case examples of nonprofit organizations from Canada. *The Journal of Applied Behavioral Science*, *50*(3), 234–260. doi:10.1177/0021886314532945

Westley, F., Zimmerman, B., & Patton, M. Q. (2006). *Getting to maybe: How the world has changed*. Toronto, Canada: Random House Canada.

What Is Political Participation? (n.d.). Retrieved from https://www.igi-global.com/dictionary/political-participation/22952

Wheeler, D., Colbert, B., & Freeman, R. E. (2003). Focusing on value: Reconciling corporate social responsibility, sustainability and a stakeholder approach in a network world. *Journal of General Management*, *28*(3), 1–27. doi:10.1177/030630700302800301

Whelan, G., Moon, J., & Grant, B. (2013). Corporations and Citizenship Arenas in the Age of Social Media. *Journal of Business Ethics*, *118*(4), 777–790. doi:10.100710551-013-1960-3

Whetstone, J. T. (2002). Personalism and moral leadership: The servant leader with a transforming vision. *Business Ethics (Oxford, England)*, *11*(4), 385–392. doi:10.1111/1467-8608.00298

Wilson, R. T. (1998). Servant leadership. *Physician Executive*, *24*(5), 6. PMID:10185648

Winker, I. (2010). *Contemporary leadership theories. Enhancing the understanding of the complexity, subjectivity and dynamic of leadership*. Springer Science and Business Media.

Winston, B. E., & Patterson, K. (2006). An integrative definition of leadership. *International Journal of Leadership Studies, 1*(2), 6-66.

Wong, P., & Page, D. (2003). *Servant leadership: An opponent-process and the revised servant leadership profile*. Virginia Beach, VA: Servant Leadership Research.

Wood, R. E., & Bandura, A. (1989). Social cognitive theory of organizational management. *Academy of Management Review*, *14*(3), 361–384. doi:10.5465/amr.1989.4279067

Woodside, A. G. (2014). Embrace perform model: Complexity theory, contrarian case analysis, and multiple realities. *Journal of Business Research*, *67*(12), 2495–2503. doi:10.1016/j.jbusres.2014.07.006

Woolley, L., Caza, A., & Levy, L. (2011). Authentic Leadership and Follower Development. *Journal of Leadership & Organizational Studies*, *18*(4), 438–448. doi:10.1177/1548051810382013

World Business Council for Sustainable Development (WBCSD). (2000). *Corporate Social Responsibility: Making Good Business Sense*. Geneva: World Business Council for Sustainable Development.

World Economic Forum and Schwab Foundation for Social Entrepreneurship. (2014). *Leadership in social enterprise. How to manage yourself and the team.* Available on line: www3.weforum.or/docs/WEF-leadership-in-social-entreprise-2014.pdf

Wright, P. M., McMahan, G. C., & McWilliams, A. (1994). Human resources and sustained competitive advantage: A resource-based perspective. *International Journal of Human Resource Management*, *5*(2), 301–326. doi:10.1080/09585199400000020

Wright, T. A. (2003). Positive organizational behavior: An idea whose time has truly come. *Journal of Organizational Behavior*, *24*(4), 437–442. doi:10.1002/job.197

Wulleman, M., & Hudon, M. (2016). Models of social entrepreneurship: Empirical evidence from Mexico. *Journal of Social Entrepreneurship*, *7*(2), 162–188. doi:10.1080/19420676.2015.1057207

Wu, T. Y., Hsu, W. L., & Cheng, B. S. (2003). Expressing anger? Or suppressing anger? Subordinates' anger responses to supervisors' authoritarian leadership behaviors in Chinese business. *Indigenous Psychological Research in Chinese Societies*, *18*, 3–49.

Yildiz, M. (2003). *Elektronik (E)-Devlet Kuram ve Uygulamasina Genel Bir Bakiş ve Değerlendirme. In Çağdaş Kamu Yönetimi I (Konular-Kuramlar-Kavramlar)* (pp. 305–327). Ankara: Nobel Yayinevi.

Yin, R. K. (1989). *Case study research: Design and method.* Sage Publications.

Yin, R. K. (2003). *Applications of case study research.* Applied social research methods series Thousand Oaks, CA: Sage Publications.

Yin, R. K. (2003). *Case study research: Design and methods.* Thousand Oaks, CA: Sage Publications.

Yoke Sin, C. (2019). *Foundation of Effective E-Government-The Singapore Experience.* Belfer Center for Science and International Affairs, Harvard Kennedy School.

Young, I. M. (2002). *Inclusion and Democracy.* Oxford University Press. doi:10.1093/0198297556.001.0001

Youssef, C. M. (2004). *Resiliency Development of Organizations, Leaders and Employees: Multulevel Theory Building and Individual- Level, Path-Analytical Emprical Testing*. University of Nebraska.

Youssef, C. M., & Luthans, F. (2007). Positive Organizational Behavior in the Workplace: The Impact of Hope, Optimism, and Resilience. *Journal of Management*, *33*(5), 774–800. doi:10.1177/0149206307305562

Yu, K. H. (2013). Institutionalization in the context of institutional pluralism: Politics as a generative process. *Organization Studies*, *34*(1), 105–131.

Yukl, G. (1994). *Leadership in Organizations* (3rd ed.). Prentice Hall International Inc.

Yukl, G. (1999). An evaluation of conceptual weaknesses in transformational and charismatic leadership theories. *The Leadership Quarterly*, *10*(2), 285–305. doi:10.1016/S1048-9843(99)00013-2

Yukl, G. (2006). *Leadership in organizations*. Upper Saddle River, NJ: Prentice-Hall.

Yukl, G. (2012). Effective Leadership Behavior: What We Know and What Questions Need More Attention. *The Academy of Management Perspectives*, *26*(4), 66–85. doi:10.5465/amp.2012.0088

Yukl, G., Gordon, A., & Taber, T. (2002). A hierarchical taxonomy of leadership behavior: Integrating a half century of behavior research. *Journal of Leadership & Organizational Studies*, *9*(1), 15–32. doi:10.1177/107179190200900102

Yu, X., Meng, X., Chen, Y., Chen, Y., & Nguyen, B. (2017). Work-Family Conflict, Organizational Ambidexterity and New Venture Legitimacy In Emerging Economies. *Technological Forecasting and Social Change*, 1–12.

Zaccaro, S. J., & Horn, Z. N. (2003). Leadership theory and practice: Fostering an effective symbiosis. *The Leadership Quarterly*, *14*(6), 769–806. doi:10.1016/j.leaqua.2003.09.009

Zacher, H., & Rosing, K. (2015). Ambidextrous leadership and team innovation. *Leadership and Organization Development Journal*, *36*(1), 54–68. doi:10.1108/LODJ-11-2012-0141

Zahra, S. A., Gedajlovic, E., Neubaum, D. O., & Shulman, J. M. (2009). A typology of social entrepreneurs: Motives, search processes and ethical challenges. *Journal of Business Venturing*, *24*(5), 519–532. doi:10.1016/j.jbusvent.2008.04.007

Zahra, S., & Nambisan, S. (2011). Entrepreneurship in global innovation ecosystems. *AMS Review*, *1*(1), 4–17. doi:10.100713162-011-0004-3

Zehir, C., & Yavuz, A. (2017). Değişim Odaklı Liderlik, Örgütsel Özdeşleşme ve Örgütsel Vatandaşlık Davranışı İlişkisi. *Doğuş Üniversitesi Dergisi*, *18*(2), 97–115. doi:10.31671/dogus.2018.34

Ženko, Z., Mulej, M., & Potočan, V. (2017). Knowledge-cum-values Management belongs to the Way out from Global Crisis. *Business Systems Research Journal*, *8*(1), 113–123. doi:10.1515/bsrj-2017-0009

Zhang, F., Wang, Y., Li, D., & Cui, V. (2016). Configurations of Innovations across Domains: An Organizational Ambidexterity View. *Journal of Product Innovation Management*, *34*(6), 821–841. doi:10.1111/jpim.12362

Zhang, Y., Huai, M., & Xie, Y. (2015). Paternalistic leadership and employee voice in china: A dual process model. *The Leadership Quarterly*, *26*(1), 25–36. doi:10.1016/j.leaqua.2014.01.002

Zheng, J., Wu, G., Xie, H., & Xu, H. (2017). Ambidextrous Leadership and Sustainability-Based Project Performance: The Role of Project Culture. *Sustainability*, *9*(2336), 1–24.

Zivkovic, S. (2015). A complexity based diagnostic tool for tackling wicked problems. *Emergence*, *17*(4), 1–9.

About the Contributors

Erdal Akdeve is an associate professor of Management at the Social Sciences University of Ankara. His main research areas are innovation, strategic management, and entrepreneurship.

Devi Akella is a Full Professor of Business Administration at College of Professional Studies, Albany State University, GA where she teaches Organizational Behavior and Human Resources Management at undergraduate and graduate levels. Dr. Akella completed her PhD in 2001 and MBA in 1997 from University of Leeds, UK. Her research interests consist of learning organizations, critical managements studies, social enterprises and teenage pregnancy.

Ana Aleksić, Ph.D., is an assistant professor at the Department of Organization and Management, Faculty of Economics and Business, University of Zagreb. Her research interests include various aspects of organizational behaviour, with a special emphasis on the role of organizational design practices in shaping behaviour within different organizational levels. She is the author and co-author of several book chapters and journal publications and has participated as a consultant on several scientific and commercial projects.

Melis Attar is an Assistant Professor in the Department of International Trade at Selçuk University, Turkey. She received a BSc. degree in Economics at Koç University, Istanbul, Turkey and has a MSc. in International Management from Essex University, UK. She completed her PhD. in Organization and Management at Selçuk University, Turkey. Her doctoral thesis is about the leadership styles and organizational ambidexterity in Turkish Banking Sector. Her research interests are on the field of leadership, organizations, organizational behavior, international management, innovation and strategic management.

Nevra Baker is an Assistant Professor in the Department of International Trade at Altinbas University, where she has been since 2017. She received a B.A. in

Tourism Administration and an MBA from Bogazici University in 2006 and 2008, respectively. She received her Ph.D. in Management from Bogazici University in 2017. From 2008 to 2010 she worked at GlaxoSmithKline Pharmaceuticals in Istanbul, eventually as a Key External Expert Specialist. Her research interests span both organizational behavior and leadership.

Nazlı Berberoglu Yilmaz is a Ph. D. student and research assistant at Ankara Yıldırım Beyazıt University Management School. They are interested in positive organizational behavior and the political connections of organizations.

Anju Bharti is an Assistant Professor in the Department of Management, Maharaja Agrasen Institute of Technology, GGSIP University, Delhi. She has a Doctorate in Management. She has 22 years of experience of working at corporate and in teaching. She has published 45 research papers in journals of National and International repute. She has co-authored and edited two books and has been a reviewer of two international journals. She has participated and presented papers in more than 32 national and international seminars and conferences.

Joseph Crawford is a leadership and behavior researcher who seeks to create empirical solutions for growing unethical behavior in leaders. He has provided conceptualizations for the authentic leader and follower and published on best practice methodology for leadership measurement.

Konur Alp Demir was born in Sakarya in 1983. He graduated from Çanakkale Onsekiz Mart University Social Sciences Institute with a master's degree and a doctorate degree from the Department of Public Administration. There are articles and papers on local governments, public administration, civil society, autonomy, municipal councils and administration. The main areas of study are public administration, local governments, urbanization and environmental issues, municipalities, provincial administrations, special provincial administrations, public administration supervision, civil society organizations, property management, governance, coastal zone management, water resource management and marine environment protection.

Çağlar Doğru, Ph.D. of Management and Organization at Ufuk University, Ankara, Turkey, completed bachelor's degree at Hacettepe University, department of business administration. He received master's degree in human resources management and doctoral degree in business administration from Gazi University, Ankara, Turkey. His research interests are concentrated on management, leadership, organization, organizational behavior, human resources management, and strategic management. He has given lessons in stated research areas, both bachelor's and master's degrees.

He has supervised many master's theses on management and organization as well. Additionally, he has numerous research articles in both national and international journals. Moreover, he has many book chapters in the international publishing houses and he presented numerous studies at many national and international conferences and published at proceedings. He serves as editor at both international books and national/international prestigious journals. Among his hobbies playing piano and swimming come first. He speaks English, French, German and Russian. He is married and has a daughter.

Niveen Labib Eid is a Business Lecturer at Birzeit University / Palestine. She has been teaching core business and marketing courses for over twelve years. While taking a critical research philosophy, Niveen has published journal articles in the areas of social enterprise management, social entrepreneurship, servant leadership, strategy, business ethics and social responsibility. She is a double Masters degree holder in two programs one in Business Administration (MBA) from Birzeit University, and another, an MSc in Organisation Studies and Human Resource Management from the University of Essex, United Kingdom. She served as the 'Chairperson' of the Business and Marketing Department at Birzeit University for two years (2016-2018). In addition, she is a board member at a number of community-based Palestinian organisations. Niveen has provided consultancy services as a business trainer and consultant for institutional clients with a concentration in Strategic Management, Business Planning and Talent Management. Among these were Palestine Investment Fund (PIF), Leaders Organization, Al Nayzak, SPARK, Najjad Zeenni Information Technology Center of Excellence (NZITCE) at Birzeit University, Rozana Association, Konrad Adenauer-Stiftung (KAS), Juhoud Association for Community and Rural Development and many others. She also has valuable knowledge and experience in running strategic planning workshops and steering the strategic planning efforts from a critical perspective. She wrote strategic plans for a number of local private and not-for-profit organizations. She has worked on economic studies, annual evaluations, and Training Need Analysis/ Assessment (TNA) assignments. She is also competent in devising business plans and the mentorship of small local entrepreneurial businesses. She has received awards in academic disciplines including receiving a HESPAL scholarship (Higher Education Scholarships for Palestinians) offered by the British Council. Furthermore, received an award from Ibn Rushd Institute titled: 'distinctive cases in University teaching', and lately received The Emerald Award for Excellence for the year 2017 for a publication about employees' commitment in arts and cultural organizations in Palestine. Her research areas of interest include: organizational culture, critical theory, social entrepreneurship, case study research, institutional theory, knowledge management, ethics and social responsibility, talent management and organizational behavior. Niveen Pursues to become a change citizen toward an academic world that emphasizes practicality, innovation and pluralism. Niveen places high value on social engagement and having researchers and businesses work for the general good. As an educator she thinks her mission revolves around creating a learning approach that emphasize a human-oriented management along with an ethical, critical, and a practical research philosophy.

Derya Fındık is currently working at the Department of Management Information Systems, Ankara Yıldırım Beyazıt University and have an academic background in science and technology policy studies. She did both master and Ph.D. at the METU Science and Technology Policy Studies Department. Her main research area has been the adoption of specific technologies as ERP, CRM, and SCM and the effect of software investment on firm performance. She also studied university-industry collaboration and firms 'innovation strategy and the impact of external collaborations on firms' innovation performance and wrote papers touching upon these issues.

Ravinder Jit is a professor of Organizational Behavior and Human Resource Management at Maharaja Agrasen Institute of Technology, Delhi, India. She has a Doctorate in Psychology. She has co-authored and edited four books and has been the editor of three refereed journals. She has presented papers in various national and international conferences and seminars. She has published 16 articles in various journals and conference proceedings. In 2016, she published an article in International Journal of Conflict Management, Vol. 27, Issue 4 – an Emerald Publication. In 2017, one of her articles was published in Vikalpa, Vol. 42, Issue 2 – a Sage Publication.

Serap Kalfaoğlu is a doctoral candidate at Faculty of Economics and Administrative Sciences in Business Management and a research assistant in Organization and Management Department.

Gülsen Kaya is an assistant professor of Political Science and Public Administration at the Social Sciences University of Ankara. She is concomitantly the vice-director of the Social Innovation Center.

Jo-Anne Kelder is a Senior Lecturer at the University of Tasmania and has a research focus on evaluation within higher education curriculum. Her work has been recognized with a series of awards and grants. Jo-Anne is currently the Senior Editor of the Journal of University Learning and Teaching Practice.

Matthew Knox is a recent graduate of the University of Tasmania with a passion for understanding how to create a better world. His current work is focused on understanding authentic followers and their role in enabling ethical leader behaviors.

Stefan Lie is an industrial designer whose work in practice, postgraduate study and academic research has led to a focus on interaction design for physical products. At UTS he is the director of Product Design and lectures in the Product Design program. His research lies in the ways that new manufacturing methods, such as 3D printing, enable interactive products to be manufactured in very low volumes for specific applications. He investigates the development of new design methodologies that will lead to such outcomes.

Yvonne Lomas-Montaudon is a university professor and researcher. Bachelor's degree in business engineering. Master's degree in logistics and supply chain management.

Nikolina Markota Vukić is a sustainability and non-financial reporting expert with a Ph.D. in CSR and social strategies. She is the co-founder and president of The Croatian Institute for CSR (IDOP) and has almost ten-year experience working in CSR, non-financial reporting and sustainability - from project management to consultancy in business strategies. During her path and consulting business sector how to integrate CSR and increase transparency, she workes as a teacher at RRiF College for Financial Management. As a complement to her passion, Nikolina is a mentor at undergraduate and graduate programmes under a sustainability and CSR topics.

Cynthia Montaudon is a full time professor at the Business School at UPAEP in Puebla, Mexico. She holds a post doctorate in Organizational Leadership, a PhD in Business, and a PhD in Strategic Planning. She also holds an MBA, a Master in Quality Engineering and Master in Communication Sciences. She is head of the Observatory on Competitiveness and New Ways of Working and develops training courses for the use of different active methodologies including Service Learning, Relations and Connections Based Learning, Challenge Based Learning, Project and Problem Based Learning to connect learning to real life experiences.

Anton Nemme is an Industrial Designer and graduate of UTS. He began his carreer in the field of product visualisation producing 3D models and renderings for some of Sydney's leading Product Design consultancies. Anton has a strong focus on teaching being involved across all the undergraduate year groups. His most recent research themes centre on the value of iteration in design projects, university Industry collaboration, applications for 3D printing and the notion of authenticity in design education.

Mislav Ante Omazić is Full Professor on the Faculty of Economics and Business, University of Zagreb at the Department of Organization and Management, where he teaches a variety of classes such as Corporate Social Responsibility, Business Ethics, Management, Project Management, and Change Management. He worked on various domestic and international projects for public, private and civil society organizations. During his research work, professor Omazić has published a number of papers in professional and scientific journals home and abroad and is the author and editor of several books.

Berto Pandolfo is an industrial designer who trained and worked in Australia and Italy. His design research interests are influenced by non-traditional or practice-oriented investigations. Specifically his research focuses on the application of both emerging and traditional methods to develop new approaches to complex form making. He applies his research within the context of batch production and commercialisation of designed objects in Australia. Among his self-produced designs are the SPK bowl, the RAK coat stand, the BLK side table and the SRV plate. Berto is senior lecturer in the Product Design program at University of Technology Sydney. He teaches in both undergraduate and postgraduate courses with a focus on the history of industrial design, technical and visual communication for industrial designers, object making and prototyping and commercialisation of design. His designs are cited in both local and international design journals and his work is exhibited in design events both locally and internationally.

Mirjana Pejić Bach is a Full Professor at the Department of Informatics at the Faculty of Economics & Business. She graduated at the Faculty of Economics & Business – Zagreb, where she also received her Ph.D. degree in Business, submitting a thesis on "System Dynamics Applications in Business Modelling" in 2003. She is the recipient of the Emerald Literati Network Awards for Excellence 2013 for the paper Influence of strategic approach to BPM on financial and non-financial performance published in Baltic Journal of Management. Mirjana was also educated at MIT Sloan School of Management in the field of System Dynamics Modelling, and at OliviaGroup in the field of data mining. She participates in a number of EU FP7 projects and is an Expert for Horizon 2020. Mirjana Pejić Bach is editor in chief of two journals, and serves as the Associate editor and member of Editorial Board in several other journals, in the field of Information and Management, as well as conferences.

P. Rajeev is a retired Economic Advisor, Government of India. He worked as an economist in the Government of India for 32 years. During his professional career he has been associated with the Ministry of Finance, Ministry of Agriculture, Department of Chemicals and Petro Chemicals, Planning Commission, Tariff Commission, Bureau of Industrial Costs and Prices and the Institute for Defence Studies and Analyses. He is the author of 25 books and about 200 articles in various newspapers and journals.

Urvashi Sharma is an Associate Professor and Course Coordinator of MBA (HRD) in the Department of Commerce, Delhi School of Economics, University of Delhi. She has a Doctorate in Management. Having 16 years of teaching experience, she has also worked in the capacity of Head of the Department in a GGSIP University College in Delhi. She has presented several papers in conferences and seminars and published papers in reputed journals. She has also been the Editor of a refereed Journal.

Rachel Taylor's research interests focus on the role of social innovation by non-profit organisations. In particular, her PhD research explored the sector-wide social innovation trends and practices within the disability sector in Australia as it grapples with the introduction of the National Disability Insurance Scheme (NDIS). Prior to joining academia, Rachel has worked in management roles in non-profit organisations and social enterprises, including within the disability sector for over four years. As well as publishing in the management discipline, Rachel strives to produce research that is both accessible and practical, and which can ultimately assist non-profit managers, together with organisational stakeholders and community members, to design social innovation initiatives in response to critical social challenges.

Nuttaneeya (Ann) Torugsa is an Associate Professor in Management at Ratchasuda College, Mahidol University in Thailand; and concurrently an Adjunct Senior Researcher at the Tasmanian School of Business and Economics, University of Tasmania in Australia. Previously, she worked for 8 years as a Research Fellow in Innovation at the Australian Innovation Research Centre and a Lecturer in Health Service Management at the Australian Institute of Health Service Management. Dr Torugsa has expertise in four inter-linked research areas including: (1) innovation management and policy in the public, non-profit and private sectors; (2) sustainable development, corporate social responsibility and strategic management; (3) health service management; and (4) innovative research designs, methods and analyses that make research intellectually relevant and meaningful for practice. She has extensive experience in analyzing large-scale data obtained from international, national and regional sources, as well as in employing various statistical and modelling techniques.

Roderick Walden undertakes research into the contemporary methods and practice of professional industrial design. His principal areas of research interest lie in professional practice, practice-based design research, design expertise and academic design. He has previously worked as a product designer and production engineer for a range of local and international manufacturers in the commercial office furniture industry including Herman Miller U.S., and Wilkhahn and König + Neurath Germany. His design consultancy, Walden Design Pty Ltd, formed in 2003, designed a range of products for Australian companies, working particularly closely with manufacturing SMEs. In his academic role, he has co-developed, through research, prototypes including a mountain bike (MTB) multi-tool (2008), leather-moulded office accessories (2008) and a fire-fighters cooling vest (2009), all with long-time colleague and friend Stefan Lie. He has conducted research into additive manufacturing technologies with the design of a DMLS MTB Crank Arm (2010) as part of the CCDP-sponsored Digifacture project, which was co-curated with Berto Pandolfo. His masters (by research) degree studied self-initiated product design and modelled connections between knowledge and practice that he is currently advancing as part of his PhD. He has co-authored research on topics including university-industry collaboration (UIC), design in business, design education and academic design practice. Roderick is passionate about his role as a teacher, the technical work of design and the meaning behind the connections designers make with other areas of knowledge to solve problems, innovate and learn.

Jovana Zoroja, PhD, is an assistant professor at the Faculty of Economics and Business, University of Zagreb, Department of Informatics. She received PhD in Information Systems at the Faculty of Economics and Business Zagreb with the dissertation thesis "Influence of the Information and Communication Technologies on the Competitiveness of the European Union Countries". She was also educated at the LSE-Summer School 2008 in London in the field of Business Development and ICT Innovation. Her main research interests are information and communication technology, e-learning, simulation games and simulation modelling. She is actively engaged in number of scientific projects (FP7-ICT, bilateral cooperation, national projects). Jovana Zoroja published several scientific papers in international and national journals and participated in many scientific international conferences and programme committees as a reviewer.

Index

F

G

H

I

J

L

M

N

O

P

Q

R

S

T

U

V

www.ingramcontent.com/pod-product-compliance
Lightning Source LLC
LaVergne TN
LVHW080348011026
849161LV00009B/2223